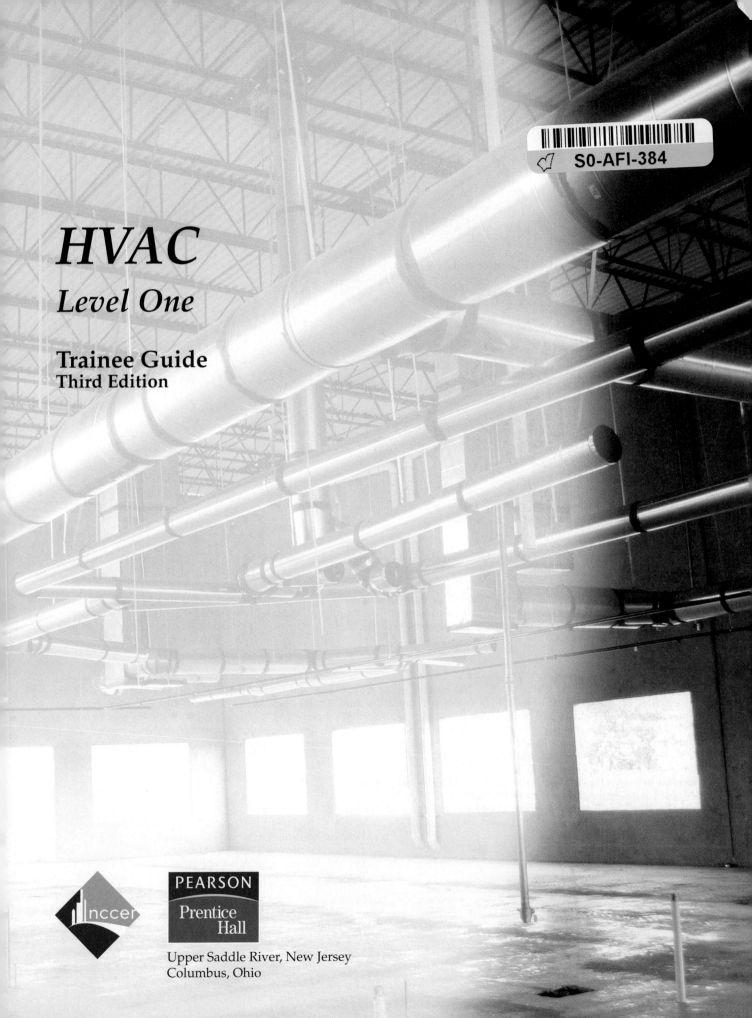

HVAC

Level One

Trainee Guide
Third Edition

nccer

PEARSON
Prentice
Hall

Upper Saddle River, New Jersey
Columbus, Ohio

National Center for Construction Education and Research
President: Don Whyte
Director of Product Development: Daniele Stacey
HVAC Project Manager: Tania Domenech
Production Manager: Jessica Martin
Product Maintenance Supervisor: Debie Ness
Editors: Brendan Coote and Bethany Harvey
Desktop Publishers: Jessica Martin and James McKay

Writing and development services provided by Topaz Publications, Liverpool, New York.

Pearson Education, Inc.
Product Manager: Lori Cowen
Project Manager: Stephen C. Robb
Design Coordinator: Diane Y. Ernsberger
Text Designer: Kristina D. Holmes
Cover Designer: Kristina D. Holmes
Copy Editor: Sheryl Rose
AV Project Manager: Janet Portisch
Operations Supervisor: Pat Tonneman
Marketing Manager: Derril Trakalo

This book was set in Palatino and Helvetica by Carlisle Publishing Services. It was printed and bound by Courier Kendallville, Inc. The cover was printed by Phoenix Color Corp.

This information is general in nature and intended for training purposes only. Actual performance of activities described in this manual requires compliance with all applicable operating, service, maintenance, and safety procedures under the direction of qualified personnel. References in this manual to patented or proprietary devices do not constitute a recommendation of their use.

Pearson Prentice Hall™ is a trademark of Pearson Education, Inc.
Pearson® is a registered trademark of Pearson plc
Prentice Hall® is a registered trademark of Pearson Education, Inc.

Pearson Education Ltd.
Pearson Education Singapore Pte. Ltd.
Pearson Education Canada, Ltd.
Pearson Education—Japan

Pearson Education Australia Pty. Limited
Pearson Education North Asia Ltd.
Pearson Educación de Mexico, S.A. de C.V.
Pearson Education Malaysia Pte. Ltd.

10 9 8 7 6 5 4
ISBN-13: 978-0-13-614416-8
ISBN-10:   0-13-614416-0

Preface

TO THE TRAINEE

Heating and air-conditioning systems (HVAC) regulate the temperature, humidity, and the total air quality in residential, commercial, industrial, and other buildings. This also extends to refrigeration systems used to transport food, medicine, and other perishable items. Other systems may include hydronics (water-based heating systems), solar panels, or commercial refrigeration. HVAC technicians and installers set up, maintain, and repair such systems. As a technician, you must be able to maintain, diagnose, and correct problems throughout the entire system. Diversity of skills and tasks is also significant to this field. You must know how to follow blueprints or other specifications to install any system. You may also need working knowledge of sheet metal practices for the installation of ducts, welding, basic pipefitting, and electrical practices.

Think about it! Nearly all buildings and homes in the United States alone use forms of heating, cooling and/or ventilation. The increasing development of HVAC technology causes employers to recognize the importance of continuous education and keeping up to speed with the latest equipment and skills. Hence, technical school training or apprenticeship programs often provide an advantage and a higher qualification for employment. NCCER's program has been designed by highly qualified subject matter experts with this in mind. Our four levels present an apprentice approach to the HVAC field, including theoretical and practical skills essential to your success as an HVAC installer or technician.

As the population and the number of buildings grow in the near future, so will the demand for HVAC technicians. According to the U.S. Bureau of Labor Statistics, employment of HVAC technicians and installers is projected to increase 18 to 26 percent by 2014. We wish you the best as you begin an exciting and promising career.

NEW WITH HVAC LEVEL ONE

Level One has been streamlined and reorganized, based on the recommendations of a standing committee of HVAC instructors and field experts who are working with NCCER to review and improve the curriculum. The new Level One, together with a reorganized and expanded Level Two, will now provide you with a broad knowledge of installation and service requirements for residential and commercial heating and cooling systems, both forced-air and hydronic. Through this new course design, you will enter the workforce with the knowledge and skills needed to perform productively in either the residential or commercial market. In addition to the general restructuring of Level One, the text, graphics, and special features were enhanced to reflect advancements in HVAC technology. A major change in Level One was the inclusion of the Air Distribution Systems module (previously a part of Level Two), which now provides an earlier exposure to this key subsystem. An introduction to HVAC-related drawings and specifications was also added to Level One.

CONTREN® LEARNING SERIES

The National Center for Construction Education and Research (NCCER) is a not-for-profit 501(c)(3) education foundation established in 1995 by the world's largest and most progressive construction companies and national construction associations. It was founded to address the severe workforce shortage facing the industry and to develop a standardized training process and curricula. Today, NCCER is supported by hundreds of leading construction and maintenance companies, manufacturers, and national associations. The Contren® Learning Series was developed by NCCER in partnership with Prentice Hall, the world's largest educational publisher.

Some features of NCCER's Contren® Learning Series are as follows:

- An industry-proven record of success
- Curricula developed by the industry for the industry
- National standardization providing portability of learned job skills and educational credits
- Compliance with Office of Apprenticeship requirements for related classroom training (*CFR 29:29*)
- Well-illustrated, up-to-date, and practical information

NCCER also maintains a National Registry that provides transcripts, certificates, and wallet cards to individuals who have successfully completed modules of NCCER's Contren® Learning Series. *Training programs must be delivered by an NCCER Accredited Training Sponsor in order to receive these credentials.*

We invite you to visit the NCCER website at **www.nccer.org** for the latest releases, training information, newsletter, Contren® product catalog, and much more. Your feedback is welcome. You may email your comments to **curriculum@ nccer.org** or send general comments and inquiries to **info@nccer.org**.

Contren® Curricula

NCCER's training programs comprise more than 40 construction, maintenance, and pipeline areas and include skills assessments, safety training, and management education.

Boilermaking
Carpentry
Cabinetmaking
Careers in Construction
Concrete Finishing
Construction Craft Laborer
Construction Technology
Core Curriculum: Introductory Craft Skills
Currículo Básico
Electrical
Electronic Systems Technician
Heating, Ventilating, and Air Conditioning
Heavy Equipment Operations
Highway/Heavy Construction
Hydroblasting
Instrumentation
Insulating
Ironworking
Maintenance, Industrial
Masonry
Millwright
Mobile Crane Operations
Painting
Painting, Industrial
Pipefitting

Pipelayer
Plumbing
Reinforcing Ironwork
Rigging
Scaffolding
Sheet Metal
Site Layout
Sprinkler Fitting
Welding

Pipeline

Control Center Operations, Liquid
Corrosion Control
Electrical and Instrumentation
Field Operations, Liquid
Field Operations, Gas
Maintenance
Mechanical

Safety

Field Safety
Orientación de Seguridad
Safety Orientation
Safety Technology

Management

Introductory Skills for the Crew Leader
Project Management
Project Supervision

Special Features of This Book

In an effort to provide a comprehensive, user-friendly training resource, we have incorporated many different features for your use. Whether you are a visual or hands-on learner, this book will provide you with the proper tools to get started in the construction industry.

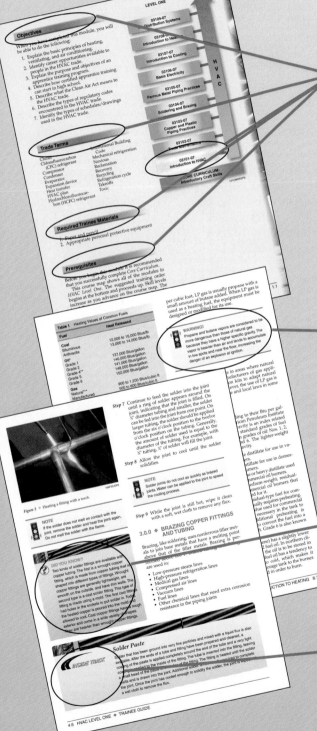

Introduction Page

This page is found in the front of each module and lists the objectives, key trade terms, required trainee materials, and prerequisites for that module. The objectives list the skills and knowledge you will need to gain in order to successfully complete the module. The list of key trade terms identifies important terms you will need to know by the end of the module. Required trainee materials lists the materials and supplies needed for the module. The prerequisites for the module are listed and illustrated in the course map. The course map also gives a visual overview of the entire course and a suggested learning sequence for you to follow.

Notes, Cautions, and Warnings

Safety features are set off from the main text in highlighted boxes and broken into three categories based on the potential danger of the issue being addressed. A note simply provides additional information on the topic area. A caution alerts you of a danger that does not present potential injury but may cause damage to equipment. A warning stresses a potentially dangerous situation that may cause injury to you or a co-worker.

Did You Know?

The Did You Know? feature introduces historical tidbits or modern information about the HVAC industry. Interesting and sometimes surprising facts about construction are also presented.

Inside Track

Inside Track features provide a head start for those entering the HVAC field by presenting technical tips and professional practices from master HVAC technicians in a variety of disciplines. Inside Tracks often include real-life scenarios similar to those you might encounter on the job site.

Step-by-Step Instructions

Step-by-step instructions are used throughout to guide you through technical procedures and tasks from start to finish. These steps show you not only how to perform a task but how to do it safely and efficiently.

Color Illustrations and Photographs

Full-color illustrations and photographs are used throughout each module to provide vivid detail. These figures highlight important concepts from the text and provide clarity for complex instructions. Each figure is denoted in the text in *italic type* for easy reference.

Think About It

Think About It features use "What if?" questions to help you apply theory to real-world experiences and put your ideas into action.

Review Questions

The review questions are provided to reinforce the knowledge you have gained. This makes them a useful tool for measuring what you have learned.

Key Trade Terms

Each module presents a list of key trade terms that are discussed within the text, defined in the glossary at the back of the module, and reinforced with a key terms quiz. These terms are denoted in the text with blue bold type upon their first occurrence. To make searches for key information easier, a comprehensive glossary of key trade terms from all modules is found at the back of this book.

Profile in Success

Profiles in Success share the apprenticeship and career experiences of and advice from successful professionals in the HVAC field.

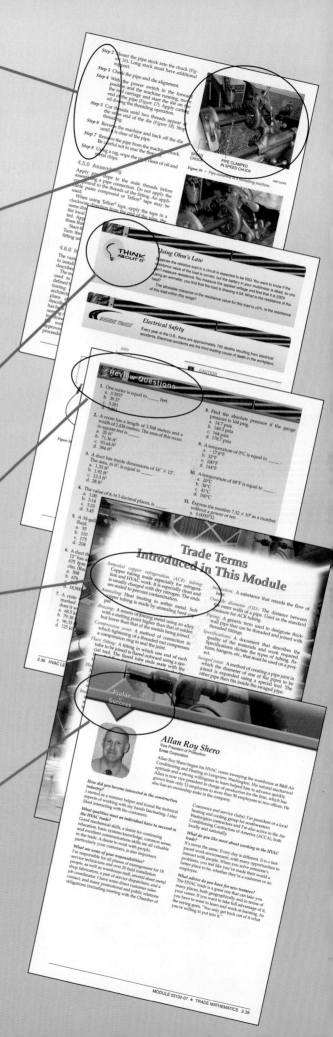

Contents

Acknowledgments

This curriculum was revised as a result of the farsightedness and leadership of the following sponsors:

ABC of Wisconsin
Comfort Systems USA
Entek Corporation
W. B. Guimarin & Co., Inc.
Hunton Trane

Newberry County Career Center
Total Comfort Service Center, Inc.
University of Florida M. E. Rinker, Sr. School
    of Building Construction

This curriculum would not exist were it not for the dedication and unselfish energy of those volunteers who served on the Authoring Team. A sincere thanks is extended to the following:

Wayne Culp
Daniel Kerkman
Dwight McDuffie
Joe Moravek

Paul Oppenheim
Jeff Plant
Troy Staton
Matthew Todd

A final note: This book is the result of a collaborative effort involving the production, editorial, and development staff at Prentice Hall and the National Center for Construction Education and Research. Thanks to all of the dedicated people involved in the many stages of this project.

PARTNERING ASSOCIATIONS

American Fire Sprinkler Association
API
Associated Builders & Contractors, Inc.
Associated General Contractors of America
Association for Career and Technical Education
Association for Skilled & Technical Sciences
Carolinas AGC, Inc.
Carolinas Electrical Contractors Association
Center for Improvement of Construction
    Management & Processes
Construction Industry Institute
Construction Users Roundtable
Design Build Institute of America
Electronic Systems Industry Consortium
Merit Contractors Association of Canada
Metal Building Manufacturers Association
NACE International
National Association of Minority Contractors

National Association of Women in Construction
National Insulation Association
National Ready Mixed Concrete Association
National Systems Contractors Association
National Technical Honor Society
National Utility Contractors Association
NAWIC Education Foundation
North American Crane Bureau
North American Technician Excellence
Painting & Decorating Contractors of America
Portland Cement Association
SkillsUSA
Steel Erectors Association of America
Texas Gulf Coast Chapter ABC
U.S. Army Corps of Engineers
University of Florida
Women Construction Owners & Executives, USA

Introduction to HVAC
03101-07

03101-07
Introduction to HVAC

Topics to be presented in this module include:

Overview

This program will pave the way for you to begin a career in one of America's most dynamic industries. Virtually every one of the tens of millions of homes and businesses in the United States has a heating system, and a large percentage have comfort cooling systems as well. Workers trained in this industry have the opportunity to install systems in new construction, service equipment in existing construction, and replace aging systems.

Working in the HVAC industry is challenging and rewarding because environmental technology is constantly changing. Technical advances in HVAC systems are made every day in advanced computerized controls, greater operating efficiency, and improved packaging.

The HVAC industry offers many opportunities for advancement. The training you are receiving can qualify you to become an installer, troubleshooter, sales technician, system design specialist, and eventually even the owner of your own HVAC service business.

Objectives

When you have completed this module, you will be able to do the following:

1. Explain the basic principles of heating, ventilating, and air conditioning.
2. Identify career opportunities available to people in the HVAC trade.
3. Explain the purpose and objectives of an apprentice training program.
4. Describe how certified apprentice training can start in high school.
5. Describe what the Clean Air Act means to the HVAC trade.
6. Describe the types of regulatory codes encountered in the HVAC trade.
7. Identify the types of schedules/drawings used in the HVAC trade.

Trade Terms

Chiller
Chlorofluorocarbon (CFC) refrigerant
Compressor
Condenser
Evaporator
Expansion device
Heat transfer
HVAC plan
Hydrochlorofluorocarbon (HCFC) refrigerant
International Building Code
Mechanical refrigeration
Noxious
Reclamation
Recovery
Recycling
Refrigeration cycle
Takeoffs
Toxic

Required Trainee Materials

1. Paper and pencil
2. Appropriate personal protective equipment

Prerequisites

Before you begin this module it is recommended that you successfully complete *Core Curriculum*.

This course map shows all of the modules in *HVAC Level One*. The suggested training order begins at the bottom and proceeds up. Skill levels increase as you advance on the course map. The local Training Program Sponsor may adjust the training order.

LEVEL ONE

03109-07
Air Distribution Systems

03108-07
Introduction to Heating

03107-07
Introduction to Cooling

03106-07
Basic Electricity

03105-07
Ferrous Metal Piping Practices

03104-07
Soldering and Brazing

03103-07
Copper and Plastic Piping Practices

03102-07
Trade Mathematics

03101-07
Introduction to HVAC

CORE CURRICULUM:
Introductory Craft Skills

H V A C

101CMAP.EPS

1.0.0 ◆ INTRODUCTION

Since ancient times, people have sought ways to make the buildings in which they live, work, and play more comfortable. Today, the Heating, Ventilating, and Air Conditioning (HVAC) industry provides the means to control the temperature, humidity, and even the cleanliness of the air in our homes, schools, offices, and factories. The members of the HVAC trade are skilled workers who install, maintain, and repair the equipment that makes this possible.

2.0.0 ◆ HEATING

Early humans burned fuel as a source of heat. That hasn't changed; what's different between then and now is the way it's done. We no longer need to huddle around a wood fire to keep warm. Instead, a central heating source such as a furnace or boiler does the job using the **heat transfer** principle; that is, heat is created in one place and carried to another place by means of air or water.

For example, in a common household furnace, fuel oil, natural gas, or propane/butane gas is burned to create heat, which warms metal plates known as heat exchangers (*Figure 1*). Air from living spaces is circulated over the heat exchangers and returned to the living spaces as heated air. This type of system is known as a forced-air system, and it is the most common type of central-heating system used in the United States.

Water is also used as a heat exchange medium. The water is heated in a boiler (*Figure 2*), then pumped through pipes to heat exchangers where the heat it contains is transferred to the surrounding air. The heat exchangers are usually baseboard heating elements located in the space to be heated. This type of system is known as a hydronic heating system, and it is more common in the Northeast and Midwest than in other parts of the country.

Natural gas and fuel oil are, by far, the most widely used heating fuels. Natural gas is currently the most popular fuel. Oil heat is more common in the Northeast than in other parts of the country. Propane/butane gas is used in many parts of the country in place of oil or natural gas. Oil and propane/butane fuels are primarily used in rural areas of the country where natural gas pipelines are not readily available.

Electricity is also used as a heat source. In an electric heating system, electricity flows through coils of heavy wire, causing the coils to become hot. Air from the conditioned space is passed over the coils and the heat from the coils is transferred to the air. Because electricity is so expensive, total electric heat is no longer common in cold climates. Total electric heat is more likely to be used in warm climates where heat is seldom required. Heat pumps are used, and in most cases are preferred over total electric heat. In some areas of the country, a main or supplementary heating system may be fueled by wood or coal. Due to increases in the cost of fossil fuels, such as oil and gas, over the last several decades, and the fact that wood is a relatively cheap and renewable resource, wood-burning stoves and furnaces have remained quite popular.

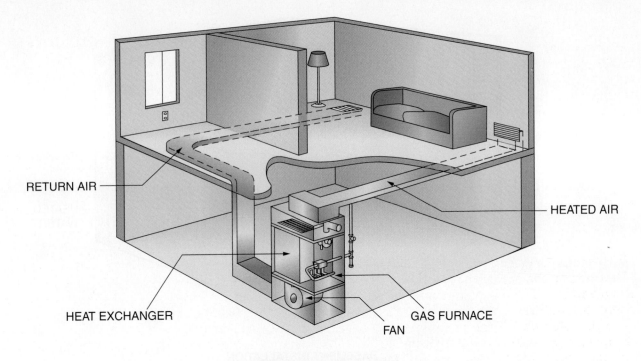

RETURN AIR

HEATED AIR

HEAT EXCHANGER

GAS FURNACE

FAN

BASEMENT INSTALLATION

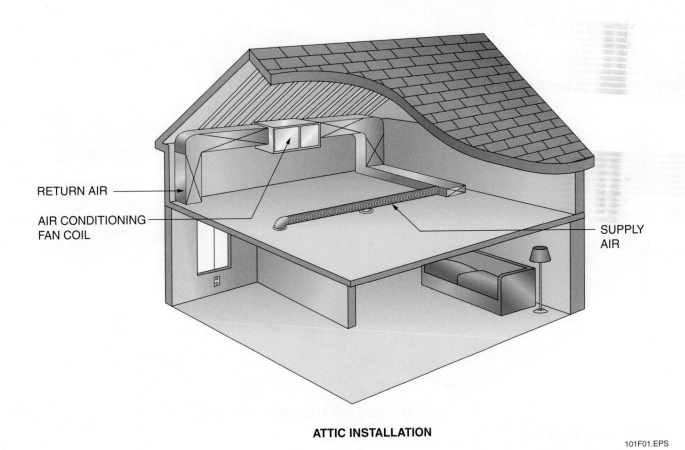

RETURN AIR

AIR CONDITIONING
FAN COIL

SUPPLY
AIR

ATTIC INSTALLATION

101F01.EPS

Figure 1 ◆ Forced-air heating.

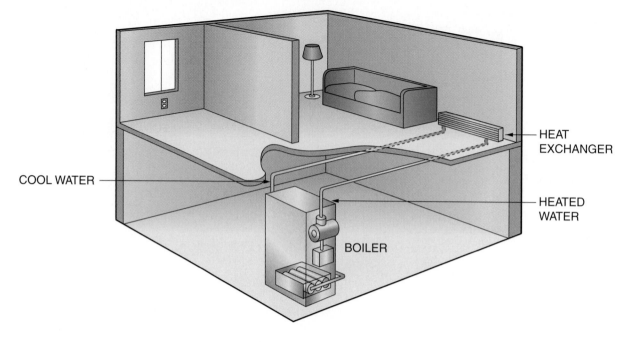

COOL WATER

HEAT EXCHANGER

HEATED WATER

BOILER

BASEMENT INSTALLATION

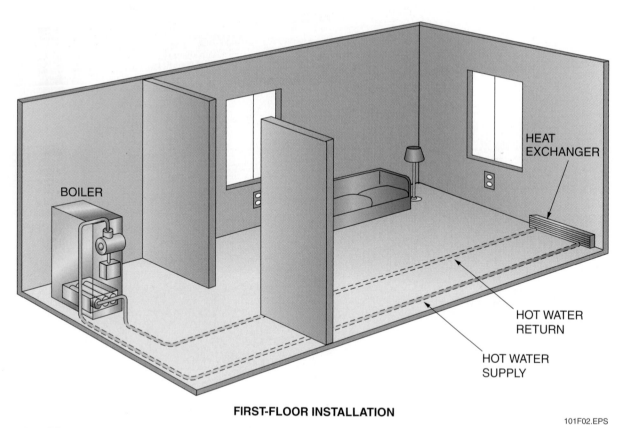

BOILER

HEAT EXCHANGER

HOT WATER RETURN

HOT WATER SUPPLY

FIRST-FLOOR INSTALLATION

101F02.EPS

Figure 2 ◆ Hot water heating.

High-Efficiency Furnaces and Boilers

Many of the newer furnaces and boilers available today are high-efficiency units with efficiency ratings ranging from 90 percent to 96.6 percent for gas. This compares to ratings of approximately 70 percent for old-style gas-burning units and 65 percent for standard oil-burning units. Although new or replacement high-efficiency equipment costs more initially, the payback savings in energy costs for a customer in the northern states could occur in as little as five years when compared to less efficient equipment. Moreover, high-efficiency units used for replacement purposes are often eligible for energy-saving cash incentives from local utilities.

A significant difference between a standard furnace and a condensing furnace is that in a condensing furnace, polyvinylchloride (PVC) piping is used to supply outdoor air for combustion and to exhaust the combustion products to the outdoors. An exhaust-vent blower is used to maintain combustion airflow through the burner section of the furnace. Because the primary and secondary heat exchangers for these furnaces are so efficient at removing heat, the exhaust gases are relatively cool and plastic exhaust-vent pipe can be used instead of metal pipe. The low temperature of the exhaust gases causes moisture to condense out of the gases in the secondary heat exchanger. The condensates are drained from the secondary heat exchanger and are usually pumped to a drain by a condensate pump.

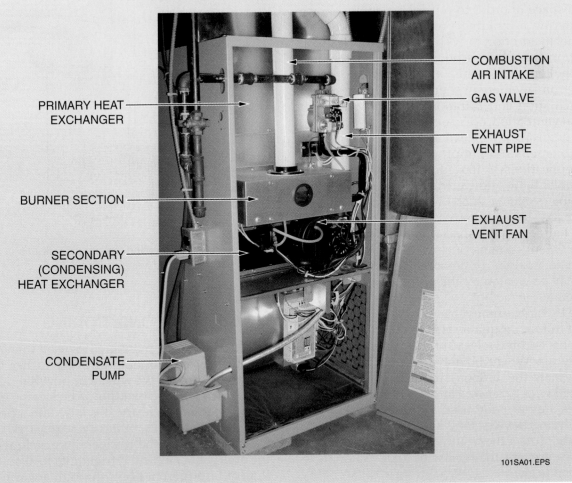

101SA01.EPS

3.0.0 ◆ VENTILATION

Ventilation is the introduction of fresh air into a closed space in order to control air quality. Fresh air entering a building provides the oxygen we breathe. In addition to fresh air, we want clean air. The air in our homes, schools, and offices contains dust, pollen, and molds, as well as vapors and odors from a variety of sources. Relatively simple air circulation and filtration methods, including natural ventilation, are used to help keep the air in these environments clean and fresh. Among the common methods of improving the air in residential applications is the addition of humidifiers, ultraviolet light, and electronic air cleaners to air-handling systems such as forced-air furnaces (*Figure 3*). Many industrial environments, on the other hand, require special ventilation and air-management systems. Such systems are needed to eliminate noxious or toxic particles and fumes that may be created by the processes and materials used at the facility.

The U.S. government has strict regulations governing indoor air quality (IAQ) in industrial environments and the release of toxic materials to the outside air. Where noxious or toxic fumes may be present, the indoor air must be constantly replaced with fresh air. Fans and other ventilating devices are normally used for this purpose. Special filtering devices may also be required; these not only protect the health of building occupants, but also prevent the release of toxic materials to the outside air.

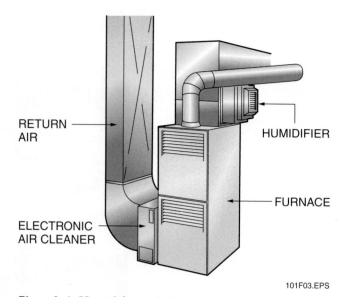

RETURN AIR

HUMIDIFIER

FURNACE

ELECTRONIC AIR CLEANER

101F03.EPS

Figure 3 ◆ Humidifier and electronic air cleaner in a residential heating system.

4.0.0 ◆ AIR CONDITIONING

Through the ages, mankind has used many methods to stay comfortable in hot weather. In this module, however, we will focus on what is known as mechanical refrigeration, which came into use in the twentieth century. It is based on a principle known as the refrigeration cycle (*Figure 4*).

Simply stated, the refrigeration cycle relies on the ability of chemical refrigerants to absorb heat. If a cold refrigerant flows through a warm space,

The Mechanical Refrigeration Cycle

Many people think that an air conditioner adds cool air to an indoor space. In reality, the basic principle of air conditioning and the mechanical refrigeration cycle is that heat is extracted from the indoor air and transferred to another location (the outdoors) by the refrigerant that flows through the cycle.

Figure 4 ◆ Basic refrigeration cycle for an air conditioner.

it will absorb heat from the space. Having given up heat to the refrigerant, the space becomes cooler. The colder the refrigerant, the more heat it will absorb, and the cooler the space will become. If the super-hot refrigerant flows to a cooler location, the outdoors for example, the refrigerant will give up the heat it absorbed from the indoors and become cool again.

A mechanical refrigeration system is a sealed system operating under pressure. The main elements of a mechanical refrigeration system are the:

- **Compressor** – Provides the force that circulates the refrigerant and creates the pressure differential necessary for the refrigeration cycle to work. A special refrigerant compressor is used.
- **Evaporator** – A heat exchanger where the heat in the warm indoor air is transferred to the cold refrigerant.
- **Condenser** – Also a heat exchanger. In the condenser, the heat absorbed by the refrigerant is transferred to relatively cooler outdoor air.
- **Expansion device** – Provides a pressure drop that lowers the boiling point and pressure of the refrigerant as it enters the evaporator. This allows the refrigerant to become a cold liquid/gas mixture and absorb heat in the evaporator.

Heat Pumps

A special type of air conditioner known as a heat pump is widely used in moderate climates to provide both cooling and heating. Heat pumps are extremely efficient; however, they are most effective in climates where the temperature generally does not fall below 25°F or 30°F. In colder parts of the country, heat pumps can be combined with a gas- or oil-fired, forced-air furnace. In such arrangements, the furnace automatically takes over heating duties when the outdoor temperature falls below the efficient range of the heat pump. Like high-efficiency furnaces and boilers, heat pumps are often eligible for energy-saving cash incentives from local power companies.

Heat pumps operate by reversing the cooling cycle. For this reason, heat pumps are sometimes called reverse-cycle air conditioners. The basic operating principle of a heat pump is that there is some heat in the air, even though the air may be very cold. In fact, the temperature would have to be –460°F for a total absence of heat to exist. In the heating mode, a special valve, known as a reversing valve, switches the compressor input and output so that the condenser operates as the evaporator and the evaporator becomes the condenser. Because of this role reversal, the coils in a heat pump are referred to as the outdoor coil and indoor coil instead of the condenser and evaporator.

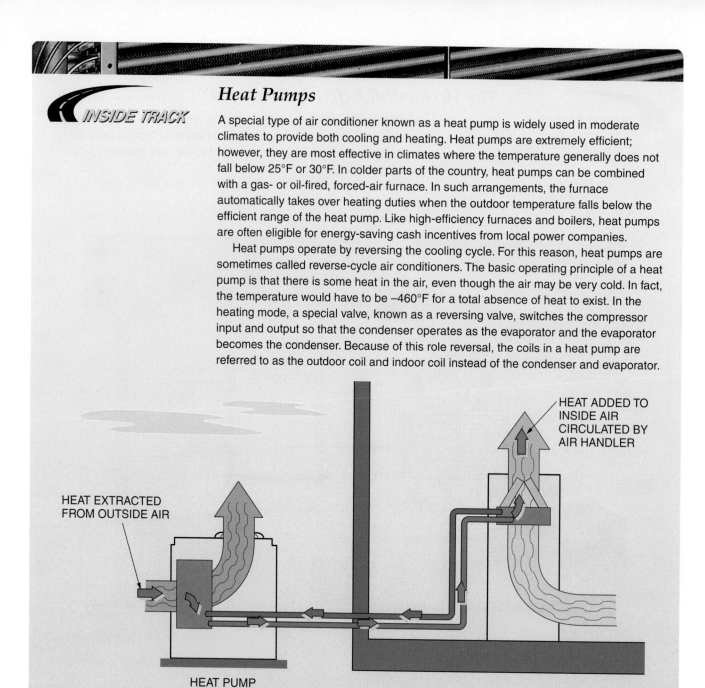

HEAT ADDED TO INSIDE AIR CIRCULATED BY AIR HANDLER

HEAT EXTRACTED FROM OUTSIDE AIR

HEAT PUMP

HEAT PUMP IN THE HEATING MODE

101SA02.EPS

The relationship between temperature and pressure is critical to mechanical refrigeration. As you study the process, you will learn that the same refrigerant can be very cold at one point in the system (the evaporator input) and very hot at another (the condenser input). These two points are often only inches apart. This is possible because of pressure changes caused by the compressor and expansion device. In addition to the circulation of refrigerant, air must also circulate. Fans at the condenser and evaporator move air across the condenser and evaporator coils.

This is a simple explanation of the refrigeration cycle. It is meant to give you a basic idea of how an air conditioner works. Later in the training program, you will explore this subject in greater detail. The relationship between temperature and pressure will also be studied in depth. It is the key

Hermetic and Semi-Hermetic (Servicible) Compressors

Hermetically sealed compressors are typically used in residential and light commercial air conditioners and heat pumps. Semi-hermetic compressors, also known as servicible compressors, are used in large-capacity refrigeration or air conditioning chiller units. Semi-hermetic compressors can be partially disassembled for repair in the field. Hermetic compressors are sealed and cannot be repaired in the field.

HERMETIC RECIPROCATING COMPRESSOR

HERMETIC ROTARY COMPRESSOR

101SA03.EPS

SEMI-HERMETIC (SERVICIBLE) COMPRESSOR

101SA04.EPS

to understanding and troubleshooting mechanical refrigeration systems.

The refrigeration cycle is the same in all refrigeration equipment, from the small air conditioner in your car to the huge system that cools the largest office building. The difference is in the size and construction of the components and piping and the amount and type of refrigerant.

5.0.0 ◆ BLUEPRINTS, CODES, AND SPECIFICATIONS

Commercial construction is a complex process involving the work of many different trades, as well as the participation of people from local government. As an HVAC technician, it is important that you be aware of the HVAC drawings that you will encounter and recognize that building codes will affect the work you do. Just as important are the specifications that answer the many questions that arise on any job site.

5.1.0 Blueprints

For a commercial project, a complete set of blueprints will contain several types of drawings or plans. HVAC drawings are generally part of the mechanical plans. Mechanical plans show what is required to install building systems such as hot water or chilled-water distribution, air distribution, refrigeration, sprinkler, and control systems. For more complex jobs, a separate **HVAC plan** is added to the set of plans.

Information about the HVAC system equipment, ductwork, and wiring can be found on the architectural, electrical, and mechanical plans of the construction drawings.

The power supplied to the various HVAC systems is typically shown on an electrical drawing. HVAC drawings (*Figure 5*) show piping for water supplies and returns, air handling equipment, AC systems, HVAC component diagrams and schematics, and more.

In addition to the detail work of installation, the HVAC drawings are also used for developing **takeoffs**, the process of itemizing and counting all materials and equipment needed for the HVAC installation.

Changes that occur during the course of a project must be marked on the drawings. Offsets in ductwork and piping are often not specified on the drawings. These offsets must be documented during the installation, in order to have an accurate set of drawings that reflects how the system was actually installed. At the end of the project, the marked-up drawings represent the as-built configuration of the building. For that reason, the as-builts are often considered to be the most valuable document for representing the completed project.

5.1.1   HVAC Schedules

Schedules are not drawings. They are tables shown on HVAC drawings and on other drawings in the overall drawing set. The schedule provides details you will need, for example, on the mechanical components and equipment shown on the mechanical/HVAC plans. Portions of a typical mechanical equipment schedule are shown in *Figure 6*.

5.1.2   Shop Drawings

Shop drawings show how a specific portion of the work is to be done, such as the fabrication and installation of duct runs. For large commercial jobs, a drafter creates shop drawings based on a design drafted by an engineer. On smaller jobs, the drafter may work from freehand sketches based on field measurements. Shop drawings, like section and detail drawings, are drawn to a larger scale than the engineer's design drawing.

5.1.3   Piping

When reading and using the project drawings, you will encounter various types of pipe. Knowing which type of piping is used on the systems you are working with is important.

The materials commonly used in piping systems include steel (coated and uncoated), black iron, copper (soft and hard), alloy steel and stainless steel, and thermoplastic such as PVC.

Steel pipe, produced with wall thicknesses that are identified by schedule and weight, is used for gas piping and, occasionally, hot water heating. Steel and stainless steel may occasionally be used for refrigerant piping.

Copper tubing is seamless and is manufactured in ¼" to 12" sizes. It is classified by wall thickness. Copper tubing is generally used in plumbing, heating, and refrigeration applications. It should not be used in ammonia systems.

PVC and CPVC are widely used types of plastic pipe. PVC is a rigid material, resistant to chemicals and corrosion and used for many types of drain, waste, and vent (DWV) systems. CPVC, a slightly different formula than PVC, adds high-temperature performance and improved impact resistance over PVC. PEX (cross-linked polyethylene) tubing is a flexible tubing that can withstand heat and high pressure. It is commonly used to carry liquid in radiant floor heating systems.

The piping material selected should be checked for design, temperature, and pressure ratings and must conform to code requirements.

5.2.0 Building Codes

As noted earlier, people from local government are involved in construction projects, primarily enforcement of building codes. The objective of a building code is to regulate the health and safety aspects of building construction in a community. Building codes regulate new construction, for example, by establishing limits on the height and

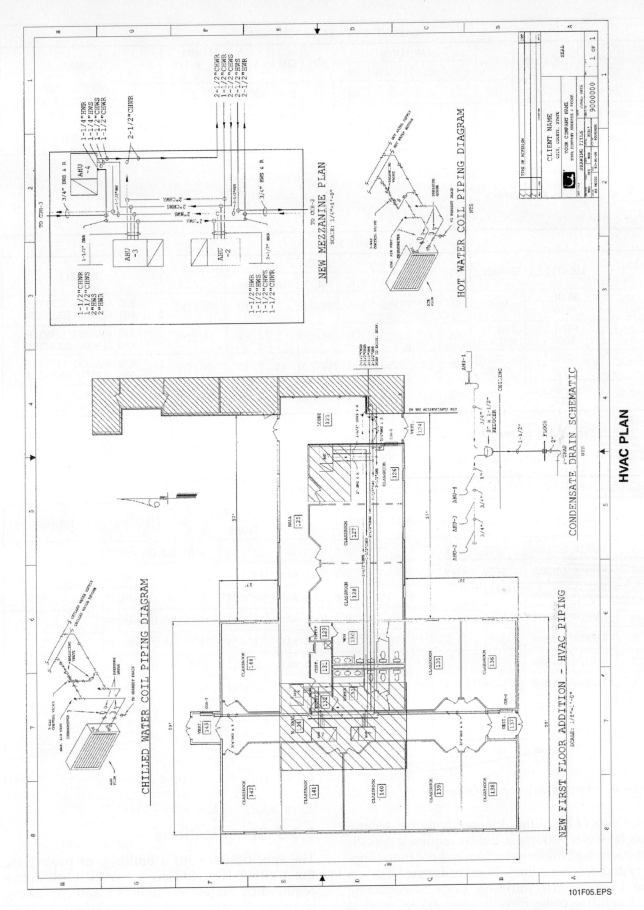

Figure 5 ◆ HVAC drawing.

CABINET UNIT HEATER SCHEDULE

UNIT HEATER NO.	LOCATION	C.F.M.	FAN MOTOR				MBH	GPM	EWT	EAT	MAX. WATER P.D.	REMARKS
			H.P.	VOLTS	PHASE	Hz						
CUH-1	124	400	1/12	115	1	60	23	2.3	180°F	60°F	2.7	McQUAY #CHF004 SEMI-RECESSED, R.H COIL
CUH-2	137	400	1/12	115	1	60	23	2.3	180°F	60°F	2.7	McQUAY #CHF004 SEMI-RECESSED, L.H. COIL
CUH-3	143	400	1/12	115	1	60	23	2.3	180°F	60°F	2.7	McQUAY #CHF004 SEMI-RECESSED, R.H COIL

NOTES:
1. 3 Speed Control
2. Front Discharge
3. With Return Air Filters

PUMP SCHEDULE

UNIT NO.	LOCATION	SERVICE.	GPM	MBH	MOTOR					TYPE	REMARKS
					RPM	H.P.	VOLTS	PHASE	Hz		
P-1	MECH. ROOM	NAVE	45	41'	1750	1½	208/230	3	60	IN-LINE	B/G #60-20T SERVICE 40% GLYCOL SOLUTION
P-2	MECH. ROOM	CHW TO AHU 1-4	65	37'	1750	1½	208/230	3	60	IN-LINE	B/G #60-20T SERVICE 40% GLYCOL SOLUTION
P-3	MECH. ROOM	RECIR. TO TANK	40	17'	1750	½	208/230	3	60	IN-LINE	B/G #60-13T SERVICE 40% GLYCOL SOLUTION
P-4	EXIST. MECH. ROOM	HW	73	31'	1750	1½	208/230	3	60	IN-LINE	B/G #60-20T HOT WATER

NOTES:
1. Starters And Disconnects By E.C.

GRILLE, REGISTER AND DIFFUSER SCHEDULE

ITEM	MANUFACTURER	MODEL NO.	QTY.	LOCATION	CFM EACH	AIR PATTERN	SIZE		FINISHES	REMARKS
							FRAME	NECK		
A	BARBER COLMAN	SFSV	8	126,127,128 144	245	4-WAY	12"× 12"	8"Ø	#7 OFF-WHITE	
B	BARBER COLMAN	SFSV	2	142	275	4-WAY	18"× 18"	10"Ø	#7 OFF-WHITE	
C	BARBER COLMAN	SFSV	4	140, 141	240	4-WAY	12"× 12"	8"Ø	#7 OFF-WHITE	
D	BARBER COLMAN	SFSV	2	139	270	4-WAY	18"× 18"	10"Ø	#7 OFF-WHITE	
E	BARBER COLMAN	SFSV	2	138	280	4-WAY	18"× 18"	10"Ø	#7 OFF-WHITE	
F	BARBER COLMAN	SFSV	2	136	250	4-WAY	12"× 12"	6"Ø	#7 OFF-WHITE	
G	BARBER COLMAN	SFSV	2	135	235	4-WAY	12"× 12"	6"Ø	#7 OFF-WHITE	
H	BARBER COLMAN	SFSV	3	134	100	4-WAY	12"× 12"	6"Ø	#7 OFF-WHITE	FIRE DAMPER SEE DETAIL A
I	BARBER COLMAN	SFSV	1	134	190	4-WAY	12"× 12"	8"Ø	#7 OFF-WHITE	FIRE DAMPER SEE DETAIL A
	MAN	SFSV	1	134	AY		12"×			FIRE DAMPER

Figure 6 ◆ Mechanical equipment schedules.

101F06.EPS

floor area of buildings, by specifying the separation between buildings, and by requiring specific set-backs for buildings and equipment from property and easement lines. Local codes are based on the **International Building Code**. Recognizing that building codes directly affect HVAC work is crucial to HVAC workers.

5.3.0 Specifications

The specifications for a building or project are the written descriptions of work and duties required by the owner, architect, and consulting engineer. Together with the working drawings, these specifications form the basis of the contract requirements for the construction of the building or project. Those who use the construction drawings and specifications must always be alert to discrepancies between the working drawings and the written specifications. These are some situations where discrepancies may occur:

- Architects or engineers use standard or prototype specifications and attempt to apply them without any modification to specific working drawings.
- Previously prepared standard drawings are changed or amended by reference in the specifications only and the drawings themselves are not changed.
- Items are duplicated in both the drawings and specifications, but an item is subsequently amended in one and overlooked in the other contract document.

In such instances, the person in charge of the project has the responsibility to ascertain whether the drawings or the specifications take precedence. Such questions must be resolved, preferably before the work begins, to avoid added costs to the owner, architect/engineer, or contractor.

5.4.0 Commissioning

System commissioning is a process by which a formal and organized approach is taken to obtaining, verifying, and documenting the installation and performance of a particular system or systems. The goal of commissioning is to make sure that a system operates as intended and at optimum efficiency. Commissioning is normally required for newly installed or retrofitted commercial and industrial systems. Because each building and its systems are different, the specific elements of the commissioning process must be tailored to fit each situation. However, the objectives for performing system commissioning are the same:

- Verify that the system design meets the functional requirements of the owner.
- Verify that all systems are properly installed in accordance with the design and specifications.
- Verify that all systems and components meet required local, state, and other required codes.
- Verify and document the proper operation of all equipment, systems, and software.
- Verify that all documentation for the system is accurate and complete.
- Train building operator and maintenance personnel to efficiently operate and maintain the installed equipment and systems.

6.0.0 ◆ CAREERS IN HVAC

Career opportunities in the HVAC trade are many and varied. There is a large existing base of HVAC systems that needs service, repair, and replacement. In addition, every time a new residential, commercial, or industrial building is constructed, it contains one or more HVAC system elements.

To get an idea of how vast the HVAC trade is, picture the town in which you live. Then think about the fact that almost every building in town contains some form of equipment to provide heating and cooling, as well as air circulation and purification. Expand that view to include the entire country and you realize that there are tens of millions of heating, air conditioning, and air-management systems. New ones are being added every day; old ones are wearing out and being repaired or replaced. From that perspective, the opportunities in the trade appear limitless.

Figure 7 provides an overview of career opportunities in the HVAC trade. For the purposes of this discussion, it is convenient to view the HVAC industry as having three segments:

- *Community-based* – Companies that sell, install, and service residential and light commercial equipment and systems such as furnaces and packaged air conditioners.
- *Commercial/industrial* – Companies that install and maintain systems for large office buildings, factories, apartment complexes, shopping malls, and so forth.
- *Manufacturing* – Companies that build and market HVAC systems and equipment.

6.1.0 Community-Based

At the community level, you might find anything from a one-person installation and service business to a firm with a hundred or more employees, including heating and air conditioning specialists, installers, sheet metal workers, and sales engineers. In such businesses, HVAC specialists may work alone or with a single partner. They typically respond to service calls from homes or small businesses. They may also install furnaces and air conditioning equipment sold by their firm's sales engineer. In other firms, one group may do installations while another group handles troubleshooting and maintenance. At this level, a technician is expected to work with a wide variety of products from many different manufacturers. Systems can consist of anything from a window air conditioner to a heating/air-conditioning system containing two or three major components.

Local or regional distributors provide equipment, parts, special tools, and other services for

COMMUNITY-BASED

• Installer 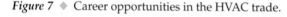 • Service Supervisor
• Service Technician • Sales Engineer
• Distributor Sales • Owner
• Inspector
• Code Enforcement Officer
• Estimator

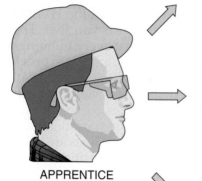

COMMERCIAL/INDUSTRIAL

• Installer • Foreman
• Air System Technician • Technical Specialist
• Service Technician • Training Specialist
• Operating Engineer
• Project Manager
• Estimator

APPRENTICE

MANUFACTURING

• Test Technician • Supervisor
• Engineering Assistant • Technical Specialist
• Assembler • Trainer
• Field Service Engineer

101F07.EPS

Figure 7 ◆ Career opportunities in the HVAC trade.

Modern Air Conditioning

Dr. Willis Carrier, founder of Carrier Corporation, is credited with the invention of modern air conditioning. In 1902, he developed a system that could control both humidity and temperature using a non-toxic, non-flammable refrigerant. Air conditioning started out as a means of solving a problem in a printing facility where heat and humidity were causing paper shrinkage. Later systems served similar purposes in textile plants. The concept wasn't applied to comfort air conditioning until about 20 years later when Carrier's centrifugal chillers began to be installed in department stores and movie theaters.

High-Efficiency Air Conditioners and Heat Pumps

Like furnaces, high-efficiency heat pumps and air conditioners are also available for both commercial and residential installations. A number of manufacturers have voluntarily partnered with the Environmental Protection Agency (EPA) in an ENERGY STAR® program to market units that exceed minimum efficiency standards. Building designs may also qualify for this program through the use of high-efficiency HVAC systems combined with thermal storage systems. For example, to reduce the energy costs of air conditioning a building, ice is generated by refrigeration systems at night using low-cost power. The ice is stored in insulated tanks and is used to aid the building's air-conditioning system during the daytime. This reduces energy consumption during peak usage hours and makes power generation by the utility company more efficient.

INSIDE TRACK

Large Commercial Chiller Unit

As the name implies, chillers use chilled water as a cooling medium. The chiller acts as the evaporator. Chillers are often combined with a cooling tower that acts as the condensing unit. Water flowing over the cooling tower absorbs the heat extracted from the indoor air. This photo shows a large commercial chiller unit.

101SA05.EPS

the firms that sell, install, and service HVAC equipment. The distributor needs salespeople who know HVAC equipment. The distributor may also provide engineering support and service training for its dealers. Distributorships are often affiliated with a single manufacturer.

NOTE

Many community-based firms are subsidiaries of nationwide firms known as consolidators. A consolidator is an umbrella organization that provides centralized management, purchasing, training, and other functions that give small, local companies the power of a large, national company.

6.2.0 Commercial/Industrial

Large commercial and industrial systems have many components and may require thousands of feet of ductwork and piping. Such systems are designed by engineers and architects. Many HVAC trade people are required for these projects and an individual is more likely to specialize. For example, where residential systems are usually controlled by a single wall thermostat, a large commercial system will often have central computer controls that are installed and serviced by a system control specialist. Others may specialize in working with large steam boilers; still others may choose to become experts in installing and servicing high-volume cooling units known as **chillers**.

Companies that install such equipment are often large construction firms that may work anywhere in the world. They are likely to do only the installation and let the building owner contract with another firm for maintenance. Many large facilities employ their own HVAC maintenance people.

6.3.0 Manufacturing

There are thousands of HVAC manufacturers. Some of the larger ones cover the entire HVAC spectrum, while others focus on a particular product or market. For example, one may make window air conditioners, another gas furnaces, and yet another might make only heavy commercial equipment. Regardless of their market, they employ HVAC specialists such as:

- Test technicians
- Engineering assistants
- Training specialists
- Instruction book writers
- Field service technicians

Working as an HVAC journeyman is satisfying and financially rewarding. If you want to go beyond the journeyman level, there are many opportunities in supervision and executive management with dealerships, construction companies, and manufacturers. Companies that install and service HVAC equipment are usually founded by someone who started as a service technician. For those who want to teach, there are opportunities to work in vocational schools and training programs such as the one in which you are now participating.

7.0.0 ◆ TYPES OF TRAINING PROGRAMS

There are two basic forms of training programs that most employers consider. The primary one is on-the-job training (OJT) to improve the competence of their employees in order to provide better customer service and for the continuity and growth of the company. The second is formal apprenticeship training, which provides the same type of training, but also conforms to federal and state requirements under the *Code of Federal Regulations (CFR), Titles 29:29* and *29:30*.

7.1.0 Standardized Training by the NCCER

The National Center for Construction Education and Research (NCCER) is a not-for-profit education foundation established by the nation's leading construction companies. NCCER was created to provide the industry with standardized construction education materials, the Contren® Learning Series (the HVAC modules are part of this series), and a system for tracking and recognizing students' training accomplishments—NCCER's National Registry.

NCCER also offers accreditation, instructor certification, and skills assessments. NCCER is committed to developing and maintaining a training process that is internationally recognized, standardized, portable, and competency-based.

Working in partnership with industry and academia, NCCER has developed a system for program accreditation that is similar to those found in institutions of higher learning. NCCER's accreditation process ensures that students receive quality training based on uniform standards and criteria. These standards are outlined in NCCER's *Accreditation Guidelines* and must be adhered to by NCCER Accredited Training Sponsors.

More than 550 training and assessment centers across the U.S. and eight other countries are proud to be NCCER Accredited Training Sponsors. Millions of craft professionals and construction managers have received quality construction education through NCCER's network of Accredited Training Sponsors and the thousands of Training Units associated with the Sponsors. Every year the number of NCCER Accredited Training Sponsors increases significantly.

A craft instructor is a journeyman craft professional or career and technical educator trained and certified to teach NCCER's Contren® Learning Series. This network of certified instructors ensures that NCCER training programs will meet the standards of instruction set by the industry. There are more than 3,350 master trainers and 33,000 craft instructors within the NCCER instructor network. More information is available at **www.nccer.org**.

The basic idea of the NCCER is to replace governmental control and credentialing of the construction workforce with industry-driven training and education programs. The NCCER departs from traditional classroom or distance learning by offering a competency-based training regimen. Competency-based training means that instead of simply attaining required hours of classroom training and set hours of OJT, you have to prove that you know what is required and successfully demonstrate specific skills. All completion information on every trainee is sent to the NCCER and kept within the National Registry. The NCCER can confirm training and skills for workers as they move from company to company, state to state, or to different offices in the same company. These are portable credentials and are recognized nationally.

In an effort to provide industry credentials and ensure national portability of skills, more than three million module completions have been

Technician Certification

A knowledgeable HVAC installation technician or service technician can obtain nationally recognized certification by taking the Air Conditioning Excellence (ACE) exams. North American Technical Excellence, Inc. (NATE) administers the exams. Although other certification programs are available, the NATE certification has become the most widely accepted in the industry.

NATE tests are divided into tests for service technicians and tests for installation technicians. NATE certification is for either the service path or the installation path, with each specialty having a core test plus a test in that specialty. Specialty tests are specific to the path you choose.

A technician who wants to obtain certification can take the core exam and any one of the specialty exams to be certified for that specialty. A technician may want to be certified in several specialties. For example, a technician living in the Middle Atlantic states may want to obtain certification in all five specialties, whereas a person living in the desert Southwest might not have a need to be certified for gas or oil heating.

EPA Refrigerant Transition and Recovery Certification

Various organizations conduct tests and issue EPA-approved certifications for the handling and recovery of refrigerants. For non-mobile or non-vehicular equipment, these certifications are for Small Appliance (Type I), High-Pressure Appliance (Type II), or Low-Pressure Appliance (Type III) equipment or a Universal certificate covering all three types.

101SA06.EPS

delivered to students and craft professionals nationwide. The National Registry provides transcripts, certificates, and wallet cards for students of the Contren® Learning Series when training is delivered through an NCCER Accredited Training Sponsor. These valuable industry credentials benefit students as they seek employment and build their careers.

7.1.1 Apprenticeship Training

As stated earlier, formal apprenticeship programs conform to federal and state requirements under *CFR Titles 29:29* and *29:30*. All approved apprenticeship programs provide OJT as well as classroom instruction. The related training requirement is fulfilled by all NCCER craft training programs. The main difference between NCCER training and registered apprenticeship programs is that apprenticeship has specific time limits in which the training must be completed. Apprenticeship standards set guidelines for recruiting and outreach, and a specific time limit for each of a variety of OJT tasks. Additionally, there are reporting requirements and audits to ensure adherence to the apprenticeship standards. Companies and employer associations

register their individual apprenticeship programs with the Office of Apprenticeship within the U.S. Department of Labor, and in some instances, with state apprenticeship councils (SAC). OJT of 2,000 hours per year and a minimum of 144 hours of classroom-related training are required. Apprenticeship programs vary in length from 2,000 hours to 10,000 hours.

7.1.2 Youth Training and Apprenticeship Programs

Youth apprenticeship programs are available that allow students to begin their apprenticeship or craft training while still in high school. A student entering the program in the 11th grade may complete as much as one year of the NCCER training program by high school graduation. In addition, programs (in cooperation with local construction industry employers) allow students to work in the craft and earn money while still in school. Upon graduation, students can enter the industry at a higher level and with more pay than someone just starting in a training program.

Students participating in the NCCER or youth apprenticeship training are recognized through official transcripts and can enter the second level or year of the program wherever it is offered. They may also have the option of applying credits at two-year or four-year colleges that offer degree or certificate programs in their selected field of study.

8.0.0 ◆ THE HVAC TECHNICIAN AND THE ENVIRONMENT

Scientific studies have shown that the chlorine in **chlorofluorocarbon (CFC)** and **hydrochlorofluorocarbon (HCFC)** refrigerants can damage the ozone layer that protects the Earth and its inhabitants from the sun's ultraviolet rays. One of the effects is an increased rate of skin cancer. Many refrigerants used in air conditioning equipment are CFCs and HCFCs.

In addition to the toxic effect of chlorine, there is evidence that the release of refrigerant to the atmosphere contributes to global warming, a condition known as the greenhouse effect. The heat trapped in the atmosphere leads to a gradual increase in the Earth's temperature. Refrigerants, especially CFCs, are viewed as major contributors to global warming.

In the 1980s the nations of the world made a commitment to phase out these chemicals. They agreed to take steps in the interim to prevent the discharge of refrigerants into the atmosphere. In 1990, the U.S. Congress passed the *Clean Air Act,* which calls for early phaseout of the most toxic refrigerants, eventual elimination of all CFCs and HCFCs, and strict control and labeling of refrigerants. As a result of the requirement to phase out CFCs and HCFCs, a number of new, environmentally friendly refrigerants have been developed. The U.S. Environmental Protection Agency (EPA) is responsible for implementing and enforcing this law, which has a significant impact on the HVAC trade. For example, it has imposed the following restrictions on refrigerants, regardless of the chlorine content:

- Anyone releasing these refrigerants to the atmosphere is subject to a stiff fine and possibly a prison term.
- Anyone handling these refrigerants must have EPA-sanctioned certification. Without it, you cannot even buy refrigerants. The training needed to obtain EPA certification is included in this curriculum.
- Records must be kept on all transactions involving these refrigerants. These include purchase, use, reprocessing, and disposal.

Before a sealed refrigeration system can be opened for repair, the refrigerant it contains must be identified. With few exceptions, all refrigerants must be **recovered** and stored in approved containers. When the repair is complete and the system is resealed, the same refrigerant may be returned to the system. It may also be used in another system belonging to the same owner. Recovered refrigerant should, however, be recycled before reuse. This removes moisture and impurities that could damage the system. Some refrigerant recovery units have a built-in **recycling** capability.

If the refrigerant is badly contaminated or no longer needed, it can be **reclaimed**. This is done at remanufacturing centers where the refrigerant is returned to the standards of purity that govern new refrigerants. Reclaimed refrigerant can be resold on the open market, but cannot be classified as "new" refrigerant.

In addition to federal regulations, there may be state regulations that apply to refrigerants. State regulations may be stricter. It is critical that everyone in the HVAC industry understands and follows the EPA regulations regarding the handling, storage, and labeling of refrigerants. Failure to do so could be very costly to both you and your employer.

Summary

The HVAC trade involves equipment used for heating, cooling, and purifying indoor air. It covers everything from the window air conditioners and furnaces used in our homes to the giant heating and cooling systems used in large office buildings and industrial complexes.

A construction project is a complex process involving many people who represent the various construction trades and local government. Without a clear set of blueprints and specifications, assigning and organizing the work efficiently would be impossible. Blueprints and specifications spell out every detail of the construction job, from preparing the site to putting on the finishing touches.

As an HVAC trainee, you must learn the basics of blueprint reading so that you can interpret the drawings that show how to accurately install an HVAC system and its accessories. However, you must also be able to read the prints used by other trades because you and your co-workers will have to coordinate your work with them.

Because of the widespread use of HVAC equipment, there are many career opportunities in the trade. Jobs are available with small local firms, large industrial and commercial contractors, and the manufacturing firms that build and market HVAC equipment. The apprentice program provides an opportunity to learn the trade through a combination of hands-on training and related classroom learning. The Youth Apprentice Program allows students to begin their training while still in high school.

One of the most serious issues affecting the trade is the damage that can be done to the Earth's ozone layer by the improper release of refrigerants into the atmosphere. There are severe penalties for improper refrigerant disposal.

Notes

1. In a common household forced-air furnace, heat is transferred from the _____ .
   a. heat exchangers to the air
   b. natural gas or oil to the conditioned space
   c. air to the heat exchangers
   d. refrigerant to the outside air

2. Water is often used in heating systems as a _____ .
   a. fuel
   b. refrigerant
   c. heat exchange medium
   d. heat exchanger

3. Ventilation is concerned with _____ .
   a. air circulation
   b. air temperature
   c. the circulation and cleanliness of indoor air
   d. the circulation and cleanliness of indoor air and air discharged to the outdoors

4. The term refrigeration cycle refers to _____.
   a. the process by which circulating refrigerant absorbs heat in one location and moves it to another location
   b. the process by which refrigerant moves through a compressor
   c. mobile refrigeration units
   d. the process by which refrigerant is recycled to remove impurities

5. Heat is transferred from the indoor air to the refrigerant at the _____ .
   a. compressor
   b. furnace
   c. evaporator
   d. condenser

6. The expansion device _____ .
   a. raises the boiling point of the refrigerant entering the evaporator
   b. lowers the boiling point of the refrigerant entering the evaporator
   c. gets bigger as the temperature increases
   d. lowers the pressure at the evaporator outlet

7. The layout of the HVAC system ductwork is shown in the _____ plans.
   a. site
   b. structural
   c. civil
   d. mechanical

8. The primary purpose of a building code is to regulate the quality of all mechanical installation on a commercial site.
   a. True
   b. False

9. In order to be resold on the open market, a refrigerant must have been _____ .
   a. reclaimed
   b. recycled
   c. recovered
   d. refurbished

10. Releasing CFC and HCFC refrigerants or certain substitutes to the atmosphere is _____ .
    a. okay, provided that the refrigerant has been recycled
    b. okay, provided that the refrigerant has been reclaimed
    c. prohibited by federal law
    d. prohibited in some states

Trade Terms Quiz

1. Anything _____ is considered poisonous.

2. _____ is the remanufacturing of used refrigerant to bring it up to the standards required of new refrigerant.

3. A class of refrigerants that contains hydrogen, chlorine, fluorine, and carbon is called _____.

4. A(n) _____ is a heat exchanger that transfers heat from the air flowing over it to the cooler refrigerant flowing through it.

5. A class of refrigerants that contains chlorine, fluorine, and carbon is called _____.

6. The removal and temporary storage of refrigerant in containers approved for that purpose is called _____.

7. A liquid metering device is also known as a(n) _____.

8. A(n) _____ is a heat exchanger that transfers heat from the refrigerant flowing inside it to the air or water flowing over it.

9. The process by which a circulating refrigerant absorbs heat from one location and transfers it to another location is called a(n) _____.

10. _____ is the transfer of heat from a warmer substance to a cooler substance.

11. _____ is circulating recovered refrigerant through filtering devices that remove moisture, acid, and other contaminants.

12. The use of machinery to provide cooling is called _____.

13. A substance harmful to your health is said to be _____.

14. A(n) _____ is added to the mechanical plans for complex jobs that require separate heating, ventilating, and air conditioning systems.

15. A high-volume cooling unit is called a(n) _____.

16. A(n) _____ is the process of itemizing and counting all material and equipment needed for the HVAC installation.

17. The _____ is a series of model construction codes that set standards that apply across the country.

18. The mechanical device that converts low-pressure, low-temperature refrigerant gas into high-temperature, high-pressure refrigerant gas in a refrigeration system is a(n) _____.

Trade Terms

Chiller	Expansion device	International Building	Recycling
Chlorofluorocarbon (CFC)	Heat transfer	Code	Refrigeration cycle
refrigerant	HVAC plan	Mechanical refrigeration	Takeoffs
Compressor	Hydrochlorofluorocarbon	Noxious	Toxic
Condenser	(HCFC) refrigerant	Reclamation	
Evaporator		Recovery	

John Tianen

HVAC Training Consultant
Tucson, AZ

John Tianen's career is a good example of how a commitment to learning can help an individual succeed. John started his HVAC career working on the assembly line for an air conditioning manufacturer. By applying himself to learning HVAC technology, he became an installation and service technician, then an instructor, and finally a designer and developer of HVAC training. Along the way, he earned bachelor's and master's degrees by taking college classes on his own time.

Following high school, John spent four years in the U.S. Air Force as an electronic systems technician. After completing military service, he worked on a production line in a factory. After a year he was hired by a major manufacturer of air conditioning equipment as a maintenance technician on the production line. His duties included maintaining the equipment used to evacuate (empty) and charge (fill) room air conditioners as they moved down the assembly line. John studied HVAC technology and soon had a job in the laboratory where central air conditioners were tested. This experience led him to establish a business installing and servicing residential air conditioners and furnaces. This additional experience enabled him to obtain a position as a service engineer in which he had worldwide service support responsibility for all split-system air conditioners and heat pumps. A subsequent promotion moved John into the technical training department, where he produced the installation and service training programs for all residential heating and cooling products. During that period, he took advantage of his company's generous tuition refund program and obtained college degrees, including a master of science in education. Now semi-retired, John has formed his own consulting company, which provides training support to the HVAC industry.

John's advice to someone just entering the trade is to make the most of apprentice training. It gives the foundation you need for a successful career. Also, take advantage of any other learning opportunities that come your way. The industry is constantly changing, and the only way to keep up with it is to make a lifelong commitment to learning. Although this sometimes requires taking classes, it can also be done by subscribing to trade journals and by participating in local, regional, and national trade organizations.

Trade Terms
Introduced in This Module

Chiller: A high-volume cooling unit. The chiller acts as an evaporator.

Chlorofluorocarbon (CFC) refrigerant: A class of refrigerants that contains chlorine, fluorine, and carbon. CFC refrigerants have a very adverse effect on the environment.

Compressor: In a refrigeration system, the mechanical device that converts low-pressure, low-temperature refrigerant gas into high-temperature, high-pressure refrigerant gas.

Condenser: A heat exchanger that transfers heat from the refrigerant flowing inside it to the air or water flowing over it.

Evaporator: A heat exchanger that transfers heat from the air flowing over it to the cooler refrigerant flowing through it.

Expansion device: Also known as the liquid metering device or metering device. Provides a pressure drop that converts the high-temperature, high-pressure liquid refrigerant from the condenser into the low-temperature, low-pressure liquid refrigerant entering the evaporator.

Heat transfer: The transfer of heat from a warmer substance to a cooler substance.

HVAC plan: Added to the mechanical plans for complex jobs that require separate heating, ventilating, and air-conditioning systems.

Hydrochlorofluorocarbon (HCFC) refrigerant: A class of refrigerants that contains hydrogen, chlorine, fluorine, and carbon. Although not as high in chlorine as chlorofluorocarbon (CFC) refrigerants, HCFCs are still considered hazardous to the environment.

International Building Code: A series of model construction codes. These codes set standards that apply across the country. This is an ongoing process led by the International Code Council (ICC).

Mechanical refrigeration: The use of machinery to provide cooling.

Noxious: Harmful to health.

Reclamation: The remanufacturing of used refrigerant to bring it up to the standards required of new refrigerant.

Recovery: The removal and temporary storage of refrigerant in containers approved for that purpose.

Recycling: Circulating recovered refrigerant through filtering devices that remove moisture, acid, and other contaminants.

Refrigeration cycle: The process by which a circulating refrigerant absorbs heat from one location and transfers it to another location.

Takeoffs: The process of itemizing and counting all the material and equipment needed for an installation.

Toxic: Poisonous.

Additional Resources

This module is intended to present thorough resources for task training. The following reference work is suggested for further study. This is optional material for continued education rather than for task training.

Career Opportunities in Heating, Air Conditioning, and Refrigeration, Latest Edition. Fairfax, VA: Air Conditioning and Refrigeration Institute (ARI).

CONTREN® LEARNING SERIES – USER UPDATE

NCCER makes every effort to keep these textbooks up-to-date and free of technical errors. We appreciate your help in this process. If you have an idea for improving this textbook, or if you find an error, a typographical mistake, or an inaccuracy in NCCER's Contren® textbooks, please write us, using this form or a photocopy. Be sure to include the exact module number, page number, a detailed description, and the correction, if applicable. Your input will be brought to the attention of the Technical Review Committee. Thank you for your assistance.

Instructors – If you found that additional materials were necessary in order to teach this module effectively, please let us know so that we may include them in the Equipment/Materials list in the Annotated Instructor's Guide.

Write: Product Development and Revision
National Center for Construction Education and Research
3600 NW 43rd St., Bldg. G, Gainesville, FL 32606

Fax: 352-334-0932

E-mail: curriculum@nccer.org

Craft _____ Module Name _____

Copyright Date _____ Module Number _____ Page Number(s) _____

Description _____

_____

_____

_____

(Optional) Correction _____

_____

_____

(Optional) Your Name and Address _____

_____

_____

Trade Mathematics
03102-07

03102-07
Trade Mathematics

Topics to be presented in this module include:

Overview

HVAC technicians use math for many purposes. Math is an essential skill required to advance in the HVAC profession and is necessary to do many tasks in your everyday work. Math is needed when cutting and fitting pipe, when sizing and installing ductwork, and when calculating electrical current flow.

In HVAC work, you must understand the metric system and various conversion units. Knowing powers of ten, square roots, and how to work basic problems in algebra and geometry is extremely useful when calculating things such as air flow and sensible load.

The HVAC technicians who know math and can solve basic math problems will have a big advantage over those workers who cannot use math to do their job. Learning trade math principles is one of the more important objectives in your training program.

Objectives

When you have completed this module, you will be able to do the following:

1. Identify similar units of measurement in both the inch-pound (English) and metric systems and state which units are larger.
2. Convert measured values in the inch-pound system to equivalent metric values and vice versa.
3. Express numbers as powers of ten.
4. Determine the powers and roots of numbers.
5. Solve basic algebraic equations.
6. Identify various geometric figures.
7. Use the Pythagorean theorem to make calculations involving right triangles.
8. Convert decimal feet to feet and inches and vice versa.
9. Calculate perimeter, area, and volume.
10. Convert temperature values between Celsius and Fahrenheit.

Trade Terms

Absolute pressure
Acceleration
Area
Atmospheric pressure
Barometric pressure
Coefficient
Constant
Exponent
Force
Length
Liter
Mass
Newton (N)
Unit
Vacuum
Variable
Volume

Required Trainee Materials

1. Paper and pencil
2. Appropriate personal protective equipment

Prerequisites

Before you begin this module it is recommended that you successfully complete *Core Curriculum;* and *HVAC Level One,* Module 03101-07.

This course map shows all of the modules in *HVAC Level One.* The suggested training order begins at the bottom and proceeds up. Skill levels increase as you advance on the course map. The local Training Program Sponsor may adjust the training order.

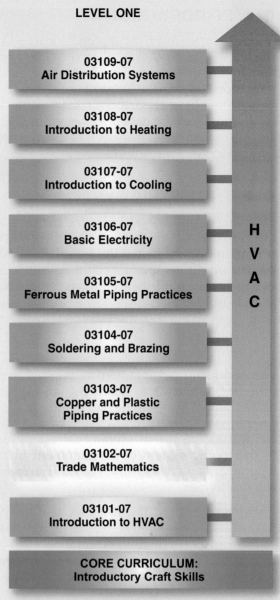

LEVEL ONE

03109-07
Air Distribution Systems

03108-07
Introduction to Heating

03107-07
Introduction to Cooling

03106-07
Basic Electricity

03105-07
Ferrous Metal Piping Practices

03104-07
Soldering and Brazing

03103-07
Copper and Plastic
Piping Practices

03102-07
Trade Mathematics

03101-07
Introduction to HVAC

HVAC

CORE CURRICULUM:
Introductory Craft Skills

102CMAP.EPS

1.0.0 ◆ INTRODUCTION

This module expands on the materials learned in *Introduction to Construction Math* in the Core Curriculum. In that module, you studied whole numbers, fractions, decimals, percentages, and the metric system. If necessary, you may want to review all or part of the material covered in *Introduction to Construction Math* before proceeding with the material covered here.

2.0.0 ◆ THE METRIC SYSTEM

Over 95 percent of the world uses the metric system of measure. Given the fact that the United States is an important part of the global economy, companies would be operating at a definite disadvantage if they (and their employees) did not know and use the metric system. Also, the government has established the *Omnibus Trade and Competitiveness Act,* which provides us (in part) with the following national policy:

- To designate the metric system of measurement as the preferred system of weights and measures for United States trade and commerce.
- To require that each federal agency, by a certain date and to the extent economically feasible, use the metric system of measurement in its procurements, grants, and other business-related activities.

If you have not used the metric system on the job, you most certainly will use it in the near future, because it is widely used on tools such as rulers and in HVAC manufacturers' equipment product literature.

2.1.0 Fundamental Units

Most work in science and engineering is based on the exact measurement of physical quantities. A measurement is simply a comparison of a quantity to some definite standard measure of dimension called a **unit.** Whenever a physical quantity is described, the units of the standard to which the quantity was compared must be specified. A number alone is insufficient to describe a physical quantity.

The importance of specifying the units of measurement for a number used to describe a physical quantity is clearly seen when you note that the same physical quantity may be measured using a variety of different units. For example, **length** may be measured in inches, feet, yards, miles, centimeters, meters, kilometers, etc.

All physical quantities can ultimately be expressed in terms of three fundamental units:

- *Length* – The distance from one point to another
- **Mass** – The quantity of matter
- *Time* – The period during which an event occurs

The three most widely used systems of measurement are:

- Meter-kilogram-second (MKS) system
- Centimeter-gram-second (CGS) system
- English or inch-pound (I-P) system

Table 1 shows the fundamental units of length, mass, and time in each of these three systems. The MKS and CGS units are both part of the metric system of measure. The inch-pound system is probably most familiar to you. Notice that time is measured in the same units (seconds) in all systems.

The existence of different sets of fundamental units contributes to a considerable amount of confusion in many calculations. Today, both the inch-pound and metric systems are widely employed in engineering and construction calculations. Therefore, it is necessary to have some degree of understanding of both systems of units.

The metric system is actually much simpler to use than the inch-pound system because it is a decimal system in which prefixes are used to denote powers of ten. The older inch-pound system requires the use of conversion factors that must be memorized and are not categorized as logically as powers of ten. For example, one mile is 5,280 feet, and 1 inch is 1/12 of a foot. *Table 2* lists some of the more common units in the inch-pound system.

The metric system prefixes are listed in *Table 3.* From this table, you can see that the use of the metric system is logically arranged, and that the name of the unit will also represent an order of magnitude (via the prefix) that foot and pound cannot.

Transferring U.S. engineering practices and equipment to the metric system was a very expensive transition for most industries. Manufacturers are currently publishing their technical manuals and instrument data sheets displaying all values, as in the past, in the inch-pound system units, but also putting the metric equivalents in parentheses behind the inch-pound units. In the future, you may find only the metric units listed, so it is a good time to become familiar with both and understand how to convert from one to the other.

Table 1 Fundamental Units

Unit	MKS	CGS	Inch-Pound
Length	Meter (m)	Centimeter (cm)	Foot (ft)
Mass	Kilogram (kg)	Gram (g)	Pound (lb)
Time	Second (sec)	Second (sec)	Second (sec)

102T01.EPS

Table 2 Common Units in the Inch-Pound System

Unit	Equivalent
12 inches (in)	1 foot (ft)
1 yard (yd)	3 ft
1 mile (mi)	5,280 ft
16 ounces (oz)	1 pound (lb)
1 ton	2,000 lbs
1 minute (min)	60 seconds (sec)
1 hour (hr)	3,600 sec
1 U.S. gallon (gal)	0.1337 cubic foot (cu ft)

102T02.EPS

Table 3 Metric System Prefixes

Prefix	Unit		
micro- (μ)	1/1,000,000	0.000001	10^{-6}
milli- (m)	1/1,000	0.001	10^{-3}
centi- (c)	1/100	0.01	10^{-2}
deci- (d)	1/10	0.1	10^{-1}
deka- (da)	10	10.	10^{1}
hecto- (h)	100	100.	10^{2}
kilo- (k)	1,000	1,000.	10^{3}
mega- (M)	1,000,000	1,000,000.	10^{6}
giga- (G)	1,000,000,000	1,000,000,000.	10^{9}

102T03.EPS

The most common metric system prefixes are mega- (M), kilo- (k), centi- (c), milli- (m), and micro- (μ). Even though these prefixes may seem difficult to understand at first, you are probably already using them regularly. For example, you have probably seen the terms megawatts, kilometers, centimeters, millivolts, and microamps.

There are four basic parameters used to describe different quantities: weight, length, **volume**, and temperature. *Figure 1* references the familiar inch-pound unit for each of these parameters and the metric unit that is becoming more and more common.

The most common measurements can be classified into four categories:

- Dimensional measurements (e.g., lengths, levels, **areas**, and volumes)
- Measurements of mass and weight
- Pressure measurements
- Temperature measurements

2.2.0 Length, Area, and Volume

In *Introduction to Construction Math,* you were concerned primarily with the development of mathematical skills for solving problems. In the field, numbers usually represent physical quantities. In order to give meaning to these quantities, measurement units are assigned to the numbers.

2.2.1 The Meter

The statement "the length of a room is 17" is completely meaningless unless we indicate the units into which the room is divided. With regard to length, a unit is simply a standard distance.

Originally, the meter was defined as 1/10,000,000 of the Earth's meridional quadrant (the distance from the North Pole to the equator). This was how the meter got its name. This distance was etched onto a metal bar that is kept in France. In 1866, the United States legalized the use of the metric system and placed an exact copy of the metal bar into the U.S. Bureau of Standards.

In October 1983, the world's scientists redefined the meter to avoid the minor potential of error associated with the metal bar in France. Now a meter is defined as being equal to the distance that light travels in 1/299,792,458 of a second. For our purposes, one meter is equal to 39.37 inches.

2.2.2 Length (Level)

Length typically refers to the long side of an object or surface. With liquids, the measurement of the level of fluid in a tank is basically a measurement of length from the surface of the fluid to the bottom of the tank. Length can be expressed in either inch-pounds or metric units (i.e., inches or centimeters).

When working in construction, instructions or plans can be in either system of measurement. It is important that you know the relationships so you can convert from one system to the other. *Table 4* shows the relationships of the most common units of length. *Figure 2* shows the comparison of inches directly to centimeters.

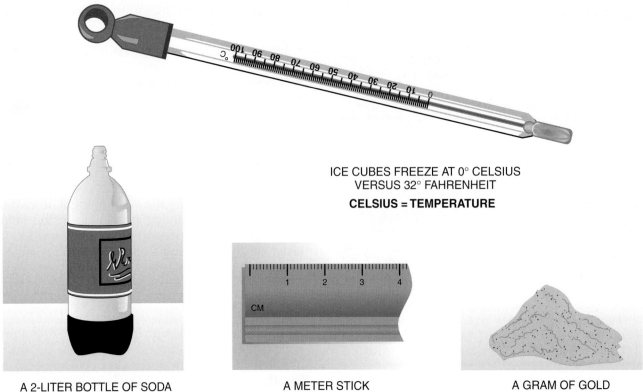

ICE CUBES FREEZE AT 0° CELSIUS
VERSUS 32° FAHRENHEIT

CELSIUS = TEMPERATURE

A 2-LITER BOTTLE OF SODA
INSTEAD OF A HALF-GALLON

LITERS = VOLUME

A METER STICK
INSTEAD OF A YARDSTICK

METERS = LENGTH

A GRAM OF GOLD
INSTEAD OF AN OUNCE

GRAMS = WEIGHT

102F01.EPS

Figure 1 ◆ Basic measurement units.

Table 4 Length Conversion Multipliers

Unit	cm	in	ft	m	km
1 millimeter	0.1	0.03937	0.003281	0.001	0.0000001
1 centimeter	1	0.3937	0.03281	0.01	0.00001
1 inch	2.54	1	0.08333	0.0254	0.0000254
1 foot	30.48	12	1	0.3048	0.0003048
1 meter	100	39.37	3.281	1	1,000
1 kilometer	10,000	39,370	3,281	1,000	1

102T04.EPS

INCHES

CENTIMETERS

102F02.EPS

Figure 2 ◆ Comparison of inches to centimeters.

The Meter

In the English system, the standard of distance is called the foot. It was originally specified as the measure of the distance from King Henry's toes to his heel. At the time, it seemed reasonable to use this as a reference as there was only one king. As time went by and King Henry died, the folly of this reference became obvious. However, the unit had been put into use and it was difficult to get people to change, just as it is today.

When seeking to define a new standard for length in developing the metric system, scientists wanted to select a reference that would be more precise and less likely to ever change. The meter became this standard.

Length

THINK ABOUT IT

If the manufacturer's product data sheet states that a fan coil unit is 1.22 meters high and 0.46 meter wide, what size opening (in feet) is needed to get this unit inside a building?

You may be called upon to convert a measurement in one system into the other system's units. The multipliers in *Table 4* may be used to make these conversions.

Example:

An installation plan requires a thermostat to be mounted 122 centimeters (cm) above the floor, but your measuring tape is calibrated in inches. How many inches above the floor should the thermostat be placed?

$$\begin{aligned}
1 \text{ cm} &= 0.3937" \\
122 \text{ cm} &= 122 \times 0.3937" \\
122 \text{ cm} &= 48.0314" \\
&= 48" \text{ (rounded off)}
\end{aligned}$$

or (changing to feet)

$$\begin{aligned}
122 \text{ cm} &= 122 \times 0.03281' \\
122 \text{ cm} &= 4.00282' \\
&= 4' \text{ (rounded off)}
\end{aligned}$$

NOTE

The *Appendix* contains listings of common conversion factors used in construction. Also, some scientific calculators have the capability of converting English units to metric units and vice versa.

2.2.3 Area

Area is the measurement of the surface of a two-dimensional object. The area of any rectangle is equal to the length multiplied by the width. In the familiar inch-pound system, the units of area are square feet (ft^2) or square inches (in^2). *Figure 3* shows the application of this concept.

The area of a circle is found using the following formula:

$$\text{Area} = \pi r^2$$

Where:

A = the area of the circle

π = a **constant** of 3.14159 (often abbreviated 3.14)

r = the radius (distance from the center to the edge of the circle)

Figure 4 shows the application of this concept.

When converting areas from one measurement system to the other, every dimension must be converted. So, to convert the dimensions of the square shown previously in *Figure 3(A)*, we must convert both the length and the width, as shown in *Figure 5*.

To convert the dimensions of the rectangle shown in *Figure 3(B)*, we must convert both the length and the width, as shown in *Figure 6*.

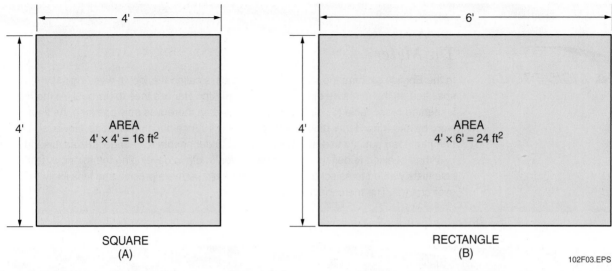

Figure 3 ◆ Measuring the area of a square and a rectangle.

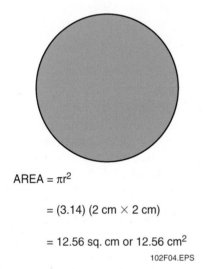

AREA = πr²

= (3.14) (2 cm × 2 cm)

= 12.56 sq. cm or 12.56 cm²

102F04.EPS

Figure 4 ◆ Measuring the area of a circle.

When converting the dimensions of a circle, only the radius has a measured dimension that must be converted. In the example shown in *Figure 7*, centimeters must be converted into inches.

2.2.4 Volume

Volume is the amount of space occupied by a three-dimensional object. The volume of a cube or rectangular prism, such as HVAC ductwork, is the product of three lengths. It has units of length × length × length, or length cubed. In the familiar inch-pound system, the most common units of volume are cubic feet (ft^3) or cubic inches (in^3).

Figure 8 shows the three-dimensional measurement of a rectangular prism. Its volume is calculated by multiplying length × width × height.

This will suffice for finding the volume of box-shaped containers. The next shape we will discuss is that of a tank or other cylindrical object, as shown in *Figure 9*.

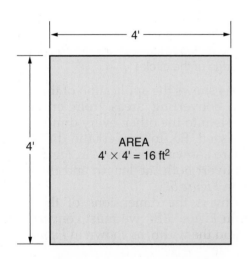

Figure 5 ◆ Conversion of units for a square.

AREA = LENGTH × WIDTH

= 4 ft × 4 ft

= 16 ft²

CONVERT TO METERS²:

1 m = 3.281 ft

THEREFORE: AREA = (4 ft ÷ 3.281) × (4 ft ÷ 3.281)

= 1.219 m × 1.219 m

= 1.486 m²

16 ft² = 1.486 m²

102F05.EPS

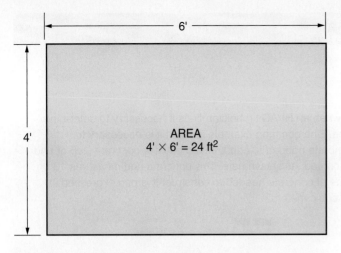

AREA = LENGTH × WIDTH

= 4 ft × 6 ft

= 24 ft²

CONVERT TO METERS²:

1 m = 3.281 ft

THEREFORE: AREA = (4 ft ÷ 3.281) × (6 ft ÷ 3.281)

= 1.219 m × 1.829 m

= 2.23 m²

24 ft² = 2.23 m²

102F06.EPS

Figure 6 ◆ Conversion of units for a rectangle.

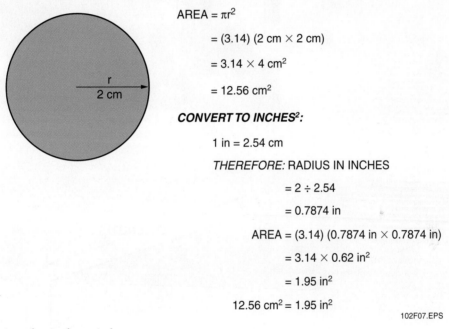

AREA = πr²

= (3.14) (2 cm × 2 cm)

= 3.14 × 4 cm²

= 12.56 cm²

CONVERT TO INCHES²:

1 in = 2.54 cm

THEREFORE: RADIUS IN INCHES

= 2 ÷ 2.54

= 0.7874 in

AREA = (3.14) (0.7874 in × 0.7874 in)

= 3.14 × 0.62 in²

= 1.95 in²

12.56 cm² = 1.95 in²

102F07.EPS

Figure 7 ◆ Conversion of units for a circle.

Since the volume of the box can be considered the area of any side times the height, likewise, the volume of a cylindrical tank, such as a refrigerant cylinder, can be simplified by considering it as the area of the circle times its depth (or height).

When converting these two volumes from one measurement system to the other, again, we must remember that every dimension must be independently converted. So, for converting the dimensions of the box, we must convert the length, width, and height, as shown in *Figure 10.*

Volume

There are many occasions when an HVAC technician finds it necessary to determine the volume of a room or area. One common example is when it is necessary to determine the volume of concrete needed to construct an outdoor concrete slab or pad upon which equipment is mounted, as shown here. This concrete pad measures 13' × 6' × 6". What was the volume of concrete needed to construct this pad expressed in cubic feet, cubic yards, and cubic meters?

102SA01.EPS

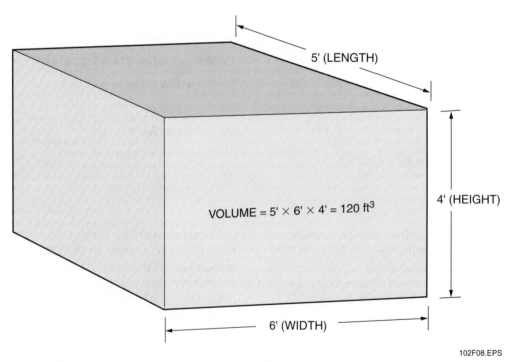

VOLUME = 5' × 6' × 4' = 120 ft³

5' (LENGTH)

4' (HEIGHT)

6' (WIDTH)

102F08.EPS

Figure 8 ◆ Volume of a rectangular prism.

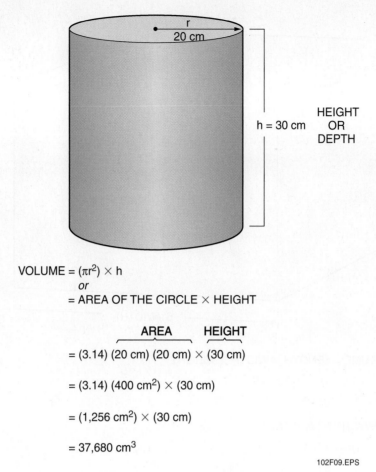

VOLUME = $(\pi r^2) \times h$
or
= AREA OF THE CIRCLE × HEIGHT

$\overbrace{}^{\text{AREA}}$ $\overbrace{}^{\text{HEIGHT}}$
= (3.14) (20 cm) (20 cm) × (30 cm)

= (3.14) (400 cm^2) × (30 cm)

= (1,256 cm^2) × (30 cm)

= 37,680 cm^3

102F09.EPS

Figure 9 ◆ Volume of a cylindrical tank or cylinder.

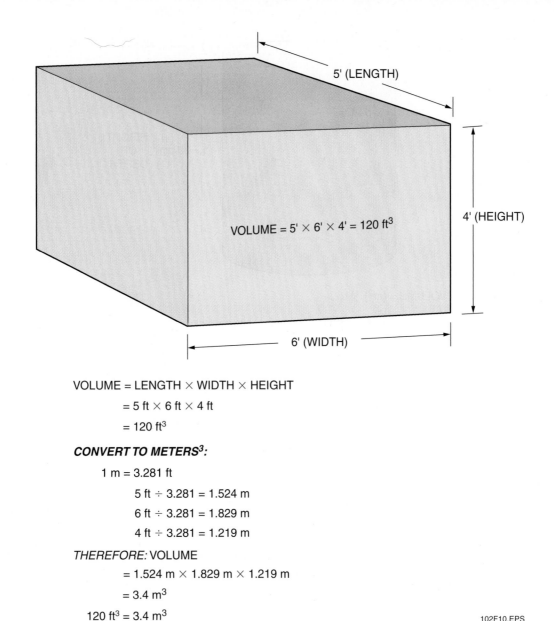

VOLUME = 5' × 6' × 4' = 120 ft³

5' (LENGTH)

4' (HEIGHT)

6' (WIDTH)

VOLUME = LENGTH × WIDTH × HEIGHT

= 5 ft × 6 ft × 4 ft

= 120 ft³

CONVERT TO METERS³:

1 m = 3.281 ft

5 ft ÷ 3.281 = 1.524 m

6 ft ÷ 3.281 = 1.829 m

4 ft ÷ 3.281 = 1.219 m

THEREFORE: VOLUME

= 1.524 m × 1.829 m × 1.219 m

= 3.4 m³

120 ft³ = 3.4 m³

102F10.EPS

Figure 10 ◆ Conversion of units for a box or rectangular duct.

In the example of the cylindrical tank, the same rule applies, as shown in *Figure 11*.

2.2.5 Wet Volume Measurements

In our discussion of volume measurements, we have been using the shape of the container to calculate the volume. This is referred to as a dry measure. In practice, we are much more used to dealing with wet measures (i.e., the amount of a fluid that would fill the volume). Common wet measures in the inch-pound system include the pint, quart, and gallon.

The metric system also uses wet measuring units. The **liter** is the most common and is about 5 percent greater than a quart.

By definition, one liter is one cubic decimeter. In other words, a cube with each side equal to one decimeter (or ten centimeters) will hold one liter of fluid. Knowing the wet measures for a substance allows easy handling and measuring of fluids, since the fluid will conform to the shape of the container. If you had to recalculate the amount each time you moved a fluid, you would soon see the advantage of using wet measures.

Table 5 shows the volume relationships between the liter and the dry volume of the cubic meter in the metric system, and the pint and gallon in the inch-pound system.

Figure 12 shows that the same metric prefixes that apply to the meter can be used with the liter.

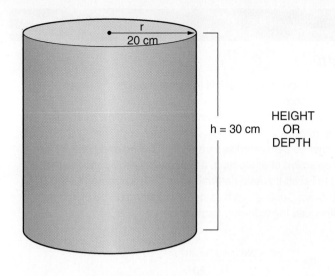

VOLUME = $(\pi r^2) \times h$

$= (3.14 \times 20^2) \times h$

$= (3.14 \times 400 \text{ cm}^2) \times 30 \text{ cm}$

$= 1{,}256 \text{ cm}^2 \times 30 \text{ cm}$

$= 37{,}680 \text{ cm}^3$

CONVERT TO INCHES³:

1 cm = 0.3937 in

20 cm = 7.874 in

30 cm = 11.811 in

THEREFORE: VOLUME

$= [(3.14)\,(7.874^2)] \times 11.811 \text{ in}$

$= (3.14 \times 62) \times 11.811 \text{ in}$

$= 194.68 \text{ in}^2 \times 11.811 \text{ in}$

$= 2{,}299.36 \text{ in}^3$

$37{,}680 \text{ cm}^3 = 2{,}299.36 \text{ in}^3$

102F11.EPS

Figure 11 ◆ Conversion of units for a cylindrical tank.

Figure 12 ◆ Common metric prefixes used with volumes.

2.2.6 Airflow in a Duct

To calculate the airflow in a duct in cubic feet per minute (cfm), the area of the duct (in square feet) and the velocity of the air flowing in the duct (in feet per minute) must be known. Expressed mathematically:

Airflow (cfm) = area (in sq ft) × velocity (in ft/min)

If the cfm and area of the duct are known, how would you solve the equation to find the velocity?

Example:

A duct has been installed that is 12" × 15" inside. Using a velometer inserted in the duct, the velocity is measured at 740 feet per minute (fpm). Calculate the actual air volume flowing through the duct.

Area = length × width
Area = 12" × 15"
Area = 180 in²

Convert to ft²:

180 in² ÷ 144 in²/ ft² = 1.25 ft²

Therefore:

Flow volume = 1.25 ft² × 740 fpm = 925 cfm

Table 5 Volume Relationships				
Unit	**Cubic Meter**	**Gallon**	**Liter**	**Pint**
Cubic meter (m³)	1	0.03785	0.001	0.0004732
U.S. gallon (gal)	264.2	1	0.264	0.125
Liter (L)	1,000	3.785	1	0.4732
U.S. pint (pt)	2,113	8	2.113	1

102T05.EPS

Calculating Airflow

One of the objectives of the HVAC design process is to select the size and type of equipment needed to deliver the correct amount of conditioned air to the building. One of the factors that must be calculated is the cooling and/or heating airflow needed in terms of cubic feet per minute (cfm).

The airflow requirement must be calculated separately for cooling and heating. In order to make a preliminary selection of equipment, it is necessary to approximate the heating and/or cooling cfm. However, the final determination on fan size and speed is based on the duct design process. Cooling cfm is generally higher than heating cfm. One method of estimating cfm uses the following formula:

Cooling:

$$cfm = \frac{\text{sensible load (Btuh)}}{1.08 \times (t_1 - t_2)}$$

Heating:

$$cfm = \frac{\text{sensible load (Btuh)}}{1.08 \times (t_2 - t_1)}$$

Where:
t_1 = room temperature
t_2 = supply air temperature
sensible load = heat gain of a structure due to several factors, such as equipment, occupants, and lighting. Sensible load is a calculation done as part of the HVAC design process.

Example 1
Cooling:

$$cfm = \frac{14,170}{1.08 \times (92 - 76)^*}$$

$$cfm = \frac{14,170}{1.08 \times 16}$$

$$cfm = \frac{14,170}{17.28}$$

$$cfm = 820$$

Example 2
Heating:

$$cfm = \frac{26,832}{1.08 \times (76 - 42)^*}$$

$$cfm = \frac{26,832}{1.08 \times 34}$$

$$cfm = \frac{26,832}{36.72}$$

$$cfm = 731$$

*Temperatures come from tables created to factor in differing locations and conditions.

2.3.0 Mass versus Weight

Mass is defined as the quantity of matter present. We often use the term weight to mean mass, but this is technically incorrect. Weight is actually the force on an object that is due to the pull of the Earth's gravity. As a body gets further away from the Earth, the effect of the Earth's gravity decreases. Therefore, as you climb a mountain, your actual weight decreases with the increasing altitude. However, the actual mass of your body has not changed at all. These two terms are used interchangeably because they are proportionally the same anywhere. However, if you are measuring very small amounts or are trying to be extremely accurate, it will be necessary to compensate for the altitude (or the distance above or below sea level). For example, Denver is almost a mile above sea level. This could cause a slight error in the measurement of weights. Since we most often use weight to determine mass, it could also be measured inaccurately.

2.3.1 Mass

Since mass is a term used more often by scientists, it is not surprising that the most common units of mass are the metric system units. The basic unit of mass is the gram. A gram (g) is a relatively small amount of matter. It is equivalent to about 1 milliliter of water. The same prefixes used with the meter and the liter in the metric system are used with the gram. The most common units are the milligram (mg), gram (g), and kilogram (kg).

2.3.2 Weight (Force)

Since weight is actually force, it is the push we exert on the surface of the Earth due to our mass and the pull of the Earth's gravity.

Weight is a force with its direction always assumed to be downward. In the inch-pound system, the most common units for weight are the pound (lb) and the ounce (oz). The inch-pound system also uses these units to represent mass (e.g., pounds-mass [lb-m] or pounds force [lb-f]).

Table 6 shows the relationships between mass and weight units in both systems. For most instances, they are interchangeable for both mass and weight.

The metric system uses the **newton (N)** as the basic unit of force. Sir Isaac Newton discovered the true importance of gravity, created the field of mathematics known as calculus, and defined the properties of matter and bodies in motion.

To understand the force of one newton, try this simple exercise. If you hold a 100-gram (small) apple in your hand, the Earth's gravity (at sea level) exerts a force of about one newton on the apple. Therefore, you are exerting an equal force (one newton) to hold the apple in the air.

Newton's First Law of Motion states that every body persists in its state of rest, or of uniform motion in a straight line, unless it is compelled to change that state by forces impressed on it. In other words, a body at rest stays at rest and a body in motion stays in motion at a constant velocity, unless something acts to influence it. That something is force.

With respect to the body in Newton's First Law, a force can be applied in a direction that aids the moving body to increase its velocity. Conversely, a force can be applied in a direction that opposes the moving body in order to decrease its velocity. **Acceleration** is the term used to represent a change in velocity with respect to time. Notice that a body at rest can be viewed as having a velocity equal to zero. If a force applied to the body causes motion at some velocity, it can be said that the body was accelerated from rest.

Figure 13 shows equal forces applied to two different bodies, which are viewed as two different surface areas.

The surface area that Body A presents to Force A is four times larger than the surface area that Body B presents to Force B. Another way of describing this is to say that Force A is spread over a larger area than Force B. As a result, the force per unit area applied to Body A is one-fourth of the force per unit area applied to Body B.

Figure 14 shows the same two bodies with the surface areas and forces specified.

Pounds per square inch (psi), the most common unit of pressure in the inch-pound system, is equivalent to newtons per square meter (N/m^2) in the metric system. One N/m^2 is also referred to as one pascal, after Blaise Pascal, a seventeenth-century French philosopher who contributed greatly to mathematics and engineering. The

Table 6	Mass and Weight Equivalences			
Unit	Kilogram	Pound	Ounce	Gram
1 kilogram (kg)	1	2.205	35.27	1,000
1 pound (lb)	0.4536	1	16	453.6
1 ounce (oz)	0.02835	0.0625	1	28.35
1 gram (g)	0.001	0.002205	0.03527	1

102T06.EPS

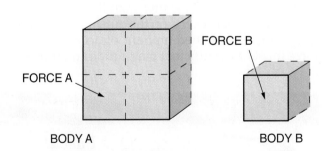

FRONT OF BODY A = 4 UNITS OF SURFACE AREA
FRONT OF BODY B = 1 UNIT OF SURFACE AREA
FORCE A = FORCE B

102F13.EPS

Figure 13 ◆ Equal forces applied to different surface areas.

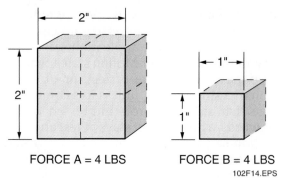

FORCE A = 4 LBS   FORCE B = 4 LBS

102F14.EPS

Figure 14 ◆ Known forces applied to known surface areas.

pascal, or Pa, is equal to a force of one newton exerted on an area of one square meter.

It would take 200,000 pascals (N/m^2) to inflate an ordinary automobile tire to about 28 psi.

A second common metric unit of pressure is the bar (b), which is equal to 100,000 Pa. The bar is used by weather forecasters. You are probably familiar with **barometric pressure** readings from watching the weather information on the evening news. You will also see the term millibar, which is 0.001 bar (b).

Occasionally, you will see the metric unit of a dyne to represent very small forces (one newton = 1,000 dynes).

In the next section, we will relate these metric units to the more familiar pressure units of the inch-pound system and review the concept of pressure measurement at the same time.

2.4.0 Pressure (Force) and Acceleration

Pressure is a quantity that is also closely related to force, but is probably more familiar to you. In a service station, you may ask the attendant for 28 pounds of air in the tires of your car. What you really want is an amount of air put into the tires that can exert 28 more pounds of force per square inch on the inside of the tires than the air pressure of the atmosphere exerts on the outside of the tires.

To cause a specified amount of acceleration, a force must have sufficient magnitude and must be applied in such a direction that it overcomes any opposing forces that are present.

The mass of a body has a direct effect on the magnitude of force required.

Newton's Second Law of Motion states that if a body is acted upon by a force, that body will be accelerated at a rate directly proportional to the force, and inversely proportional to its mass. In other words, a body with a large mass requires more force to obtain a specified amount of acceleration than does a body with less mass. Mass can simply be described as the amount of matter contained in a body. The relationship of force, mass, and acceleration is expressed as follows:

$$\text{Acceleration} = \frac{\text{force}}{\text{mass}}$$

or

$$\text{Force} = \text{mass} \times \text{acceleration}$$

Two bodies of the same size but made of different materials can have different masses. Two bodies can also have different masses if one body has been made larger by adding more of the material to it.

The term acceleration describes the process of a body at rest becoming a body in motion at some velocity. A body at a certain velocity travels a specific distance per unit of time.

The basic unit of force in the metric system, the newton (N), can now be defined as the amount of force required to accelerate 1 kilogram at the rate of one meter per second per second. Expressed mathematically:

$$\text{Force (1 newton)} = \text{mass (1 kilogram)} \times \text{acceleration} \text{ (1 meter/sec/sec)}$$

or

$$1 \text{ newton} = 1 \text{ kilogram} \times 1 \frac{\text{meter}}{\text{sec}^2}$$

The definition of pressure is force per unit area, or:

$$\text{Pressure} = \frac{\text{force}}{\text{area}}$$

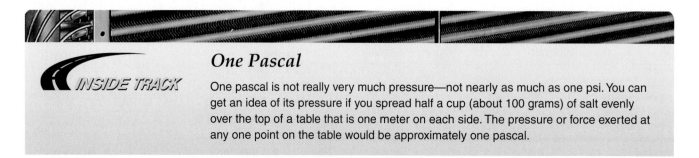

The force per unit area that is applied to Body A (shown in *Figure 14*) is:

$$\text{Pressure} = \frac{4 \text{ lbs}}{4 \text{ square inches}} = 1 \text{ lb per square inch}$$

The force per unit area that is applied to Body B (shown in *Figure 14*) is:

$$\text{Pressure} = \frac{4 \text{ lbs}}{1 \text{ square inch}} = 4 \text{ lbs per square inch}$$

Pressure represents the force applied perpendicular to a surface. The pressure applied to Body A in *Figure 14* is one pound per square inch (1 psi). The pressure applied to Body B is four pounds per square inch (4 psi).

As shown, applying equal forces to bodies of different surface areas results in applying different pressures to the bodies.

When converting these two pressures to their metric system equivalents, remember that each dimension must be converted. As we deal with more complex measurements, it becomes much easier to use conversion factors directly for the more common units of measure. *Table 7* shows the more common conversion factors used to convert the newton or dyne directly for general forces. *Table 8* shows the conversion factors used to convert the pascal or newton/meter^2 or dyne/cm^2 for very small pressures. Since the pascal is such a small unit and it takes so many pascals to define even a low pressure, pressure is normally represented in kilopascals (kPa).

To convert the 1 psi for Body A in *Figure 14*:

$$1 \text{ psi} \times 6{,}895 \text{ Pa} = 6{,}895 \text{ Pa}$$

To convert the 4 psi for Body B in *Figure 14*:

$$4 \text{ psi} \times 6{,}895 \text{ Pa} = 27{,}580 \text{ Pa or } 27.580 \text{ kPa}$$

As you can see, the pascal is indeed small in comparison to the pound per square inch.

For example, a person weighing 160 pounds is wearing ice skates. The total surface area of the two skate blades that are in contact with the ice is:

$$\text{Total area} = \text{length} \times \text{width} \times 2 \text{ blades}$$

or

$$\text{Total area} = 12" \times \tfrac{1}{8}" \times 2$$

or

$$\text{Total area} = 3 \text{ square inches}$$

Therefore, the person exerts a pressure on the surface of the ice equal to:

$$\text{Pressure} = \frac{\text{force}}{\text{unit area}}$$

or

$$\text{Pressure} = \frac{160 \text{ lbs}}{3 \text{ in}^2}$$

or

$$\text{Pressure} = 53.3333 \text{ psi or } 367{,}733 \text{ Pa}$$

Now suppose that the person weighing 160 pounds is wearing snow skis. The total surface area of the two skis that are in contact with the ice is:

$$\text{Total area} = 6' \times 3" \times 2 \text{ skis}$$
$$\text{Total area} = 72" \times 3" \times 2$$

or

$$\text{Total area} = 432 \text{ square inches}$$

Table 7	Force Conversion Factors		
Unit	**Kilogram-Force**	**Pound-Force**	**Gram-Force**
Newton	9.807	4.448	0.009807
Dyne	980,700.0	444,800.0	980.7

102T07.EPS

Table 8	Pressure Conversion Factors			
Unit	**Kilogram-Force per cm²**	**Pound-Force per in² (psi)**	**Pound-Force per ft²**	**Kilogram-Force per m²**
Kilopascal (kPa)	98.040	6.895	0.04788	0.009807
Pascal (Pa)	98,040	6,895	47.88	9.807
Newton/meter² (N/m²)				
Dyne/cm²	980,400	68,950	478.8	98.07

102T08.EPS

Pressure

Both liquids and gases are capable of exerting pressure. Gases are compressed under pressure and expand when the pressure is lowered. Liquids are generally considered to be incompressible. Vessels (drums) of refrigerant provide a good example of liquids/gases exerting pressure. The pressure inside a refrigerant drum depends on the type of refrigerant in the drum and the ambient temperature surrounding the drum. For every refrigerant, there is a specific temperature/pressure relationship. These values are listed on temperature/pressure charts, such as the one shown here. For example, a drum of R-22 refrigerant sitting in the back of an open truck with the sun shining on it may have a drum temperature of 105°F. Using a temperature/pressure chart for R-22, we would find that this corresponds to a pressure of about 210 psig. This means that the drum has a pressure of about 210 pounds pushing outward against its walls for each square inch of drum surface area. At 32°F, the pressure is only 57.5 psig.

102SA02.EPS

Therefore, the person exerts a pressure on the surface of the ice equal to:

$$\text{Pressure} = \frac{160 \text{ lbs}}{432 \text{ in}^2}$$

or

$$\text{Pressure} = 0.3703 \text{ psi or } 2,553 \text{ Pa}$$

Suppose you are the person weighing 160 pounds and you see a friend break through the ice and fall into the lake. You should immediately lie down flat on the ice to distribute your weight over as large a surface area as possible. For example, if you are approximately 6' (72") in height and on average 18" wide, the surface area now becomes:

$$\text{Total area} = 72" \times 18"$$

$$\text{Total area} = 1,296 \text{ square inches}$$

You would exert a pressure on the surface of the ice equal to:

$$\text{Pressure} = \frac{160 \text{ lbs}}{1,296 \text{ in}^2}$$

$$\text{Pressure} = 0.1234 \text{ psi or only } 851 \text{ Pa}$$

That is a reduction in pressure of over 2,000 times your pressure when standing upright wearing ice skates.

2.4.1 Absolute Pressure

The standard **atmospheric pressure** exerted on the surface of the earth is 14.696 psi taken at sea level with the air at 70°F. For most practical applications, this is often rounded off to 14.7 psi. Atmospheric pressure is also expressed as 29.9213 inches of mercury, which is equal to 14.696 psi. The rounded off value for inches of mercury (in Hg) is 29.92.

Actual weather conditions usually cause a slight variation in the atmospheric pressure. The actual atmospheric pressure is known as the barometric pressure. It is usually ignored in measuring hydraulic machinery pressure, but it cannot be ignored in power plant work and when dealing with the low pressures generated by fans and blowers.

Most gauges measure the difference between the actual pressure in the system being measured and the atmospheric pressure. Pressure measured by gauges is called gauge pressure (psig). The total pressure that exists in the system is called the **absolute pressure** (psia). Absolute pressure is equal to gauge pressure plus the atmospheric pressure, either the standard, 14.7, or the actual pressure, if measured. In other words:

$$psia = gauge\ pressure\ (psig) + 14.7$$

Example:

A steam boiler pressure gauge reads 295 psig. Find the absolute pressure.

$$\begin{aligned} psia &= psig + 14.7 \\ &= 295 + 14.7 \\ &= 309.7 \end{aligned}$$

Boiler pressure is sometimes specified in terms of atmospheres, where one atmosphere is equal to 14.696 (14.7) psi (see *Table 9*).

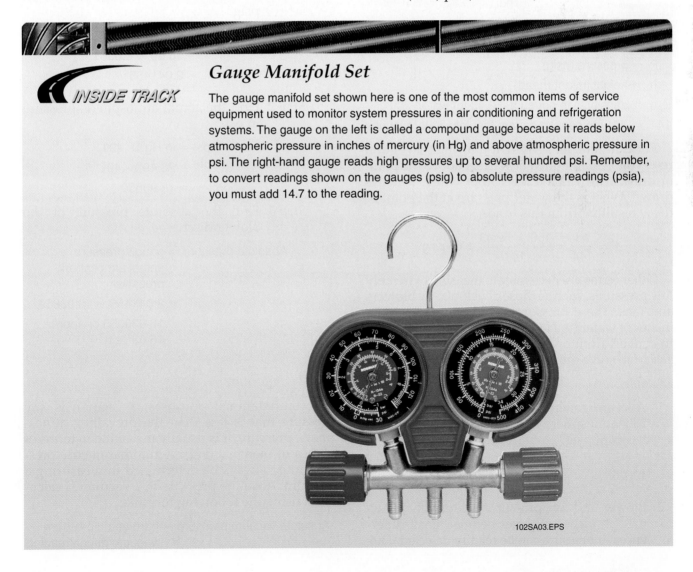

INSIDE TRACK

Gauge Manifold Set

The gauge manifold set shown here is one of the most common items of service equipment used to monitor system pressures in air conditioning and refrigeration systems. The gauge on the left is called a compound gauge because it reads below atmospheric pressure in inches of mercury (in Hg) and above atmospheric pressure in psi. The right-hand gauge reads high pressures up to several hundred psi. Remember, to convert readings shown on the gauges (psig) to absolute pressure readings (psia), you must add 14.7 to the reading.

102SA03.EPS

Table 9   Pressure Conversions

Multiply	By	To Obtain
Atmospheres	14.7	Pounds per square inch
Pounds per square inch	0.0680	Atmospheres
Pounds per square inch	27.68	Inches of water (H_2O)
Pounds per square inch	2.31	Feet of water (H_2O)
Pounds per square inch	2.04	Inches of mercury (Hg)
Inches of water	0.0361	Pounds per square inch
Feet of water	0.433	Pounds per square inch
Inches of mercury	0.491	Pounds per square inch

102T09.EPS

2.4.2   Static Head Pressure

Municipal water systems usually define pressure in terms of the static head or height in feet from the use point to elevated water reservoirs. Gauge pressure can be converted to static head pressure using the formula:

$$P = \frac{hd}{144}$$

Where:

P   = pressure (psig)

h   = height in feet (head)

d   = density in lb per cu ft (62.43 for water)

144   = used to convert square feet into square inches

For example, let's say that the water level in a reservoir is maintained at 150 feet above the point of draw, which is the level from which water is drawn. What is the water pressure at the point of draw resulting from this head?

$$P = \frac{hd}{144} = \frac{150 \times 62.43}{144} = 65 \text{ psig}$$

Other terms are necessary to measure extremely low pressures, such as those developed by blowers and fans. These pressures are often measured in inches of water (in H_2O). Sometimes it is necessary to convert from one measure to another. Consult *Table 9* to solve the following problems.

Examples:

1. A steam generator operates at a pressure of 320 atmospheres. What is this pressure in terms of psig?

   $$\text{psig} = 320 \times 14.7$$
   $$= 4{,}704 \text{ psig}$$

2. Blower and fan pressures are usually measured in inches of water because small differences in pressure can be readily detected. Due

to the low pressures involved, the exact atmospheric pressure is usually measured to determine the actual discharge pressure. Calculate the discharge pressure in psia if the measured discharge pressure of a blower is 56.55 in H_2O and the barometric pressure is 28.49.

Step 1   Find the blower discharge pressure in psig:

$$\text{psig} = \text{in } H_2O \times 0.0361$$
$$= 56.55 \times 0.0361$$
$$= 2.041455$$

Step 2   Find the actual atmospheric pressure in psi.

$$\text{psi} = \text{in Hg} \times .491$$
$$= 28.49 \times .491$$
$$= 13.98859$$

Step 3   Find the absolute pressure of the blower discharge.

Absolute pressure = gauge pressure + actual atmospheric pressure

$$= 2.041455 + 13.98859$$
$$= 16.030055 \text{ or approx.}$$
$$16.03 \text{ psia}$$

2.4.3   Vacuum

For computational purposes, a **vacuum** is any pressure that is less than the prevailing atmospheric pressure. It is usually measured in terms of inches of mercury (Hg). Actual barometric pressure must always be determined in measuring a vacuum. Absolute pressure in a vacuum is calculated using the following formula:

Absolute vacuum pressure = barometric pressure − vacuum gauge reading

Creating and Measuring a Vacuum

Any air and/or moisture (noncondensibles) trapped in an air conditioning or refrigeration system must be removed before the system can be charged with refrigerant. This requires that a vacuum be drawn on the system using a vacuum pump, such as the one shown here. The vacuum pump creates a pressure differential between the system and the pump. This causes air and moisture vapor trapped in the system at a higher pressure to move into a lower pressure (vacuum) area created in the vacuum pump. When the vacuum pump lowers the pressure (vacuum) in the system enough, as determined by the ambient temperature of the system, liquid moisture trapped in the system will boil and change into a vapor. Like free air, this water vapor is then pulled out of the system, processed through the vacuum pump, and exhausted to the atmosphere. The level of vacuum present in the system can be measured using a vacuum gauge, such as the one shown here.

VACUUM PUMP

102SA04.EPS

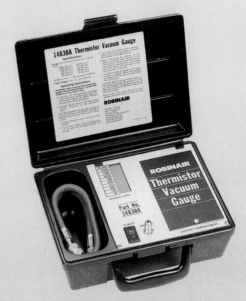

VACUUM GAUGE

102SA05.EPS

Example:

A vacuum gauge attached to a line reads 17.2 in Hg. The barometric pressure reads 29.85 in Hg. What is the absolute pressure in the line?

Absolute vacuum pressure = barometric pressure −
vacuum gauge reading

= 29.85 − 17.2

= 12.65 in Hg

2.5.0 Temperature Scales

Temperature is the intensity level of heat and is usually measured in degrees Fahrenheit or degrees Celsius. Temperature is measured in degrees on a temperature scale. In order to establish the scale, a substance is needed that can be placed in reproducible conditions. The substance used is water. The point at which water freezes at atmospheric pressure is one reproducible condition and the point at which water boils at atmospheric pressure is another. The four temperature scales commonly used today are the Fahrenheit scale, Celsius scale, Rankine scale, and Kelvin scale (see *Figure 15*). On the Fahrenheit scale, the freezing temperature of water is 32°F and the boiling temperature is 212°F. On the Celsius scale, the freezing temperature of water is 0°C and the boiling temperature is 100°C. The temperatures at which these fixed points occur were arbitrarily chosen by the inventors of the scales.

The Rankine scale and the Kelvin scale are based on the theory that at some extremely low temperature, no molecular activity occurs. The

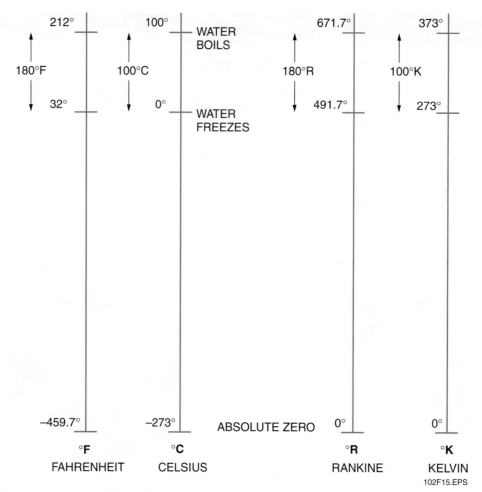

Figure 15 ◆ Comparison of temperature scales.

temperature at which this condition occurs is called absolute zero, the lowest temperature possible. Both the Rankine and Kelvin scales have their zero degree points at absolute zero. On the Rankine scale, the freezing point of water is 491.7°R and the boiling point is 671.7°R. The increments on the Rankine scale correspond in size to the increments on the Fahrenheit scale; for this reason, the Rankine scale is sometimes called the absolute Fahrenheit scale. Both the Rankine and Fahrenheit scales are part of the English system of measurement.

On the Kelvin scale, the freezing point of water is 273°K and the boiling point is 373°K. The increments on the Kelvin scale correspond to the increments on the Celsius scale; for this reason, the Kelvin scale is sometimes called the absolute Celsius scale. Both the Kelvin and Celsius scales are part of the metric system of measurement.

In the construction industry, the scales of primary importance are the Fahrenheit scale and the Celsius scale. The Rankine scale and the Kelvin scale are used primarily in scientific applications.

2.5.1 Temperature Conversions

Because both the Fahrenheit and Celsius scales are used in our industry, it sometimes becomes necessary to convert between the two. You should be familiar with these conversions.

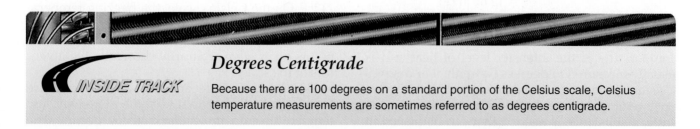

INSIDE TRACK

Degrees Centigrade

Because there are 100 degrees on a standard portion of the Celsius scale, Celsius temperature measurements are sometimes referred to as degrees centigrade.

INSIDE TRACK

Digital Thermometers

One advantage of some digital thermometers is that they can provide temperature readings in both Celsius and Fahrenheit.

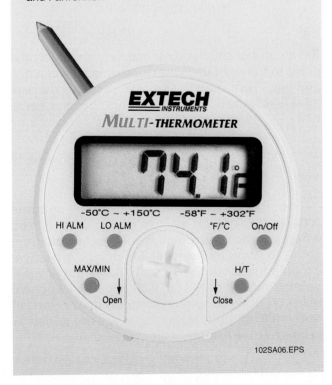

102SA06.EPS

On the Fahrenheit scale, there are 180 degrees between the freezing temperature and boiling temperature of water. On the Celsius scale, there are 100 degrees between the freezing temperature and the boiling temperature of water. The relationship between the two scales can be expressed as follows:

$$\frac{\text{Fahrenheit range (freezing to boiling)}}{\text{Celsius range (freezing to boiling)}} = \frac{180°}{100°} = \frac{9}{5}$$

Therefore, one degree Fahrenheit is ⅝ of one degree Celsius and conversely, one degree Celsius is ⅝ of one degree Fahrenheit. Thus, to convert a Fahrenheit temperature to a Celsius temperature, it is necessary to subtract 32° (since 32° corresponds to 0°), then multiply by ⅝. To convert a Celsius temperature to a Fahrenheit temperature, it is necessary to multiply by ⅝, then add 32°C.

$$°C = \tfrac{5}{9}\,(°F - 32°)$$
$$°F = (\tfrac{9}{5} \times °C) + 32°$$

You will need to practice these calculations to become more comfortable with using them. *Figure 16* shows two examples.

NOTE

The *Appendix* contains temperature conversion formulas.

EXAMPLE: 77°F to °C

$$°C = \frac{5}{9}\,(77° - 32°)$$

$$°C = \frac{5}{9}\,(45°)$$

$$°C = 25°C$$

EXAMPLE: 90°C to °F

$$°F = \left(\frac{9}{5} \times 90°\right) + 32°$$

$$°F = 162° + 32°$$

$$°F = 194°F$$

102F16.EPS

Figure 16 ◆ Sample temperature unit conversions.

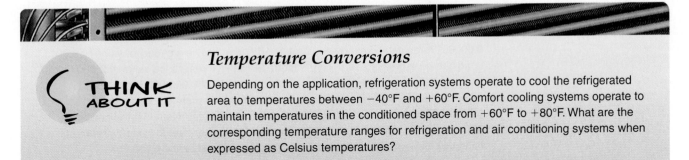

Temperature Conversions

THINK ABOUT IT

Depending on the application, refrigeration systems operate to cool the refrigerated area to temperatures between −40°F and +60°F. Comfort cooling systems operate to maintain temperatures in the conditioned space from +60°F to +80°F. What are the corresponding temperature ranges for refrigeration and air conditioning systems when expressed as Celsius temperatures?

3.0.0 ◆ SCIENTIFIC NOTATION

The mathematics related to all technical trades commonly uses numbers in the millions and larger, as well as numbers of less than one on down to a millionth or even lower. The complete number expressed in basic units can be used in calculations, but this is very cumbersome and increases the chance for error. For example, suppose we want to multiply 10 megohms times 50 picofarads in order to find the time constant of a circuit. The calculation would be:

10,000,000 ohms × 0.00000000005 farad = 0.0005

A method called scientific notation can be used to simplify calculations by expressing all numbers as a power of ten. In scientific notation, numbers are expressed as a base number times ten raised to some power (or **exponent**). The base number consists of an integer (usually between the numbers one and ten) followed by a decimal. For example, in the base number 3.6099, 3 is the integer and .6099 is the decimal. The power of ten is the exponent; it tells how many times ten is used as a factor in multiplying. For example, 10^2 indicates 10×10; 10^3 is $10 \times 10 \times 10$, etc. *Table 10* shows a list of the powers of ten that occur most often.

There are two basic rules for converting numbers to powers of ten.

Rule 1:

For a number larger than one, move the decimal point to the left until the number is between one and ten. Then count the number of places the decimal was moved, and use that number as a positive (+) power of ten. For example:

$$725 = 7.25 \times 10^2$$
$$3,000 = 3 \times 10^3$$
$$500,000 = 5 \times 10^5$$
$$10,000,000 = 10 \times 10^6$$

Rule 2:

For a number smaller than one, move the decimal point to the right until the number is between one and ten. Then count the number of places the decimal was moved, and use that number as a negative (−) power of ten. For example:

$$0.005 = 5 \times 10^{-3}$$
$$0.000008 = 8 \times 10^{-6}$$
$$0.00000000005 = 5 \times 10^{-11}$$

The process is reversed by performing opposite actions. The power of ten exponent tells you where the decimal point would be if you were going to write the number longhand. A positive exponent tells you how many places the decimal point is moved to the right. A negative exponent tells you how many places the decimal is moved to the left. In both instances, zeros are added as needed. Some examples are:

$$7.25 \times 10^2 = 725$$
$$3 \times 10^3 = 3,000$$
$$5 \times 10^5 = 500,000$$
$$10 \times 10^6 = 10,000,000$$
$$5 \times 10^{-3} = 0.005$$
$$8 \times 10^{-6} = 0.000008$$
$$5 \times 10^{-11} = 0.00000000005$$

Table 10	Common Powers of Ten
$10^0 = 1$	$10^{-1} = 0.1$
$10^1 = 10$	$10^{-2} = 0.01$
$10^2 = 100$	$10^{-3} = 0.001$
$10^3 = 1000$	$10^{-4} = 0.0001$
$10^4 = 10,000$	$10^{-5} = 0.00001$
$10^5 = 100,000$	$10^{-6} = 0.000001$
$10^6 = 1,000,000$	$10^{-7} = 0.0000001$

102T10.EPS

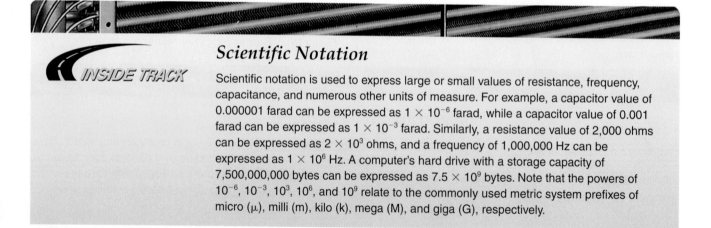

Scientific Notation

Scientific notation is used to express large or small values of resistance, frequency, capacitance, and numerous other units of measure. For example, a capacitor value of 0.000001 farad can be expressed as 1×10^{-6} farad, while a capacitor value of 0.001 farad can be expressed as 1×10^{-3} farad. Similarly, a resistance value of 2,000 ohms can be expressed as 2×10^3 ohms, and a frequency of 1,000,000 Hz can be expressed as 1×10^6 Hz. A computer's hard drive with a storage capacity of 7,500,000,000 bytes can be expressed as 7.5×10^9 bytes. Note that the powers of 10^{-6}, 10^{-3}, 10^3, 10^6, and 10^9 relate to the commonly used metric system prefixes of micro (μ), milli (m), kilo (k), mega (M), and giga (G), respectively.

3.1.0 Using Scientific Notation with a Calculator

Scientific calculators are capable of performing calculations using data entered in scientific notation. Some calculator manufacturers refer to this capability as the scientific or engineering display format. The procedure for entering calculation data using the scientific notation format may differ depending on the calculator being used and should be done according to the instructions for the specific calculator.

The procedure for entering a number in scientific notation using one popular type of calculator is done by entering the base number, then pressing the [+/−] key if the number is negative. Following this, the enter exponent [EE] key is pressed, then the power of ten is entered. If negative, the [+/−] key is also pressed. Addition, subtraction, multiplication, and division of the numbers entered in scientific notation are all performed in the usual manner.

4.0.0 ◆ POWERS AND ROOTS

Many mathematical formulas used in electronics and construction require that the power or root of a number be found. As in scientific notation, a power (or exponent) is a number written above and to the right of another number, which is called the base. For example, the expression y^x means to take the value of y and multiply it by itself x number of times. The xth root of a number y is another number that when multiplied by itself x times returns a value of y. Expressed mathematically:

$$\left(x\sqrt{y}\right)^x = y$$

4.1.0 Square and Square Roots

The need to find squares and square roots is common in trade mathematics. A square is the product of a number or quantity multiplied by itself. For example, the square of 6 means 6 × 6. To denote a number as squared, simply place the exponent 2 above and to the right of the base number. For example:

$$6^2 = 6 \times 6 = 36$$

The square root of a number is the divisor which, when multiplied by itself (squared), gives the number as a product. Extracting the square root refers to a process of finding the equal factors which, when multiplied together, return the original number. The process is identified by the radical symbol [$\sqrt{\ }$]. This symbol is a shorthand way of stating that the equal factors of the number

under the radical sign are to be determined. Finding the square roots is necessary in many calculations, including calculating loads and determining the airflow in a duct.

For example, $\sqrt{16}$ is read as the square root of 16. The number consists of the two equal factors 4 and 4. Thus, when 4 is raised to the second power or squared, it is equal to 16. Squaring a number simply means multiplying the number by itself.

The number 16 is a perfect square. Numbers that are perfect squares have whole numbers as the square roots. For example, the square roots of perfect squares 4, 25, 36, 121, and 324 are the whole numbers 2, 5, 6, 11, and 18, respectively.

Squares and square roots can be calculated by hand, but the process is very time consuming and subject to error. Most people find squares and square roots of numbers using a calculator. To find the square of a number, the calculator's square key [x^2] is used. When pressed, it takes the number shown in the display and multiplies it by itself. For example, to square the number 4.235, you would enter 4.235, press the [x^2] key, then read 17.935225 on the display.

Similarly, to find the square root of a number, the calculator's square root key [$\sqrt{\ }$] or [$\sqrt{x}$] is used. When pressed, it calculates the square root of the number shown in the display. For example, to find the square root of the number 17.935225, enter 17.935225, press the [$\sqrt{\ }$] or [$\sqrt{x}$] key, then read 4.235 on the display. Note that on some calculators, the [$\sqrt{\ }$] or [$\sqrt{x}$] key must be pressed before entering the number.

4.2.0 Other Powers and Roots

It is sometimes necessary to find powers and roots other than squares and square roots. This can easily be done with a calculator. The powers key [y^x] raises the displayed value to the xth power. The order of entry must be y, [y^x], then x. For example, to find the power 2.86^3 you would enter 2.86, press the [y^x] key, enter 3, press the [=] key, then read 23.393656 on the display.

The root key [$x\sqrt{y}$] is used to find the xth root of the displayed value y. The order of entry is y [INV] [$x\sqrt{y}$]. The [INV] or [2nd F] key, when pressed before any other key that has a dual function, will cause the second function of the key to be operated. For example, to find the cube root of 1,500 ($^3\sqrt{1,500}$), you would enter 1,500, press the [INV] [$x\sqrt{y}$] keys, enter 3, press the [=] key, then read 11.44714243 on the display. Note that on some calculators, the [$x\sqrt{y}$] key must be pressed before entering the number.

Remembering that the xth root of a number y is another number that when multiplied by itself

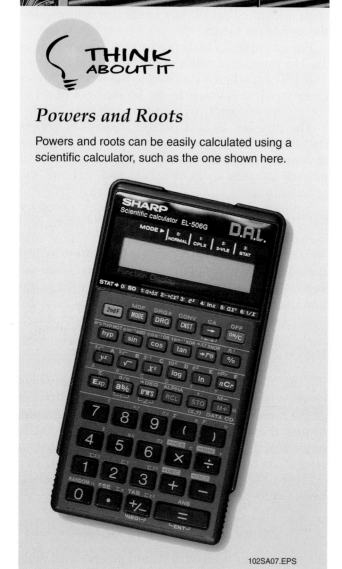

THINK ABOUT IT

Powers and Roots

Powers and roots can be easily calculated using a scientific calculator, such as the one shown here.

102SA07.EPS

x times returns the value of *y*, you can easily check your answer by using the [y^x] function to raise the answer on the display to a power of 3. For example, $11.44714243^3 = 1500$.

5.0.0 ◆ INTRODUCTION TO ALGEBRA

Algebra is the mathematics of defining and manipulating equations containing symbols instead of numbers. The symbols may be either constants or **variables**. They are connected to each other with mathematical operators such as $+$, $-$, $\times$, and $\div$. Knowing how to do calculations in algebra is an important skill for HVAC technicians. Basic algebra is necessary to get through the training program and is essential for doing airflow, electrical current, and voltage calculations. Algebraic calculations are used in many types of troubleshooting.

As in all fields of mathematics, an understanding of algebra requires the knowledge of some basic rules and definitions. These definitions and rules will be introduced where required to promote the understanding necessary to become proficient in algebra.

5.1.0 Definition of Terms

This section defines basic algebraic terms, including operators, equations, variables, constants, and coefficients.

5.1.1 *Mathematical Operators*

Mathematical operators define the required action using a symbol. Common operators include the following:

$+$ Addition

$-$ Subtraction

$\times$ Multiplication

$\cdot$ Multiplication

$\div$ Division

5.1.2 *Equations*

An equation is a collection of numbers, symbols, and mathematical operators connected by an equal sign ($=$). Some examples of equations are:

$$2 + 3 = 5$$
$$P = EI$$
$$\text{Volume} = L \times W \times H$$

5.1.3 *Variables*

A variable is an element of an equation that may change in value. For example, let's examine the simple equation for the area of a rectangle:

$$\text{Area} = L \times W$$

If the area is 12 and the length (L) is 6, the equation would read as follows:

$$12 = 6 \times W$$

In this case, it is easy enough to see that the width (W) is equal to 2 ($12 = 6 \times 2$). What is the width if the length is equal to 3?

$$12 = 3 \times W$$

In this case, the width is equal to 4 ($12 = 3 \times 4$). Therefore, in these two equations ($12 = 6 \times W$

and $12 = 3 \times W$), W must be considered a variable because it may change, depending on the value of L.

5.1.4 Constants

A constant is an element of an equation that does not change in value. For example, consider the following equation:

$$2 + 5 = 7$$

In this equation, 2, 5, and 7 are constants. The number 2 will always be 2, 5 will always be 5, and 7 will always be 7, no matter what the equation. Constants also refer to accepted values that represent one element of an equation and do not change from situation to situation. One of the most common constants that you will be dealing with is pi (π). It has an approximate value of 3.14 and represents the ratio of the circumference to the diameter in a circle.

5.1.5 Coefficients

A **coefficient** is a multiplier. Consider the following equation:

$$Area = L \times W$$

In this equation, L is the coefficient of W. It can also be written as LW, without the multiplication sign. No multiplication symbol is required when the intended relationship between symbols and letters is clear. For example:

- 2L means two times L (2 is the coefficient of L)
- IR means I times R (I is the coefficient of R)

5.2.0 Sequence of Operations

Complicated equations must be solved by performing the indicated operations in a prescribed sequence. This sequence is: multiply, divide, add, and subtract (MDAS). For example, the following equation can result in a number of answers if the MDAS sequence is not followed:

$$3 + 3 \times 2 - 6 \div 3 = ?$$

To come up with the correct result, this equation must be solved in the following order:

Step 1 Multiply:

$$3 + \underline{3 \times 2} - 6 \div 3$$

Step 2 Divide:

$$3 + 6 - \underline{6 \div 3}$$

Step 3 Add:

$$\underline{3 + 6} - 2$$

Step 4 Subtract:

$$9 - 2$$

Result:

$$7$$

5.3.0 Solving Algebraic Equations

Some equations may include several variables. Solving these equations means simplifying them as much as possible and, if necessary, separating the desired variable so that it is on one side by itself, with everything else on the other side. Problems such as these are known as algebraic expressions. When an algebraic expression appears in an equation, the MDAS sequence also applies. For example:

$$P = R - [5(3A + 4B) + 40L]$$

The parentheses represent multiplication, so they are worked on first. When working with multiple sets of parentheses or brackets, always begin by eliminating the innermost symbols first, then working your way to the outermost symbols. Thus, in the above equation, the expressions within parentheses (3A + 4B) with the coefficient of 5 are multiplied first, giving:

$$P = R - [15A + 20B + 40L]$$

The brackets also represent multiplication, so they are worked on next. The minus sign is the same as a coefficient of -1, so each term within the brackets is multiplied by -1, giving:

$$P = R - 15A - 20B - 40L$$

At this point, the equation has been simplified as much as possible. However, we will see what happens when we apply this same equation to a real-life situation. Suppose you have just installed ductwork in five identical apartments in a complex, and you want to determine the profit on the job. If we wrote this equation out longhand, it would look like this:

Profit (P) is equal to the payment received (R) minus five apartments times three pieces of one type of ductwork in each apartment at a certain cost per piece (A) plus four pieces of a second type of ductwork in each apartment at a certain cost per piece (B) plus forty hours of labor times an hourly rate (L).

It makes a lot more sense to simply write it algebraically:

$$P = R - [5(3A + 4B) + 40L]$$

or

$$P = R - 15A - 20B - 40L$$

Now, we will plug in numbers for the known values. Say that R = $1,500, A = $10, B = $15, and L = $15. This results in:

P = 1,500 − (15 × 10) − (20 × 15) − (40 × 15)

Multiplying, we get:

P = 1,500 − 150 − 300 − 600

Result:

P = $450

5.3.1 Rules of Algebra

There are a few simple rules that, once memorized, will help you to simplify and solve almost any equation you encounter as an HVAC technician.

Rule 1:

If the same value is added to or subtracted from both sides of an equation, the resulting equation is valid. For example, consider the following equation:

5 = 5

If 3 is added to each side of the equation, the resulting equation remains valid (both sides are still equal to each other).

5 + 3 = 5 + 3
8 = 8

In the same way, if we subtract the same number from both sides of an equation, the resulting equation is valid. For example, consider the following equation:

5 = 5

If 4 is subtracted from both sides of the equation, the resulting equation remains valid.

5 − 4 = 5 − 4
1 = 1

Moving variables from one side of an equation to another is done in the same way as when moving constants. Recall that when an equation is solved for one particular variable, that means that the variable should be on one side of the equation by itself. For example, consider the following pressure equation used frequently in the trade when working with system pressures in air conditioning and steam boiler systems:

Absolute pressure = gauge pressure + 14.7

To solve this equation for gauge pressure, 14.7 must be moved to the other side of the equation with absolute pressure. To do so, we will subtract 14.7 from both sides of the equation:

Absolute pressure − 14.7 =
gauge pressure + 14.7 − 14.7

It should be clear that the +14.7 and −14.7 on the right cancel each other out and we are left with:

Absolute pressure − 14.7 = gauge pressure

or

Gauge pressure = absolute pressure − 14.7

The equation has been solved for gauge pressure. If we wanted to take this new equation and solve it for absolute pressure again, we would simply add 14.7 to each side:

Gauge pressure + 14.7 =
absolute pressure − 14.7 + 14.7

Again, the +14.7 and −14.7 on the right cancel each other out and we are left with:

Gauge pressure + 14.7 = absolute pressure

or

Absolute pressure = gauge pressure + 14.7

Rule 2:

If both sides of an equation are multiplied or divided by the same value, the resulting equation is valid. For this rule, we will examine the equation for Ohm's law, which you will sometimes use in your work as an HVAC technician to calculate voltages, current, and resistance for equipment input power. This equation is as follows:

E = IR

Where:

E = voltage

I = current

R = resistance

If you know the voltage (E) and the current (I), but need to find the resistance (R), how do you rearrange the equation? To solve this equation for R, I must be moved to the other side of the equation with E. To do so, we will divide both sides by I:

$$\frac{E}{I} = \frac{IR}{I}$$

The two on the right cancel each other out and we are left with:

$$\frac{E}{I} = R$$

or

$$R = \frac{E}{I}$$

Fan Airflow versus Speed

THINK ABOUT IT

The performance of all fans and blowers is governed by three rules called the fan rules. One of these rules states that the amount of air delivered by a fan in cubic feet per minute (cfm) varies directly with the speed of the fan as measured in revolutions per minute (rpm). Expressed mathematically:

$$\text{New cfm} = \frac{\text{new rpm} \times \text{existing cfm}}{\text{existing rpm}}$$

Given the equation above for calculating a new cfm, how would you solve the equation to determine the new rpm, assuming you know the new cfm, the existing cfm, and the existing rpm?

The equation has been solved for resistance. If we wanted to take this new equation and solve it for E again, we would simply multiply each side by I:

$$I \times R = \frac{E}{I} \times I$$

The two on the right cancel each other out and we are left with:

$$IR = E$$

or

$$E = IR$$

Rule 3:

Like terms may be added and subtracted in a manner similar to constant numbers. For example, given the equation:

$$2A + 3A = 15$$

The like terms (2A and 3A) may be added directly:

$$5A = 15$$

Dividing both sides of the equation by 5, we have:

$$\frac{5A}{5} = \frac{15}{5}$$

$$A = 3$$

These rules may be used repeatedly until an equation is in the desired form. For example, take a look at the following pressure equation:

$$P = \frac{hd}{144}$$

Where:

P = pressure
h = height
d = density

Solve the equation for density (d).

Step 1 Remove the fraction by multiplying both sides by 144.

$$P \times 144 = \frac{hd}{144} \times 144$$

$$144P = hd$$

Step 2 Divide both sides by h.

$$\frac{144P}{h} = d$$

or

$$d = \frac{144P}{h}$$

The equation has been solved for density (d).

6.0.0 ◆ INTRODUCTION TO GEOMETRY

Geometry is the study of various figures. It consists of two main fields: plane geometry and solid geometry.

Plane geometry is the study of two-dimensional figures such as squares, rectangles, triangles, circles, and polygons. Solid geometry is the study of figures that occupy space, such as cubes, spheres, and other three-dimensional objects. The focus of this section is on the elements of plane geometry.

6.1.0 Lines

A line that forms a right angle (90 degrees) with one or more lines is said to be perpendicular to those lines (*Figure 17*). The distance from a point to a line is the measure of the perpendicular line drawn from that point to the line. Two or more straight lines that are the same distance apart at all perpendiculars are said to be parallel. Parallel lines do not intersect.

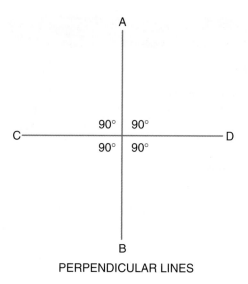

PERPENDICULAR LINES

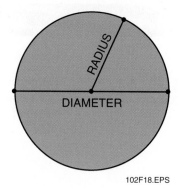

DIAMETER

RADIUS

102F18.EPS

Figure 18 ◆ Circle.

EQUAL DISTANCE AT ALL PERPENDICULARS

PARALLEL LINES

102F17.EPS

Figure 17 ◆ Perpendicular and parallel lines.

6.2.0 Circles

As shown in *Figure 18,* a circle is a finite curved line that connects with itself and has these other properties:

- All points on a circle are the same distance (equidistant) from the point at the center.
- The distance from the center to any point on the curved line, called the radius (r), is always the same.
- The shortest distance from any point on the curve through the center to a point directly opposite is called the diameter (d). The diameter is equal to twice the radius (d = 2r).
- The distance around the outside of the circle is called the circumference. It can be determined by using the equation: circumference = πd, where π is a constant equal to approximately 3.14 and d is the diameter.
- A circle is divided into 360 parts with each part called a degree; therefore, one degree = $\frac{1}{360}$ of a circle. The degree is the unit of measurement commonly used in construction layout for measuring the size of angles.

6.3.0 Angles

Two straight lines meeting at a point, called the vertex, form an angle (*Figure 19*). The two lines are the sides, or rays, of the angle. The angle is the

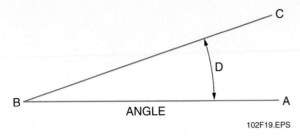

Figure 19 ◆ Angle.

amount of opening that exists between the rays and is measured in degrees. There are two ways commonly used to identify angles. One is to assign a letter to the angle, such as angle D shown in *Figure 19*. This is written: ∠D. The other way is

to name the two end points of the rays and put the vertex letter between them (e.g., ∠ABC). When you show the angle measure in degrees, it should be written inside the angle, if possible. If the angle is too small to show the measurement, you may put it outside of the angle and draw an arrow to the inside.

There are several types of angles, as shown in *Figures 20* and *21*.

- *Right angle* – This angle has rays that are perpendicular to each other (*Figure 20*). The measure of this angle is always 90 degrees.
- *Straight angle* – This angle does not look like an angle at all. The rays of a straight angle lie in a straight line, and the angle measures 180 degrees.
- *Acute angle* – An angle less than 90 degrees.
- *Obtuse angle* – An angle greater than 90 degrees, but less than 180 degrees.
- *Adjacent angles* – When three or more rays meet at the same vertex, the angles formed are said to be adjacent (next to) one another. In *Figure 21*, the angles ∠ABC and ∠CBD are adjacent angles. The ray BC is said to be common to both angles.
- *Complementary angles* – Two adjacent angles that have a combined total measure of 90 degrees. In *Figure 21*, ∠DEF is complementary to ∠FEG.
- *Supplementary angles* – Two adjacent angles that have a combined total measure of 180 degrees. In *Figure 21*, ∠HIJ is supplementary to ∠JIK.

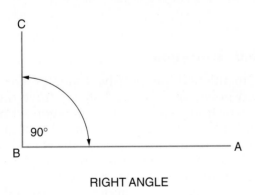

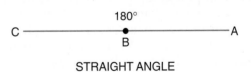

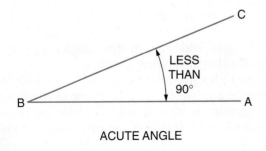

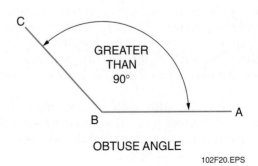

Figure 20 ◆ Angles.

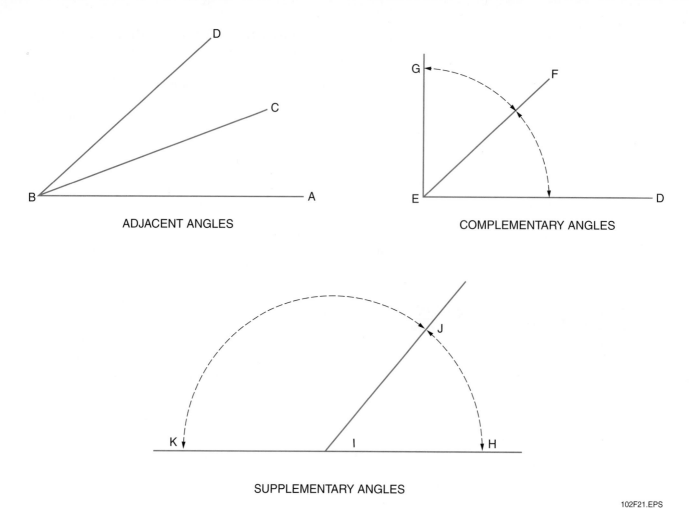

ADJACENT ANGLES

COMPLEMENTARY ANGLES

SUPPLEMENTARY ANGLES

102F21.EPS

Figure 21 ◆ Adjacent, complementary, and supplementary angles.

6.4.0 Polygons

A polygon is formed when three or more straight lines are joined in a regular pattern. Some of the most familiar polygons are shown in *Figure 22*. As shown, they have common names that generally refer to their number of sides. When all sides of a polygon have equal length and all internal angles are equal, it is called a regular polygon.

Each of the boundary lines forming the polygon is called a side of the polygon. The point at which any two sides of a polygon meet is called a vertex of the polygon. The perimeter of any polygon is equal to the sum of the lengths of each of the sides. The sum of the interior angles of any polygon is equal to $(n - 2) \times 180$ degrees, where n is the number of sides.

For example, the sum of the interior angles for a square is 360 degrees [$(4 - 2) \times 180$ degrees = 360 degrees] and for a triangle is 180 degrees [$(3 - 2) \times 180$ degrees = 180 degrees].

6.5.0 Triangles

As mentioned previously, triangles are three-sided polygons. *Figure 23* shows three different types of triangles. A regular polygon with three equal sides is called an equilateral triangle. Two types of irregular triangles are the isosceles (having two sides of equal length) and the scalene (having all sides of unequal length). An important fact to remember about triangles is that the sum of the three angles of any triangle equals 180 degrees. As shown, all three sides of an equilateral triangle are equal. In such a triangle, the three angles are also equal. The isosceles triangle has two equal sides with the angles opposite the equal sides also being equal.

Triangles are also classified according to their interior angles (*Figure 24*). If one of the three interior angles is 90 degrees, the triangle is called a right triangle. If one of the three interior angles is greater than 90 degrees, the triangle is called an

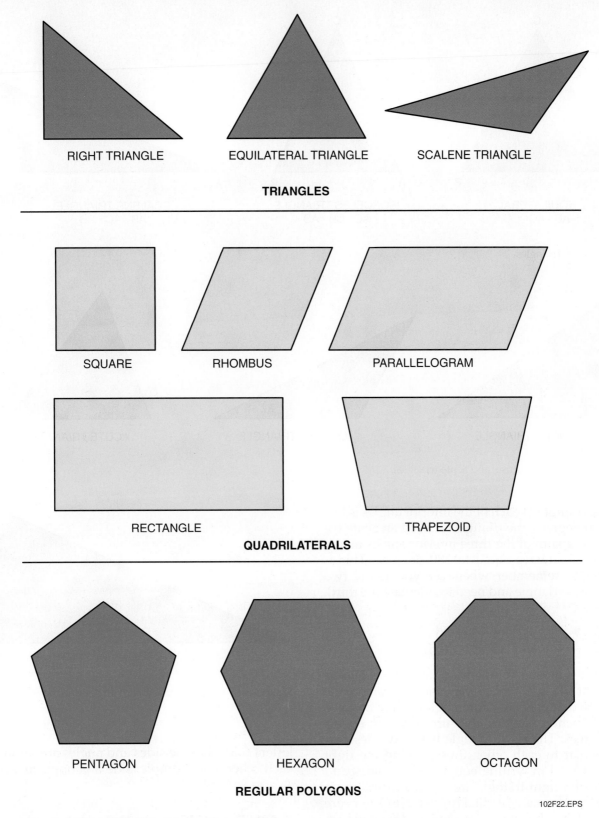

RIGHT TRIANGLE   EQUILATERAL TRIANGLE   SCALENE TRIANGLE

TRIANGLES

SQUARE   RHOMBUS   PARALLELOGRAM

RECTANGLE   TRAPEZOID

QUADRILATERALS

PENTAGON   HEXAGON   OCTAGON

REGULAR POLYGONS

102F22.EPS

Figure 22 ◆ Common polygons.

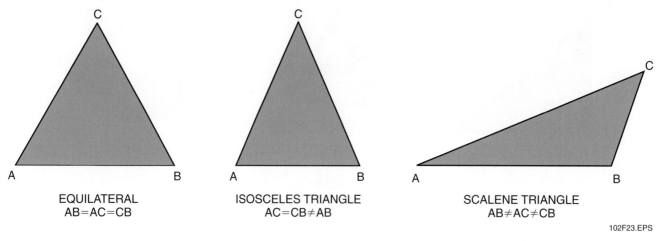

EQUILATERAL
AB=AC=CB

ISOSCELES TRIANGLE
AC=CB≠AB

SCALENE TRIANGLE
AB≠AC≠CB

102F23.EPS

Figure 23 ◆ Triangles.

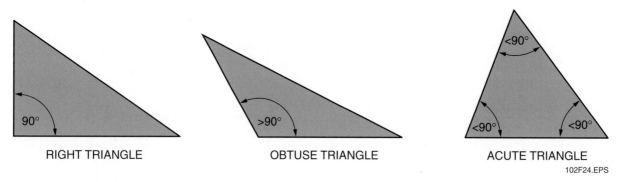

RIGHT TRIANGLE

OBTUSE TRIANGLE

ACUTE TRIANGLE

102F24.EPS

Figure 24 ◆ Right, obtuse, and acute triangles.

obtuse triangle. If each of the interior angles is less than 90 degrees, the triangle is called an acute triangle. The sum of the three interior angles of any triangle is always equal to 180 degrees. This is helpful to remember whenever you know two angles of a triangle and need to calculate the third.

7.0.0 ◆ WORKING WITH RIGHT TRIANGLES

One of the most common triangles you will use is the right triangle. Since it has one right angle, the other two angles are acute angles. They are also complementary angles, the sum of which equals 90 degrees. The right triangle has two sides perpendicular to each other, thus forming the right angle. To aid in writing equations, the sides and angles of a right triangle are labeled as shown in *Figure 25*. Normally, capital (uppercase) letters are used to label the angles, and lowercase letters are used to label the sides. The third side, which is always opposite the right angle (C), is called the hypotenuse. It is always longer than either of the other two sides. The other sides can be remembered as *a* for altitude and *b* for base. Note that the

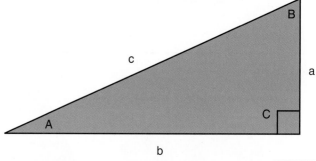

102F25.EPS

Figure 25 ◆ Common labeling of angles and sides in a right triangle.

letters that label the sides and angles are opposite each other. For example, side a is opposite angle A, and so forth.

7.1.0 Right Triangle Calculations Using the Pythagorean Theorem

If you know the lengths of any two sides of a right triangle, you can calculate the length of the third side using a rule called the Pythagorean theorem.

Trigonometry

Fifty years before Pythagoras, Thales (a Greek mathematician) figured out how to measure the height of the pyramids using a technique that later became known as trigonometry. Trigonometry recognizes that there is a relationship between the size of an angle and the lengths of the sides in a right triangle.

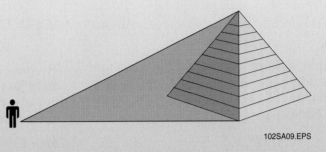

102SA09.EPS

3/4/5 Rule

The 3/4/5 rule is based on the Pythagorean theorem and has been used in building construction for centuries. It is a simple method for laying out or checking 90° angles (right angles) and requires only the use of a tape measure. The numbers 3/4/5 represent dimensions in feet that describe the sides of a right triangle. Right triangles that are multiples of the 3/4/5 triangle are commonly used (e.g., 9/12/15, 12/16/20, 15/20/25, etc.). The specific multiple used is determined by the relative distances involved in the job being layed out or checked.

An example of the 3/4/5 rule using the multiples of 15/20/25 is shown here. In order to square or check a corner as shown in the example, first measure and mark 15'-0" down the line in one direction, then measure and mark 20'-0" down the line in the other direction. The distance measured between the 15'-0" and 20'-0" points must be exactly 25'-0" to ensure that the angle is a perfect right angle.

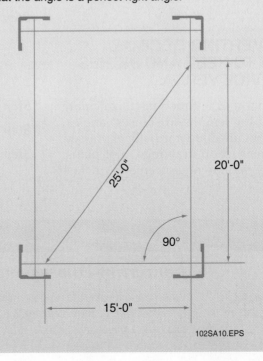

102SA10.EPS

It states that the square of the hypotenuse (c) is equal to the sum of the squares of the remaining two sides (a and b). Expressed mathematically:

$$c^2 = a^2 + b^2$$

You may rearrange to solve for the unknown side as follows:

$$a = \sqrt{c^2 - b^2}$$
$$b = \sqrt{c^2 - a^2}$$
$$c = \sqrt{a^2 + b^2}$$

For example, assume you had a right triangle with an altitude (side a) equal to 8' and a base (side b) equal to 12'. To find the length of the hypotenuse (side c), proceed as follows:

$$c = \sqrt{a^2 + b^2}$$
$$c = \sqrt{8^2 + 12^2}$$
$$c = \sqrt{64 + 144}$$
$$c = \sqrt{208}$$
$$c = 14.422$$

To determine the actual length of the hypotenuse using the formula above, it is necessary to calculate the square root of the sum of the sides squared. Fortunately, this is easy to do using a scientific calculator. On many calculators, you simply key in the number and press the square root $[\sqrt{\ }]$ key. On some calculators, the square root does not have a separate key. Instead, the square root function is the inverse of the $[x^2]$ key, so you have to press [INV] or [2nd F], depending on your calculator, followed by $[x^2]$, to obtain the square root.

8.0.0 ◆ CONVERTING DECIMAL FEET TO FEET AND INCHES AND VICE VERSA

When using trigonometric functions to calculate numerical values for lengths, distances, angles, etc., the answers obtained are normally expressed as a decimal. Construction drawings, plot plans, etc. often express similar measurements in feet and inches. For this reason, it is often necessary to convert between these two measurement systems. Conversion tables are available in many trade-related reference books that can be used for this purpose. However, in case conversion tables are not readily available, you should become familiar with the methods for making such conversions mathematically.

8.1.0 Converting Decimal Feet to Feet and Inches

To convert values given in decimal feet into equivalent feet and inches, use the following procedure. For our example, we will convert 45.3646' to feet and inches.

Step 1 Subtract 45' from 45.3646' = 0.3646'.

Step 2 Convert 0.3646' to inches by multiplying 0.3646' by 12 = 4.3752".

Step 3 Subtract 4" from 4.3752" = 0.3752".

Step 4 Convert 0.3752" into eighths of an inch by multiplying 0.3752" by 8 = 3.0016 eighths or, when rounded off, ⅜. Therefore, 45.3646' = 45'-4⅜".

8.2.0 Converting Feet and Inches to Decimal Feet

To convert values given in feet and inches (and inch-fractions) into equivalent decimal feet values, use the following procedure. For our example, we will convert 45'-4⅜" to decimal feet.

Step 1 Convert the inch-fraction ⅜" to a decimal. This is done by dividing the numerator of the fraction (top number) by the denominator of the fraction (bottom number). For example, ⅜" = 0.375.

Step 2 Add the 0.375 to 4" to obtain 4.375".

Step 3 Divide 4.375" by 12 to obtain 0.3646'.

Step 4 Add 0.3646' to 45' to obtain 45.3646'. Therefore, 45'-4⅜" = 45.3646'.

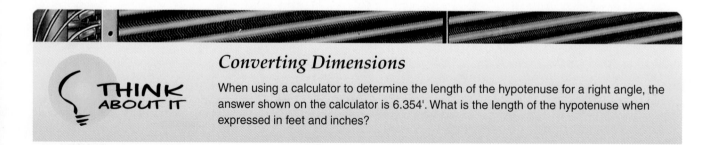

Converting Dimensions

When using a calculator to determine the length of the hypotenuse for a right angle, the answer shown on the calculator is 6.354'. What is the length of the hypotenuse when expressed in feet and inches?

THINK ABOUT IT

Summary

This module built on the knowledge gained in *Introduction to Construction Math*. It covered mathematics important to the HVAC technician, with an emphasis placed on the metric system. It also introduced basic concepts concerning scientific notation, powers and roots, algebra, geometry, and trigonometry.

HVAC technicians use math functions in their day-to-day work activities. For example, it may be necessary to perform any of the following calculations during a service call:

- Calculate the area of rectangular or round ductwork to determine airflow volume.
- Calculate the current load that an electrical device will place on an electrical service.
- Determine the required offset angle for piping or ductwork.
- Convert temperature readings from Fahrenheit to Celsius.

Notes

1. One meter is equal to _____ feet.
   a. 0.3937
   b. 39.37
   c. 3.281
   d. 3,281

2. A room has a length of 3.568 meters and a width of 2.438 meters. The area of this room in square feet is _____ .
   a. 20 ft²
   b. 71.36 ft²
   c. 93.64 ft²
   d. 384 ft²

3. A duct has inside dimensions of 16" × 12". The area, in ft², is equal to _____ .
   a. 1.33 ft²
   b. 1.92 ft²
   c. 13.3 ft²
   d. 28 ft²

4. The value of π, to 2 decimal places, is _____.
   a. 1.00
   b. 3.14
   c. 3.33
   d. 3.45

5. A 55-gallon drum can hold _____ liters of fluid.
   a. 55
   b. 110
   c. 175
   d. 208

6. A duct that has inside dimensions of 16" × 12" has air moving through it at a speed of 675 fpm. The volume of airflow is _____ cfm. (Round to decimal point.)
   a. 508
   b. 676
   c. 898
   d. 18,900

7. A crate containing a condensing unit is marked as weighing 175 pounds. What does it weigh in kilograms?
   a. 68.5 kilograms
   b. 79.38 kilograms
   c. 96.33 kilograms
   d. 125 kilograms

8. Find the absolute pressure if the gauge pressure is 164 psig.
   a. 14.7 psia
   b. 149.3 psia
   c. 164 psia
   d. 178.7 psia

9. A temperature of 0°C is equal to _____ .
   a. −17.8°F
   b. 32°F
   c. 100°F
   d. 144°F

10. A temperature of 68°F is equal to _____ .
    a. 20°C
    b. 36°C
    c. 41°C
    d. 100°C

11. Express the number 7.32×10^5 as a number without a power of ten.
    a. 0.0000732
    b. 0.00732
    c. 73,200
    d. 732,000

12. Express the number 0.000064 as a number between 1 and 10, times a power of ten.
    a. 6.4×10^{-3}
    b. 6.4×10^{-4}
    c. 6.4×10^{-5}
    d. 6.4×10^{-6}

13. The square root of 12,500, which is expressed mathematically as $\sqrt{12,500}$, is equal to _____ .
    a. 1.11803
    b. 11.1803
    c. 111.803
    d. 1118.03

14. 6^2 is the same as _____ .
    a. 4
    b. 8
    c. 12
    d. 36

15. The number 3.21413^3 is equal to _____ .
    a. 3.3203993
    b. 33.203993
    c. 333.20993
    d. 3332.0993

16. Given the equation $P = E^2/R$, solve for R.
    a. $R = E^2/P$
    b. $R = E^2 \times P$
    c. $R = P/E^2$
    d. $R = \sqrt{PE}$

17. The distance from the center of a circle to any point on the curved line is called the
    _____ .
    a. circumference
    b. radius
    c. diameter
    d. cord

18. A triangle in which all sides are of unequal length is called a(n) _____ triangle.
    a. equilateral
    b. isosceles
    c. right
    d. scalene

19. Which of the following is *not* a form of the Pythagorean theorem?
    a. $c^2 = a^2 + b^2$
    b. $a = \sqrt{c^2 - b^2}$
    c. $b = \sqrt{c^2 - a^2}$
    d. $c = \sqrt{a^2 - b^2}$

20. Convert 6.875' to feet and inches.
    a. 6'-2¼"
    b. 6'-8¼"
    c. 6'-10½"
    d. 6'-11½"

1. A(n)_____ is a definite standard measure of a dimension.

2. A(n) _____ is a standard unit of volume in the metric system.

3. A(n) _____ is a small figure or symbol placed above and to the right of another figure or symbol to show how many times the latter is to be multiplied by itself.

4. The actual atmospheric pressure at a given place and time is called _____.

5. The rate of change of velocity is known as _____.

6. _____ is the amount of space contained in a given three-dimensional shape.

7. A push or pull on a surface is called _____.

8. A(n) _____ is an element in an equation with a fixed value.

9. The standard pressure exerted on the Earth's surface is called _____.

10. _____ is the total pressure that exists in a system.

11. An element of an equation that may change in value is called a(n) _____.

12. _____ is the quantity of matter present.

13. A multiplier is also called a(n) _____.

14. Any pressure that is less than the prevailing atmospheric pressure is referred to as a(n) _____.

15. _____ is the distance from one point to another.

16. The unit of force required to accelerate one kilogram at a rate of one meter per second is a(n) _____.

17. The amount of surface contained in a given size plane or two-dimensional shape is referred to as _____.

Trade Terms

Absolute pressure	Coefficient	Length	Unit
Acceleration	Constant	Liter	Vacuum
Area	Exponent	Mass	Variable
Atmospheric pressure	Force	Newton (N)	Volume
Barometric pressure			

Allan Roy Shero

Vice President of Production
Entek Corporation

Allan Roy Shero began his HVAC career sweeping the warehouse at B&B Air Conditioning and Heating in Longview, Washington. His natural mechanical aptitude and a strong willingness to learn helped him to advance quickly. Allan is now vice president in charge of production for the firm, which has grown from only 12 employees to more than 50 employees in two offices. He also has an ownership stake in the company.

How did you become interested in the construction industry?
I started as a summer helper and found the technical aspects of working with my hands fascinating. I also liked interacting with the customers.

What qualities must an individual have to succeed in the HVAC trade?
Good mechanical skills, a desire for continuing education, basic systems knowledge, common sense, and excellent communications skills are all valuable in the trade. A desire to work with people, particularly your customers, is also important.

What are some of your responsibilities?
I'm responsible for all phases of management for 18 service technicians and over 20 field installation people, as well as warehouse staff, several sheet metal shop fabricators, a pair of service dispatchers, and a job coordinator. I have some direct customer sales contact, and many promotional and public relations obligations (including meeting with the Chamber of

Commerce and service clubs). I'm president of a local heating and cooling group for southwestern Washington contractors and I'm also active in the Air Conditioning Contractors of America (ACCA), both locally and nationally.

What do you like most about working in the HVAC trade?
It's never the same. Every day is different. It is a fast-paced work environment, with many opportunities to interact with people. When you solve someone's problems, you feel like you've made their world a better place to be, whether they're a customer or an employee.

What advice do you have for new trainees?
The HVAC trade is a great one that can take you many places, both geographically and in terms of your career. If you want to take full advantage of it, you have to want to learn and work at learning. As the saying goes, "You only get back out of it what you're willing to put into it."

Trade Terms
Introduced in This Module

Absolute pressure: The total pressure that exists in a system. Absolute pressure is expressed in pounds per square inch absolute (psia). Absolute pressure = gauge pressure + atmospheric pressure.

Acceleration: The rate of change of velocity; also, the process by which a body at rest becomes a body in motion.

Area: The amount of surface in a given plane or two-dimensional shape.

Atmospheric pressure: The standard pressure exerted on the Earth's surface. Atmospheric pressure is normally expressed as 14.7 pounds per square inch (psi) or 29.92 inches of mercury.

Barometric pressure: The actual atmospheric pressure at a given place and time.

Coefficient: A multiplier (e.g., the numeral 2 as in the expression 2b).

Constant: An element in an equation with a fixed value.

Exponent: A small figure or symbol placed above and to the right of another figure or symbol to show how many times the latter is to be multiplied by itself (e.g., $b^3 = b \times b \times b$).

Force: A push or pull on a surface. In this module, force is considered to be the weight of an object or fluid. This is a common approximation.

Length: The distance from one point to another; typically refers to a measurement of the long side of an object or surface.

Liter: A standard unit of volume in the metric system. It is equal to one cubic decimeter.

Mass: The quantity of matter present.

Newton (N): The amount of force required to accelerate one kilogram at a rate of one meter per second.

Unit: A definite standard measure of a dimension.

Vacuum: Any pressure that is less than the prevailing atmospheric pressure.

Variable: An element of an equation that may change in value.

Volume: The amount of space contained in a given three-dimensional shape.

Conversion Factors

COMMON MEASURES

WEIGHT UNITS
- 1 ton = 2,000 pounds
- 1 pound = 16 dry ounces

LENGTH UNITS
- 1 yard = 3 feet
- 1 foot = 12 inches

VOLUMES
- 1 cubic yard = 27 cubic feet
- 1 cubic foot = 1,728 cubic inches
- 1 gallon = 4 quarts
- 1 quart = 2 pints
- 1 pint = 2 cups
- 1 cup = 8 fluid ounces

AREA UNIT
- 1 square yard = 9 square feet
- 1 square foot = 144 square inches

102A01.EPS

WEIGHT UNITS

1 kilogram	=	1,000 grams
1 hectogram	=	100 grams
1 dekagram	=	10 grams
1 gram	=	1 gram
1 decigram	=	0.1 gram
1 centigram	=	0.01 gram
1 milligram	=	0.001 gram

LENGTH UNITS

1 kilometer	=	1,000 meters
1 hectometer	=	100 meters
1 dekameter	=	10 meters
1 meter	=	1 meter
1 decimeter	=	0.1 meter
1 centimeter	=	0.01 meter
1 millimeter	=	0.001 meter

VOLUME UNITS

1 kiloliter	=	1,000 liters
1 hectoliter	=	100 liters
1 dekaliter	=	10 liters
1 liter	=	1 liter
1 deciliter	=	0.1 liter
1 centiliter	=	0.01 liter
1 milliliter	=	0.001 liter

102A03.EPS

PREFIX	SYMBOL	NUMBER	MULTIPLICATION FACTOR
giga	G	billion	$1,000,000,000 = 10^9$
mega	M	million	$1,000,000 = 10^6$
kilo	k	thousand	$1,000 = 10^3$
hecto	h	hundred	$100 = 10^2$
deka	da	ten	$10 = 10^1$
			BASE UNITS $1 = 10^0$
deci	d	tenth	$0.1 = 10^{-1}$
centi	c	hundredth	$0.01 = 10^{-2}$
milli	m	thousandth	$0.001 = 10^{-3}$
micro	μ	millionth	$0.000001 = 10^{-6}$
nano	n	billionth	$0.000000001 = 10^{-9}$

102A02.EPS

U.S. TO METRIC CONVERSIONS

WEIGHTS

1 ounce	=	28.35 grams
1 pound	=	435.6 grams or
		0.4536 kilograms
1 (short) ton	=	907.2 kilograms

LENGTHS

1 inch	=	2.540 centimeters
1 foot	=	30.48 centimeters
1 yard	=	91.44 centimeters or
		0.9144 meters
1 mile	=	1.609 kilometers

AREAS

1 square inch	=	6.452 square centimeters
1 square foot	=	929.0 square centimeters
		or 0.0929 square meters
1 square yard	=	0.8361 square meters

VOLUMES

1 cubic inch	=	16.39 cubic centimeters
1 cubic foot	=	0.02832 cubic meter
1 cubic yard	=	0.7646 cubic meter

LIQUID MEASUREMENTS

1 (fluid) ounce	=	0.095 liter or 28.35 grams
1 pint	=	473.2 cubic centimeters
1 quart	=	0.9263 liter
1 (US) gallon	=	3,785 cubic centimeters or
		3.785 liters

TEMPERATURE MEASUREMENTS

To convert degrees Fahrenheit to degrees Celsius, use the following formula: $C = 5/9 \times (F - 32)$.

102A04.EPS

METRIC TO U.S. CONVERSIONS

WEIGHTS

1 gram (g)	=	0.03527 ounces
1 kilogram (kg)	=	2.205 pounds
1 metric ton	=	2,205 pounds

LENGTHS

1 millimeter (mm)	=	0.03937 inches
1 centimeter (cm)	=	0.3937 inches
1 meter (m)	=	3.281 feet or 1.0937 yards
1 kilometer (km)	=	0.6214 miles

AREAS

1 square millimeter	=	0.00155 square inches
1 square centimeter	=	0.155 square inches
1 square meter	=	10.76 square feet or
		1.196 square yards

VOLUMES

1 cubic centimeter	=	0.06102 cubic inches
1 cubic meter	=	35.31 cubic feet or
		1.308 cubic yards

LIQUID MEASUREMENTS

1 cubic centimeter (cm^3) =		0.06102 cubic inches
1 liter (1,000 cm^3)	=	1.057 quarts, 2.113 pints, or
		61.02 cubic inches

TEMPERATURE MEASUREMENTS

To convert degrees Celsius to degrees Fahrenheit, use the following formula: $F = (9/5 \times C) + 32$.

102A05.EPS

CONTREN® LEARNING SERIES – USER UPDATE

NCCER makes every effort to keep these textbooks up-to-date and free of technical errors. We appreciate your help in this process. If you have an idea for improving this textbook, or if you find an error, a typographical mistake, or an inaccuracy in NCCER's Contren® textbooks, please write us, using this form or a photocopy. Be sure to include the exact module number, page number, a detailed description, and the correction, if applicable. Your input will be brought to the attention of the Technical Review Committee. Thank you for your assistance.

Instructors – If you found that additional materials were necessary in order to teach this module effectively, please let us know so that we may include them in the Equipment/Materials list in the Annotated Instructor's Guide.

Write: Product Development and Revision
           National Center for Construction Education and Research
           3600 NW 43rd St., Bldg. G, Gainesville, FL 32606

Fax: 352-334-0932

E-mail: curriculum@nccer.org

Craft _____   Module Name _____

Copyright Date _____   Module Number _____   Page Number(s) _____

Description _____

_____

_____

_____

(Optional) Correction _____

_____

_____

(Optional) Your Name and Address _____

_____

_____

Copper and Plastic Piping Practices

03103-07

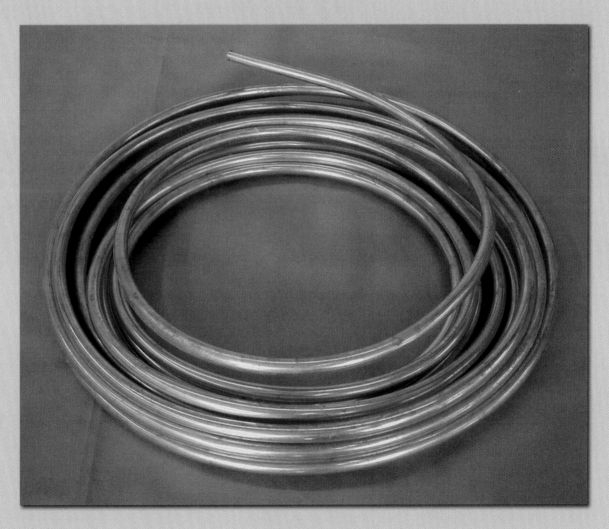

03103-07
Copper and Plastic Piping Practices

Topics to be presented in this module include:

Overview

Copper piping is used extensively in HVAC work, primarily for refrigerant piping. In this work, using the correct type and size of copper piping and correctly joining that piping are critical to the performance of the system. Solder and braze joints must be perfect. Because these systems are under pressure, even a tiny leak will cause the loss of refrigerant. This will reduce system performance and eventually could result in serious damage to the system.

Plastic piping is used in cooling, refrigeration, and heating systems. In cooling and refrigeration systems, it is used to make condensate drains. In chilled-water systems, plastic piping may be used to carry the water that circulates through the system. Plastic piping is usually joined with cement. As with copper, these joints must be correctly made to avoid leaks. One major use of plastic pipe in furnaces is to carry supply air to support combustion and to carry the byproducts of combustion to the outdoors.

Objectives

When you have completed this module, you will be able to do the following:

1. State the precautions that must be taken when installing refrigerant piping.
2. Select the right tubing for a job.
3. Cut and bend copper tubing.
4. Safely join tubing by using flare and compression fittings.
5. Determine the kinds of hangers and supports needed for refrigerant piping.
6. State the basic requirements for pressure-testing a system once it has been installed.
7. Identify types of plastic pipe and state their uses.
8. Cut and join lengths of plastic pipe.

Trade Terms

Annealed copper refrigeration (ACR) tubing
Annealing
Brazing
Compression joint
Flare fitting
Halocarbon refrigerants
Inside diameter (ID)
Insulation
Outside diameter (OD)
Piping
Specifications
Swaged joint
Sweating
Tubing

Required Trainee Materials

1. Paper and pencil
2. Appropriate personal protective equipment

Prerequisites

Before you begin this module, it is recommended that you successfully complete *Core Curriculum*; *HVAC Level One*, Modules 03101-07 and 03102-07.

This course map shows all of the modules in the first level of the *HVAC* curriculum. The suggested training order begins at the bottom and proceeds up. Skill levels increase as you advance on the course map. The local Training Program Sponsor may adjust the training order.

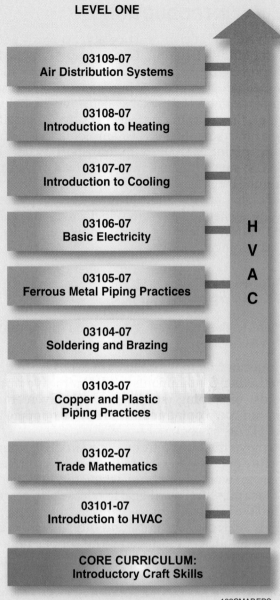

LEVEL ONE

03109-07
Air Distribution Systems

03108-07
Introduction to Heating

03107-07
Introduction to Cooling

03106-07
Basic Electricity

03105-07
Ferrous Metal Piping Practices

03104-07
Soldering and Brazing

03103-07
Copper and Plastic Piping Practices

03102-07
Trade Mathematics

03101-07
Introduction to HVAC

CORE CURRICULUM:
Introductory Craft Skills

HVAC

103CMAP.EPS

1.0.0 ◆ INTRODUCTION

An HVAC technician must be able to work with several kinds of **piping**. Copper pipe is used to transport refrigerant in residential and smaller commercial air conditioning systems. In large commercial and industrial systems, welded steel piping is more common. Steel is also used with special refrigerants, such as ammonia. Steel and plastic piping are used to transport water in HVAC systems that use water as a heat exchange medium. Black iron, brazed copper, and flexible plastic pipe are used to transport natural gas for heating systems. This module focuses on handling, cutting, bending, and joining copper and plastic piping.

2.0.0 ◆ INSTALLATION PRECAUTIONS

Adhere to the following guidelines when piping an HVAC system:

- Protect refrigerant piping from exposure to dirt and other contaminants, and don't leave open ends of piping stock exposed. Even tiny amounts of contaminants can cause damage in an HVAC system.
- Remove the charge from a pressurized system before opening any joints. It is not only dangerous to work on piping when the system is under pressure, it is illegal to release refrigerant into the atmosphere.
- Only ACR copper piping and fittings should be used in refrigeration work. The copper piping and fittings used in plumbing (Types M and DWV) are usually unsuitable for HVAC work.
- Use as few fittings as possible. Fewer fittings mean less chance for leaks and pressure drops.
- Exercise caution in making every solder connection. Use the correct solder and recommended techniques.

3.0.0 ◆ MATERIALS

Residential and small commercial refrigeration systems using **halocarbon (halogenated hydrocarbon) refrigerants** normally use copper **tubing**. However, welded steel pipe and fittings may be specified where diameters above three inches are required. Aluminum, stainless steel, and plastic tubing are used in some systems.

Refrigerant line sets with pre-insulated suction lines can be ordered in various lengths and diameters from local distributors. These line sets can also be special-ordered with the liquid line insulated. For maximum cleanliness, line sets that have been purged and charged with dry nitrogen and then tightly plugged are also available.

Large commercial refrigeration systems are not as simple to design and operate. Line sizes are larger and evaporator refrigerant temperatures vary widely. Structural location distances are more variable and some systems have refrigerant pressures at or near vacuum conditions. Oil return problems at low temperatures are more critical; therefore, precharged line sets are rarely used in these systems.

3.1.0 Selecting Copper Tubing and Fittings

The size (diameter) of the tubing to be used depends on the heating or cooling capacity of the HVAC system. Soft copper tubing is suitable for small systems. It ranges in size from ⅛" to 1⅜" in diameter. It is easily bent with hand tools and comes in rolls of 25' and 50'; larger rolls may be special-ordered. Because of its length, soft copper needs fewer connections. This saves time on installation and decreases the chances of a refrigerant leak. On small-capacity jobs, soft copper may be joined with **flare fittings**. This eliminates the need for soldering.

Hard copper tubing has a thicker wall and can therefore withstand higher system pressures. It comes in 20' lengths and should not be bent. Sections must be joined by soldering, or **brazing**, and bends must be made using fittings such as tees and elbows.

The term tubing generally applies to thin-wall materials that are joined together by methods other than threads cut into the tube wall. Piping is the term applied to thick pipe-wall material (iron and steel) into which threads can be cut and to which fittings can be joined by threading them onto the pipe.

Another distinction between tubing and piping is the method of designating size. Piping uses **inside diameter (ID)** and **outside diameter (OD)** measurements, depending on the pipe *size*. Tubing uses inside and outside diameter measurements, depending on the pipe *use*.

Tubing sizes used for plumbing and heating (hydronic piping) are expressed in terms of inside diameter (ID) (*Figure 1*). Air conditioning and refrigeration tubing sizes are expressed in terms of outside diameter (OD).

3.2.0 Copper Pipe Sizing

Pipe sizing varies depending on the type of copper pipe. Types K, L, M, and DWV use nominal, or

Precut Tubing Packages

Interconnecting tubing packages are available to connect the evaporator and condensing units of split-system air conditioners. The copper tubing in such packages is factory-sealed with plugs to keep out moisture and other contaminants. When using a tubing package, leave the plugs in place until just before making connections. This will minimize the entry of contaminants. However, after all connections are made, the system must still be evacuated to eliminate trace contaminants.

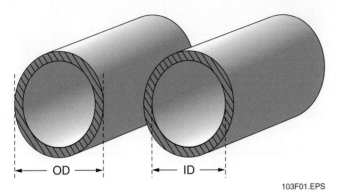

Figure 1 ◆ Tubing sizes.

103F01.EPS

Table 1   Copper Tubing Color Codes

Type	Color Code
K	Green
L	Blue
M	Red
DWV	Yellow
ACR	Blue

103T01.EPS

standard, sizing. This means that the outside diameter (OD) is always ⅛" larger than the nominal size (see *Figure 1*). The nominal size is the approximate measurement in inches of the inside diameter (ID) for most copper pipes. For example, the OD of a ¾" nominal type M pipe measures ⅞". This allows the same fittings to be used with the different wall thicknesses and IDs of different types of copper pipe. The ID is usually close to the stated size. A pipe with a ½" ID has approximately ½" between the inside walls of the pipe.

Type ACR pipe uses actual OD sizing. A ⅞" OD ACR copper pipe is actually the same OD as ¾" K, L, or M copper pipe. This means that you can use the same fittings with these pipes.

3.3.0 Labeling (Markings)

The labels on copper pipe contain very important information. Manufacturers must permanently mark Types K, L, M, and DWV to show the tube type, the name or trademark of the manufacturer, and the country of origin. Manufacturers also print this information on the hard tubes in a color that identifies the tube type. Manufacturers label most soft copper by stamping the information into the product. *Table 1* shows five types of copper tubing and the corresponding color code for labeling.

DID YOU KNOW?

Copper is one of the most plentiful metals. It has been in use for thousands of years. In fact, a piece of copper pipe used by the Egyptians more than 5,000 years ago is still in good condition. When the first Europeans arrived in the New World, they found the Native Americans using copper for jewelry and decoration. Much of this copper was from the region around Lake Superior, which separates northern Michigan and Canada.

One of the first copper mines in North America was established in 1664 in Massachusetts. During the mid-1800s, new deposits of copper were found in Michigan. Later, when miners went west in search of gold, they uncovered some of the richest veins of copper in the United States.

4.0.0 ◆ COPPER TUBING

The tubing used in all domestic refrigeration systems is special **annealed** (softened by heat treatment) copper. It is known as **annealed copper refrigeration (ACR) tubing**; that is, it has been manufactured specifically for use in air conditioning and refrigeration work.

ACR tubing is pressurized with nitrogen gas to keep out air, moisture, and impurities and to protect against oxides that are normally formed during brazing (hard soldering). Nitrogen should also be fed through the tubing during the brazing process to prevent harmful oxidation. The ends of new stock are plugged to keep out moisture and dirt. These plugs should be replaced after cutting a length of tubing for use.

4.1.0 Types of Copper Tubing

Copper is divided into three classifications based on the wall thickness. Type K (heavy wall) and Type L (medium wall) tubing are ACR approved. Type M (thin wall) is not used on pressurized refrigerant lines since it does not have the wall thickness to meet safety standards. It is, however, used on water lines and condensate drains.

Type K is meant for special use where heavy corrosion might be expected. Type L is most frequently used for normal refrigeration applications. Both types are available in soft- or hard-drawn types.

Soft copper tubing is available in sizes from ⅛" OD to 1⅜" OD. Line sizes from ¼" to ¾" OD are common. This tubing may be soldered or used with flared or other mechanical fittings. It is easily bent or shaped, but it must be held in place by clamps or other hardware as it cannot support its own weight.

Hard-drawn tubing is also widely used in commercial refrigeration and air conditioning systems. It comes in straight lengths of 20' and in sizes from ⅜" OD to 6" OD. The lengths are charged with nitrogen and plugged at each end to maintain a clean, moisture-free internal condition. It is intended for use with formed fittings to make the necessary bends or changes in direction. It is more self-supporting and therefore needs fewer supports than soft copper tubing.

4.2.0 Measuring

It is extremely important to measure pipe carefully. Several methods of measuring copper pipe are described in the following list (see *Figure 2*):

- *End-to-end* – Measure the full length of the pipe.
- *End-to-center* – Use for pipe that has a fitting joined on one end only; pipe length is equal to the measurement minus the end-to-center dimension of the fitting.
- *Center-to-center* – Use with a length of pipe that has fittings joined on both ends; pipe length is equal to the measurement minus the sum of the end-to-center dimensions of the fittings.

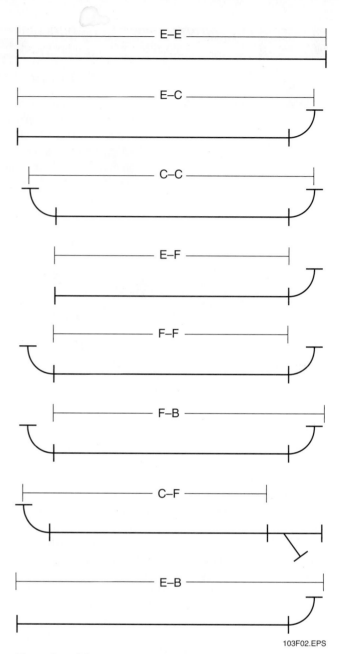

Figure 2 ◆ Measuring copper pipe.

103F02.EPS

- *End-to-face* – Use for pipe that has a fitting joined on one end only; pipe length is equal to the measurement.
- *Face-to-face* – Use for same situation as center-to-center measurement; pipe length is equal to the measurement.
- *Face-to-back* – Use with a length of pipe that has fittings joined on both ends; pipe length is equal to the measurement plus the distance from the face to the back of one sweated-on fitting.
- *Center-to-face* – Use with a length of pipe that has fittings on both ends; pipe length is equal to the measurement from the center of one of the fittings to the face of the opposite fitting, plus twice the insertion length.

- *End-to-back* – Use for pipe that has a fitting joined on one end only; pipe length is equal to the measurement plus the length of the sweated-on fitting.

4.3.0 Cutting Tubing

Copper tubing is cut with one of two tools:

- Handheld tube cutter
- Hacksaw and sawing fixture

The tube cutter is the preferred method because it produces a cleaner joint and leaves no metal particles.

NOTE
When copper is used, building codes require that soft copper tubing be protected and that its ends should be capped after cutting.

4.3.1 Handheld Tube Cutter

To use the tube cutter (*Figure 3*), place it on the tube at the desired cutting point. Tighten the knob, forcing the cutting wheel against the tube. The cut is

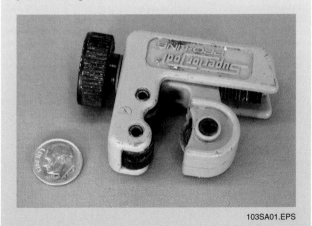

INSIDE TRACK

Midget Tubing Cutters

A variety of small tubing cutters are available for use in spaces where a turning radius for a standard tubing cutter is restricted. These cutters are used on small-diameter hard and soft copper, aluminum, brass, and plastic tubing.

103SA01.EPS

CUTTING WHEEL   ROLLERS   CLAMPING SCREW

Nº 20

FOLDING PIPE REAMER

103F03.EPS

Figure 3 ◆ Tubing cutter.

made by rotating the cutter around the tube under constant pressure. Afterward, the built-in deburring blade (reamer) is used to remove any burrs (sharp protruding edges) from inside the tube. There are a variety of models available, and the cutting ranges vary from ⅛" OD to as much as 4⅛" OD.

4.3.2 Hacksaw and Sawing Fixture

For larger size hard-drawn tubing, a hacksaw and sawing fixture may be used. The fixture helps square the ends and allows more accurate cuts. The blade of the hacksaw should have at least 32 teeth per inch. Avoid getting saw cuttings inside the tubing. File the end to produce a smooth surface, and carefully clean the inside of the tubing with a cloth.

NOTE
Portable power tools with abrasive wheels to cut, clean, and buff the tubing are available. They may be used for production pipe cutting.

4.4.0 Bending

Soft-drawn tubing should be bent around corners, as opposed to cutting and using fittings. Smaller-diameter tubing can easily be bent by hand, but care must be taken to avoid flattening the tube with a bend that is too sharp. For large tubing, the minimum bending radius to which a tube may be curved is up to ten times the tube's diameter. For smaller tubing, it may be up to five times the tube's diameter.

Tube-bending springs (*Figure 4*) are available to be placed on the outside of the tube to prevent it

Deburring and Cleaning

Always point the tubing down when deburring (reaming) the inside of the tubing. This will prevent any chips that are produced from falling down inside the tubing. Metal particles inside an air conditioning system can cause a variety of problems. To make sure that particles are removed, run a brush or cloth swab through the pipe after any cutting and deburring operations are completed. If you can't be sure that all particles can be removed, don't deburr the pipe.

If the tubing will be brazed or soldered, the outside of the tubing should be cleaned with fine emery cloth to remove any oxidation. Some tubing cutters are equipped with cleaner attachments that clean the exterior of the tubing while it is being cut. These attachments can save time when a large number of tubing cuts must be made.

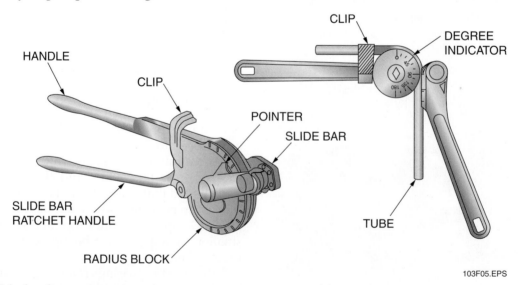

103F04.EPS

Figure 4 ◆ Bending spring.

from collapsing during hand bending. Tube-bending equipment (*Figure 5*) is also available for making accurate and reliable bends in both soft and hard tubing. Various sizes of forming attachments are available to bend tubing of diameters up to ⅞" at any angle up to 180 degrees.

4.5.0 Joining

There are two accepted methods of joining tubing. Because the copper tubing used for refrigerant piping is too thin to thread, mechanical coupling or heat bonding methods are used. Mechanical

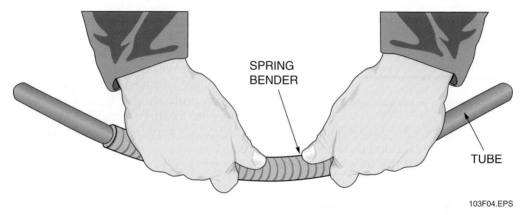

103F05.EPS

Figure 5 ◆ Tube-bending equipment.

 INSIDE TRACK

Soft Solder

Heat bonding of fitting connections using soft solder should not be employed on the high-pressure side of a refrigeration or air conditioning system. The combination of vibration and high pressure on this side of the system can cause failure of the soldered joints.

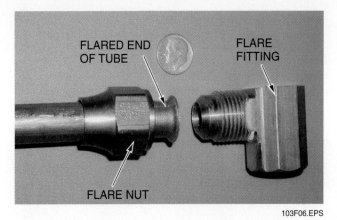

Figure 6 ◆ Flared connection and flare nut.

coupling includes flared and compression fittings that are semipermanent (can be taken apart). Heat bonding is accomplished by soldering or brazing; while these methods are considered permanent, the joints can be separated by reheating them.

> **NOTE**
>
> Standard compression fittings are still available, but not in common use. Newer compression technology such as the RIDGID® ProPress® System is available.

4.5.1 Flared Connections

Soft-drawn copper fittings should be leakproof and easily dismantled with the right tools. The flared connection (*Figure 6*) is a popular method of joining soft copper tubing.

A special flaring tool, shown in *Figure 7*, expands the tube's end into the shape of a cone, or flare. Two kinds of flare fittings are popular:

- The single-thickness flare forms a 45-degree cone that fits up against the face of a flare fitting (*Figure 8*). Under excessive pressure or expansion, larger size tubing with a single flare may be weak. In these applications, a double-thickness flare is preferable. Double-thickness flare connections can also be taken apart and reassembled more easily without damage.
- The double-thickness flare and the method of forming it are shown in *Figure 9*. In one operation the single-thickness flare is formed, then the lip is folded back onto itself and compressed.

The flare fittings used in refrigeration and air conditioning consist of a variety of elbows, tees, and unions (*Figure 10*). They are drop-forged

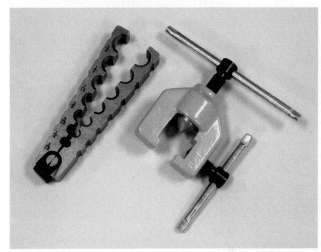

Figure 7 ◆ Flaring tool.

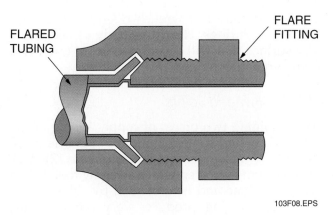

Figure 8 ◆ Single-thickness (angle) flare.

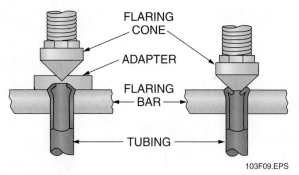

Figure 9 ◆ Double-thickness flare.

103F09.EPS

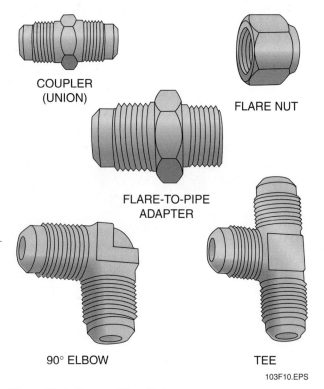

COUPLER
(UNION)

FLARE NUT

FLARE-TO-PIPE
ADAPTER

90° ELBOW

TEE

103F10.EPS

Figure 10 ◆ Types of flare fittings.

brass and are accurately machined to form the 45-degree flare face. Fittings are based on the size of tubing to be used. Flare nuts are hexagon-shaped. A flare nut wrench (*Figure 11*) is used with these fittings. The fitting body usually has a flat surface that will accommodate an open-end flare wrench.

4.5.2 Swaged Joints

When joining two pieces of copper tubing of the same diameter, some technicians feel it is more reliable to make a **swaged** joint (*Figure 12*). This type of joint is created with a swaging tool. When properly done and heat bonded by soldering or brazing, swaging produces very secure joints, thereby reducing leak hazards; however, it does

Tube Flaring

Before flaring tubing, slide the flare nut onto the tube with the threaded portion facing the end of the tube. Make sure the end of the tube is clamped at the correct position in the flaring tool. This is necessary in order to obtain a proper seal when the tube is subsequently connected to a flare fitting. Also, check that the inside edge of the tube is deburred. If not, the burr can cause an improper flare that will result in a leaky joint. A drop of the proper refrigerant oil on the flaring cone of the tool makes the flaring process easier.

103F11.EPS

Figure 11 ◆ Flare nut wrench.

take more time. Three types of swaging tools are available:

- *Combination flaring (swaging) tool* – This tool is similar to a standard flaring tool. The swaging die is substituted for the flaring die. The clamping block holds the tubing and the swaging die is forced into the end of the tube.
- *Tube expander* – With this tool, very precise, repeatable swaging can be accomplished. The die is inserted into the tube and then expanded by a hand press to enlarge the inside diameter of the tube.
- *Hammer-driven swaging tool* – This tool is inserted into the tube and driven by a hammer to expand the end of the tube.

4.5.3 Compression Joints

Compression joints (*Figure 13*) are sometimes used for joining refrigerant tubing. This method is popular because it takes less time than making flared or heat-bonded connections. A seal is

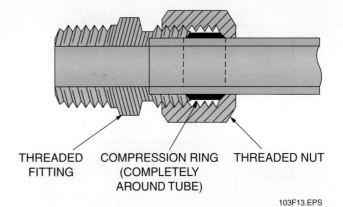

THREADED   COMPRESSION RING   THREADED NUT
FITTING   (COMPLETELY
AROUND TUBE)

103F13.EPS

Figure 13 ◆ Compression fitting.

obtained by connecting the tubing to the threaded fitting with a coupling nut and compression ring. The joint is formed by compressing the ring between the nut and the threaded fitting, forming a gas-tight seal. Compression couplings are available in types similar to those for flare fittings and in sizes from ¼" to 1⅛" OD. Adapters are available to join these couplings to other fittings.

4.5.4 Elbow Fittings

Elbows produce a large pressure drop in a piping system. With equal velocities, the amount of pressure drop depends on the sharpness of the turn. Long-radius elbows, rather than short-radius elbows, should be used whenever possible. When laying out offsets, 45-degree ells are recommended over 90-degree ells (*Figure 14*).

4.5.5 Tee Fittings

Tee fittings can cause a condition known as bullheading, which results in turbulence, greatly adding to the pressure drop. Bullheading occurs when tees are installed in such a position that the pressure input is applied to the leg that is at right

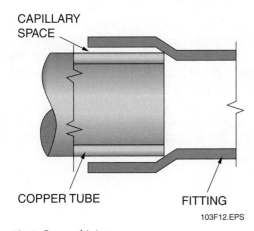

CAPILLARY
SPACE

COPPER TUBE   FITTING

103F12.EPS

Figure 12 ◆ Swaged joint.

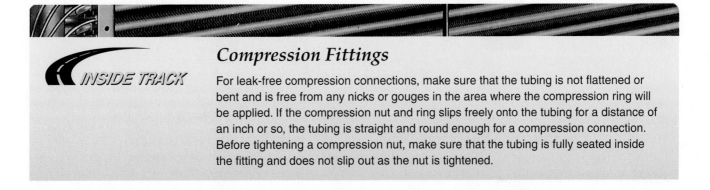

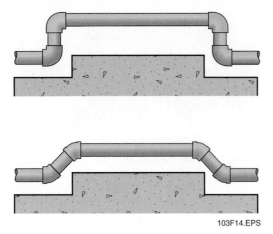

Figure 14 ◆ Elbow installations.

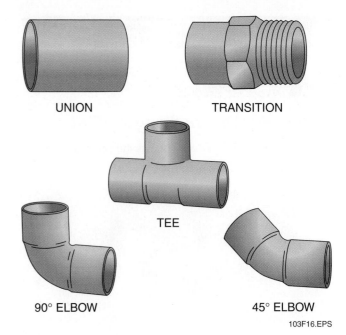

UNION   TRANSITION

TEE

90° ELBOW   45° ELBOW

103F16.EPS

Figure 16 ◆ Sweat fittings.

angles to the straight-through section of the tee. Tees should be installed so that the pressure input is applied to the straight-through section (*Figure 15*). Bullheading may also cause hammering in the line. If more than one tee is installed in the line, a straight piece of pipe 10 diameters in length between tees is recommended to reduce turbulence.

4.5.6 Sweat Fittings

As stated earlier, copper tubing is often joined by soldering or brazing techniques in which a soft metal is melted in the joint between two tubes. This type of heat bonding is also known as **sweating**. There are special fittings known as sweat fittings made for soldering or brazing copper tubes (*Figure 16*). Sweat fittings may be made of copper or brass. They are made slightly larger than the tubes to be joined, leaving only enough room for

solder to flow into the joint. Some sweat fittings, like the transition fitting shown in *Figure 16*, allow copper tubing to be joined with threaded pipe.

4.5.7 Ridgid® ProPress® System

The ProPress® System (*Figure 17*) is a flameless, solderless pipe-joining system that mechanically joins copper using specially designed fittings. Among the most critical components of the ProPress® System are the highly engineered pressing tools made by RIDGID®, which can make connections in as little as four seconds. The RIDGID® pressing tools automatically deliver the same

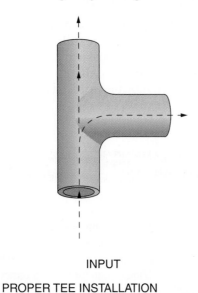

INPUT

PROPER TEE INSTALLATION

INPUT

IMPROPER TEE INSTALLATION

103F15.EPS

Figure 15 ◆ Tee installation.

Sweat Fittings Used in Refrigerant Piping

If 90-degree elbows must be used, always use long-radius elbows to reduce pressure losses. Two 45-degree elbows should be used instead of a short-radius 90-degree elbow. Always use refrigeration (ACR) fittings instead of plumbing fittings. Before heat-bonding the connections of copper tubes and fittings, always clean the inside connection surface of all fittings and the exterior of the tubing with a fine emery cloth to remove surface oxidation that can interfere with the bonding process.

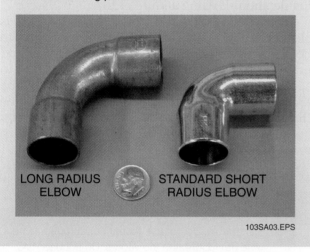

LONG RADIUS ELBOW   STANDARD SHORT RADIUS ELBOW

103SA03.EPS

Saddle Tees

Saddle tees are used to tap into pipes. Saddle tees are not used on refrigerant lines. A common use is to tap into a water line to supply a humidifier. The saddle tee can be clamped to pipes up to 1½" in diameter. The tee handle is attached to a pin. As the handle is turned, the pin penetrates the pipe, allowing water to flow into the copper tube that is attached by the flare fitting.

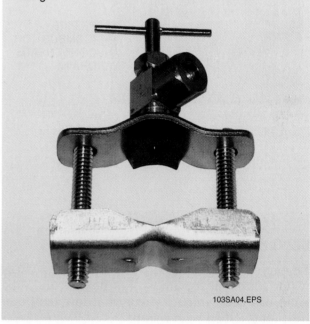

103SA04.EPS

103F17.EPS

Figure 17 ◆ ProPress® fittings and ProPress® tool.

force and consistency to every connection with every pull of the trigger. These tools use interchangeable jaws and rings to make ½" to 4" connections on copper. RIDGID® pressing tools and jaws were designed in conjunction with ProPress® fittings to ensure compatibility and consistency.

5.0.0 ◆ PLASTIC PIPE

Plastic pipe is used to vent condensing furnaces, drain condensate from furnaces and air conditioners, and transport water in water-cooled HVAC systems. There are several types of plastic pipe, each with a different application. *Figure 18* illustrates plastic pipe used in a typical condensing furnace.

5.1.0 ABS (Acrylonitrilebutadiene Styrene) Pipe

ABS pipe is rigid and has good impact strength at low temperatures. It is used for water, vents, and drains. In an unpressurized system, it can withstand heat up to 180°F.

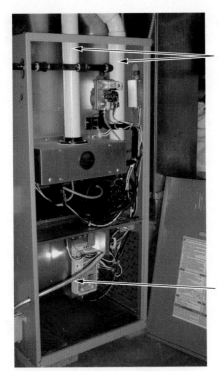

PVC PIPE
(INLET AIR AND
EXHAUST)

PE PIPE
(COMBUSTION
CONDENSATE
DRAIN)

103F18.EPS

Figure 18 ◆ Plastic pipe used in a typical condensing furnace.

5.2.0 PE (Polyethylene) Pipe

PE pipe is used in cold-water systems such as water source heat pumps. It is flexible and, like ABS, has good impact strength at low temperatures. It is joined with clamps.

5.3.0 PVC (Polyvinyl Chloride) Pipe

PVC is a rigid pipe with high impact strength. It can be used in high-pressure systems at low temperatures. PVC pipe is usually joined with cement. It can also be threaded and joined to steel pipe with a transition fitting.

5.4.0 CPVC (Chlorinated Polyvinyl Chloride) Pipe

The applications and joining methods for CPVC are very similar to those for PVC. However, CPVC can be used for hot water (up to 180°F) in a pressurized system (up to 100 psi).

5.5.0 PEX (Crosslinked Polyethylene) Tubing

PEX is a heat transfer tubing that is used in radiant floor heating systems. It is usually orange in color and is extremely resistant to chemicals. PEX tubing is rated for 200°F at 80 psi, 180°F at 100 psi, or 73.4°F at 160 psi. A version of PEX tubing, identified as hePEX, is sealed with a special polymer barrier that prevents oxygen diffusion through the sides of the tubing. This barrier reduces corrosion to ferrous components in the system.

5.6.0 Joining Plastic Pipe

Plastic pipe may be cut with a tubing cutter, a hacksaw, or a special cutter known as a plastic tubing shear (*Figure 19*). The cut must be square to ensure a good joint. Once cut, the pipe is deburred inside and out using a knife or file.

To join PVC or CPVC pipe, the pipe end is cleaned, then primer and cement are applied to both the pipe and fitting. The pipe is then inserted into the fitting and turned one quarter turn to spread the cement. Some PVC and CPVC pipe (Schedule 80) can be threaded; however, the same die should not be used for both plastic and metal because a die used for metal will not be sharp enough to cut plastic.

 WARNING!

PVC cleaner and cement are toxic and flammable. Do not use them near an open flame, and do not breathe the vapors. Do not use PVC cement in an enclosed space without the appropriate ventilation or breathing apparatus. Refer to the applicable material safety data sheet (MSDS). Avoid getting it on your skin.

103F19.EPS

Figure 19 ◆ Tubing shear.

When you join CPVC or PVC pipe and fittings, use only CPVC or PVC cement or all-purpose cement conforming to *ASTM F-493* standards, or the joint may fail.

Use special care when assembling CPVC and PVC systems in extremely low temperatures (below 40°F) or extremely high temperatures (above 100°F). In extremely hot environments, make sure both surfaces to be joined are still wet with cement when you put them together. If the cement has dried, the two surfaces may not adhere. Adapters are available to connect CTS CPVC pipe to CPVC Schedule 40 and 80 pipe for systems requiring piping diameters larger than 2".

To join CPVC or PVC pipe and fittings with solvent cement, follow these steps (see *Figure 20*). Note that these steps illustrate typical instructions. When joining pipe, always follow the cement manufacturer's instructions.

Step 1 Cut the pipes to the lengths you need, then test fit the pipes and fittings. The pipes should fit tightly against the bottom of the hubbed sockets on the fittings.

Step 2 Clean the surfaces you are joining by lightly scouring the ends of the pipe with emery paper, then wiping with a clean cloth. This ensures that the primer you apply later will soften the plastic before you apply solvent cement.

Step 3 Clean the socket interior and the spigot area of all dirt with a rag or brush.

Step 4 Mark the pipe and fitting with a felt-tipped pen to show the proper position for alignment. Also mark the depth of the fitting sockets on the pipes to make sure that you fit them back in completely when joining.

Step 5 Apply the primer to the surfaces you are joining. The primer will soften the plastic in preparation for the solvent-weld process.

Step 6 While the primer is still wet, apply a heavy, even coat of cement to the pipe end. Use the same applicator without additional cement to apply a thin coat inside the fitting socket. Too much cement can clog waterways. Do not allow excess cement to puddle in the fitting and pipe assembly.

Step 7 Immediately insert the tubing into the fitting socket, rotating the tube one-quarter to one-half turn while inserting. This motion ensures an even distribution of cement in the joint. Properly align the fitting.

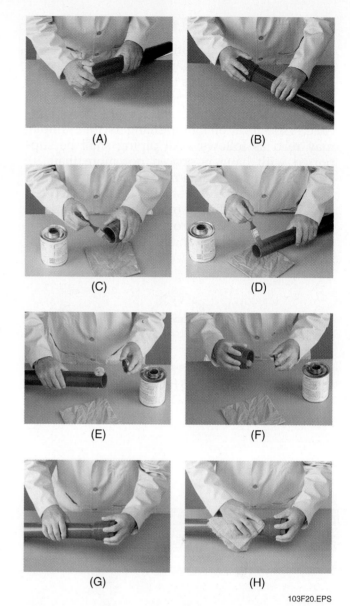

(A) (B) (C) (D) (E) (F) (G) (H)

103F20.EPS

Figure 20 ◆ Joining CPVC or PVC pipe and fittings with solvent cement.

Step 8 Hold the assembly firmly until the joint sets up. An even bead of cement should be visible around the joint. If this bead does not appear all the way around the socket edge, you may not have applied enough cement. In this case, remake the joint to avoid the possibility of leaks. Wipe excess cement from the tubing and fitting surfaces. Always follow the cement manufacturer's instructions.

The procedure for joining ABS pipe with cement is the same as for PVC or CPVC, except that no primer is required.

6.0.0 ◆ HANGERS AND SUPPORTS

Air conditioning and heating installations planned by architects and engineering consultants include plans and **specifications** that completely describe the proposed system. Specifications are based upon codes or ordinances and require strict adherence. A specification for pipe hangers, for example, may read as follows: "All piping shall be supported with hangers spaced not more than 10' apart (on center). Hangers shall be the malleable iron split-ring type and shall be as manufactured by XYZ Hangers, Inc. or other approved vendor."

Hangers (*Figure 21*) provide horizontal or vertical support of pipes and piping. The principal purpose of hangers and brackets is to keep the piping in alignment and to prevent it from bending or distorting. Horizontal hangers can be attached to wooden structures with lag screws or large nails.

Pipe hangers are used primarily to support pipe, but they may also be used as vibration isolators. If vibration problems are not anticipated, ordinary plumbing practices may be used.

When fastening pipes to beams and other metal structures, beam clamps or C-clamps (*Figure 22*) are used. Other horizontal support clamps and brackets are shown in *Figure 22*. Vertical hangers called pipe risers (*Figure 23*) consist of a friction

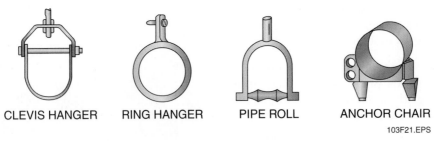

CLEVIS HANGER   RING HANGER   PIPE ROLL   ANCHOR CHAIR

103F21.EPS

Figure 21 ◆ Hangers.

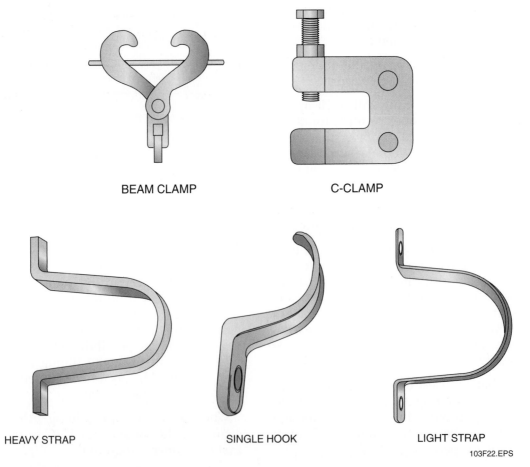

BEAM CLAMP   C-CLAMP

HEAVY STRAP   SINGLE HOOK   LIGHT STRAP

103F22.EPS

Figure 22 ◆ Clamps.

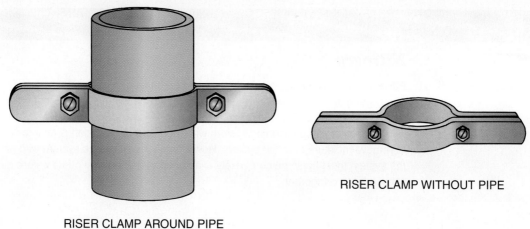

RISER CLAMP WITHOUT PIPE

RISER CLAMP AROUND PIPE

103F23.EPS

Figure 23 ◆ Riser clamps.

clamp that can be attached to structural site components to support the vertical load of the pipe. Special fasteners are used to attach hangers to masonry, concrete, or steel.

7.0.0 ◆ INSULATING

Under certain temperature and humidity conditions, condensation will form on refrigerant and cold-water piping and may drip into equipment or occupied areas. Refrigeration piping can pick up heat from the air, causing an air conditioning system to lose efficiency. Similarly, heat can escape from hot-water piping. To prevent these conditions, some piping is insulated.

Insulation is a material that prevents the transfer of heat. Cork, glass fibers, mineral wool, and polyurethane foams are examples of insulating materials. Insulation should be fire-resistant, moisture-resistant, and vermin-proof.

ACR tubing can be purchased with factory-installed insulation. If it is necessary to install the insulation at the job site, it should be added before the tubing is connected. That way, the insulation can be slid onto the tubing. The inside of the insulation is usually powdered to allow it to slip on easily.

If the insulation cannot be installed before the tubing is connected, it must be slit lengthwise to fit onto the pipe. Slit seams and connecting seams must then be sealed with adhesive. Do not use tape. Insulation should not be stretched, because its effectiveness will be reduced.

Some pipes are always insulated, while others are insulated only under certain conditions. Local building codes and job specifications will usually describe insulation requirements. Piping insulation requirements will be covered in detail in a later module.

8.0.0 ◆ PRESSURE TESTING

When the installation is complete, it must be inspected to make sure that the work has been done in accordance with the job specifications and applicable codes. Then, the system must be leak-tested to be sure that all connections are secure.

Testing under pressure is the best method of leak testing. This test is done using nitrogen. Nitrogen is inexpensive and, if there is a leak, you do not risk losing expensive refrigerant and violating laws regarding the release of refrigerants into the atmosphere. The test pressure will vary based on the size of the system, but should never exceed the maximum pressure stated on the nameplate of any component.

 WARNING!
Oxygen, air, acetylene, or other gases should never be used to pressure test a refrigerant system. Oxygen will cause an explosion if it comes into contact with refrigerant oil. Acetylene is highly flammable. The only gas other than refrigerant that should be introduced into an HVAC system is nitrogen.

 NOTE
Although nitrogen is more common, some manufacturers permit the use of CO_2 for purging and testing.

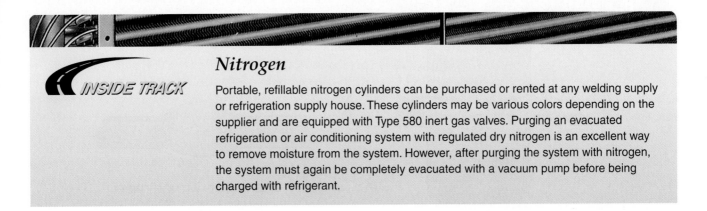

Nitrogen

INSIDE TRACK

Portable, refillable nitrogen cylinders can be purchased or rented at any welding supply or refrigeration supply house. These cylinders may be various colors depending on the supplier and are equipped with Type 580 inert gas valves. Purging an evacuated refrigeration or air conditioning system with regulated dry nitrogen is an excellent way to remove moisture from the system. However, after purging the system with nitrogen, the system must again be completely evacuated with a vacuum pump before being charged with refrigerant.

8.1.0 Technique

Exercise caution when using nitrogen to build up pressure for testing at code specifications. First, the refrigerant cylinder should be disconnected to prevent nitrogen from backing into it. Second, there must be a hand shutoff valve, pressure regulator, pressure gauge, and pressure relief valve in the charging line (*Figure 24*). The relief valve should be adjusted to open at one or two psi above the test pressure.

WARNING!

Make sure that nitrogen tanks are secured in a cart or other stabilizing device to prevent the tanks from tipping or being knocked over. If the tank valve is broken off when a tank falls, the tank cylinder can be propelled like a rocket by the escaping nitrogen gas.

A soap solution or commercial test liquid can be used to check for leaks when using nitrogen. This is done by wiping the outer surface of the joints with the solution and watching for any bubbles, which would indicate leaks. If no leaks are found, the nitrogen is usually left in the system for 24 hours. After no leaks are found, the technician pulls a vacuum in the system. If the vacuum is maintained over a specified period of time, the installation is approved. After refrigerant has been charged into a system, electronic leak detectors (*Figure 25*) are available to test for refrigerant leaks. They will detect tiny amounts of refrigerant and sound an audible alarm when a leak is detected.

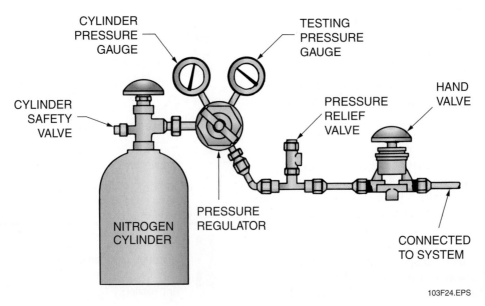

103F24.EPS

Figure 24 ◆ Nitrogen hookup.

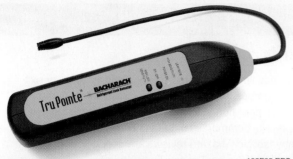

Figure 25 ◆ Electronic leak detector.

103F25.EPS

9.0.0 ◆ PIPING CODES

Most cities, counties, and states have adopted codes that may be based on national standards, but which may be subject to local interpretation and reflect local conditions. In earthquake-prone areas, for example, there may be special requirements for installing and testing equipment and piping. In flood-prone areas, other special requirements will apply. The job specifications should detail these requirements. Code violation corrections are expensive and are the responsibility of the installer. Standards serve as guidelines to improve the performance or reliability of a system or component, but codes and ordinances are specific and mandatory rules with which installers must comply.

10.0.0 ◆ SAFETY

HVAC systems operate under high pressure and are powered by electricity. Therefore, those working with HVAC systems must know and practice applicable safety precautions in order to protect themselves and their co-workers from injury and to prevent equipment damage.

Specific safety precautions will be covered as you progress through the HVAC training program. Here are a few general precautions that you should follow:

- Do not attempt any work that you have not been specifically trained to do, and then do it only under the direct supervision of your instructor or supervisor. For example, do not operate system valves unless you know exactly what the result will be. In general, before you work independently on an HVAC system, you must know the temperature and pressure conditions that exist at every point in the system, and how those conditions can be affected by malfunctions or by changes in valve positions.
- Disconnecting piping in a system that is under pressure can cause serious injury, as well as release of refrigerant into the atmosphere. The only proper way to reduce the pressure and avoid the release of refrigerant is to remove the refrigerant charge using approved recovery methods and equipment.
- Wear goggles and gloves when working with refrigerants, and don't work with refrigerants in poorly ventilated areas. Also, be aware that some refrigerants become toxic when exposed to an open flame.
- Unless it is absolutely necessary to work with electrical power applied, always shut off power and use approved lockout/tagout procedures to avoid electrical shock.

 WARNING!

Oxygen, compressed air, or acetylene must never be used when pressure testing for leaks. Oxygen will explode when exposed to oil; acetylene is highly flammable; and compressed air will contaminate the system.

Confined Space Hazards

A confined space is any space that has limited or restricted openings for entry or exit and in which rotating machinery, dangerous levels of toxic air contaminants, flammable or explosive air mixtures, excessive temperatures, lack of oxygen, or excess oxygen could exist. HVAC equipment, especially commercial or industrial systems, may be located in confined spaces. In many facilities, hazard reviews are conducted to identify non-permit and permit-required confined spaces. The facility safety office must always be contacted to obtain permission to enter any confined space. OSHA regulations specify that a permit-required confined space must be tested for hazardous atmosphere with a calibrated instrument before a worker is allowed to enter the space. In addition, the atmosphere must be continuously monitored while work is in progress in order to detect atmospheric hazards, including oxygen deficiency. Additionally, continuous communication must be maintained between workers inside and outside the confined space. HVAC workers entering any confined space should always observe proper precautions for ventilation, working in excessive temperatures, and working around operating machinery. No worker should enter any such space without appropriate rescue equipment and a means to communicate with a person stationed outside the space. The person stationed outside the space must be able to operate the necessary rescue equipment and summon help in the event of an emergency.

- Only nitrogen at the recommended pressure should be used to pressure test a system. If testing pressures are not given on the unit nameplate, never exceed national or local codes or the manufacturer's recommendations.

 WARNING!

If it is necessary to braze tubing in a confined space, make sure to connect a hose or pipe to the tubing being brazed so that the tubing is vented outside the confined space. Otherwise, the oxygen in the space will be displaced by the nitrogen gas, resulting in a hazardous oxygen deficiency.

Summary

An HVAC service technician must be able to work with several kinds of piping, including copper, steel, and plastic. Copper is the most common type used in the HVAC industry. It provides the passageways for refrigerant in most systems. Only ACR copper tubing is used for refrigerant because it is designed to withstand the pressures in HVAC systems.

Soft copper is available in rolls of 25' and 50'; diameters range from ⅛" to 1⅜". Soft copper tubing can easily be cut, bent, and joined. Hard copper tubing typically comes in 20' lengths with diameters up to 6".

Refrigerant tubing and piping must be protected from exposure to dirt and moisture. Pipe ends must remain capped, and a tubing cutter should be used, rather than a hacksaw, because it leaves no metal particles. If a hacksaw is used, the pipe must be thoroughly cleaned inside and out before it is joined.

Copper pipe can be joined with flare or compression fittings. However, it is more common to solder or braze the connections. Piping must be properly supported and insulated. The requirements for pipe hangers and insulation are usually specified in the job specifications or by local building codes.

Once the piping installation is complete, it must be inspected and tested. A common method of leak testing involves pressurizing the system with a charge of nitrogen and then using a soap solution to check for leaks.

Anyone working with pressurized systems and refrigerants must follow special safety practices to avoid injury to personnel or damage to equipment. Special safety practices also apply when working with HVAC systems in confined spaces.

Notes

1. Only _____ copper piping and fittings should be used in HVAC work.
   a. Type M
   b. ACR
   c. Type DWV
   d. soft-drawn

2. The term tubing applies to _____ .
   a. thin-wall piping that is not threaded
   b. soft copper pipe only
   c. hard copper pipe only
   d. small-diameter plastic piping

3. In an HVAC system, copper tubing with an outside diameter up to _____ is used to circulate refrigerant through the system.
   a. ¾"
   b. 1"
   c. 1⅜"
   d. 6"

4. The preferred tool for cutting copper tubing is a _____ .
   a. pipe threader
   b. hatchet
   c. hacksaw
   d. tubing cutter

5. A non-heat joining method for copper tubing is a _____ .
   a. soldered joint
   b. flare fitting
   c. brazed joint
   d. sweat connection

6. The type of plastic piping that can be threaded is _____ .
   a. PVC
   b. ABS
   c. PE
   d. Type K

7. The _____ is used to secure a pipe hanger to a beam.
   a. ring hanger
   b. pipe roll
   c. trapeze hanger
   d. C-clamp

8. If in doubt about which pipes to insulate, you should _____ .
   a. insulate all of them
   b. follow the job specifications and/or local codes
   c. insulate none of them
   d. insulate the refrigerant and hot-water lines

9. When nitrogen is used to pressure-test a system, the charging line must include a hand shutoff valve, pressure regulator, pressure gauge, and _____ .
   a. leak detector
   b. gauge manifold set
   c. refrigerant bottle
   d. pressure relief valve

10. Oxygen and acetylene should never be used for pressure testing because they _____ .
    a. do not give accurate test results
    b. are highly explosive under some circumstances
    c. are not readily available
    d. are too costly

Trade Terms Quiz

1. A connection created in which the diameter of one of the pipes to be joined is expanded using a special tool is called a(n) _____.

2. The distance between the outer walls of a pipe is referred to as the _____.

3. _____ is a class of refrigerants that includes most of the refrigerants used in residential and small commercial air conditioning systems.

4. A(n) _____ is a method of connection in which tightening of a threaded nut compresses a compression ring to seal the joint.

5. Copper tubing made especially for refrigeration and HVAC work is called _____.

6. Thin-wall pipe that can be easily bent is called _____.

7. _____ are documents that describe the quality of the material and work required.

8. A substance that retards the flow of heat is called _____.

9. _____ refers to the distance between the inner walls of a pipe.

10. _____ is a means of joining metal using an alloy with a melting point higher than that of solder, but lower than that of the metals being joined.

11. A method of joining pipe in which solder is applied to the joint and heated until the solder flows into the joint is called _____.

12. A fitting in which one end of each tube to be joined is flared outward using a special tool is called a(n) _____.

13. _____ is the use of heat treating to soften metal.

Trade Terms

Annealed copper refrigeration (ACR) tubing
Annealing
Brazing
Compression joint
Flare fitting
Halocarbon refrigerants
Inside diameter (ID)
Insulation
Outside diameter (OD)
Piping
Specifications
Swaged joint
Sweating
Tubing

Daniel Kerkman, Inside Sales and Technical Support

Climatic Control, Madison, WI
Instructor, ABC of Wisconsin

When he was a young man working as a maintenance mechanic in a resort hotel, Dan Kerkman's boss challenged him to learn about servicing HVAC systems in order to be more effective in his job and to create more opportunities for himself. Dan accepted the challenge, and that was the beginning of a long and rewarding career.

How did you choose a career in HVAC?
Actually, I think it chose me. I was working in maintenance at a hotel in Florida when the general manager suggested that I learn how to service and maintain HVAC systems. He said I could basically name my own salary if I was able to add that responsibility to my other duties. I started taking HVAC training courses right away and eventually was able to take on the responsibility for maintaining the hotel's HVAC systems.

What kinds of work have you done in your career?
Most of my career has involved installing and servicing a variety of residential and commercial equipment. I spent 16 years as a field service technician, servicing all kinds of HVAC and refrigeration equipment. In addition to my job with Climatic Control, I teach the HVAC Level 3 and Level 4 courses for the ABC of Wisconsin apprentice program. Also, for the last 3 years I developed and managed the HVAC program at the ABC of Wisconsin skill competition.

What do you like most about your job?
I like the variety. There are so many facets to HVAC work, and there's always something new to learn. Most HVAC systems have microprocessor controls these days. That means advances in computer technology filter down to the HVAC world, which makes it interesting and challenging. Also, manufacturers are constantly working to improve the efficiency of heating and cooling equipment, and to develop ways of improving indoor air quality.

What factors have contributed most to your success?
I am detail-oriented, and I think that helps a lot. When you make service calls, you need to do the job right the first time. Your company's reputation – and yours for that matter – will suffer if you have a lot of callbacks to fix problems that should have been fixed the first time. Don't forget, if you're making a call to fix a heating problem and it reoccurs, you will have some cold, unhappy customers.

What types of training have you been through?
I started with a two-year HVAC tech school in Florida. Since then, I've taken many night school courses conducted by manufacturers and distributors. I've also taken many refrigeration courses from the RSES (Refrigeration Service Engineers Society) and have also taught those courses.

What advice would you give to those who are new to the HVAC field?
One way to rise to the top is to become certified in different HVAC skill areas by taking the NATE (North American Training Excellence) tests*. Also, take advantage of training courses sponsored by manufacturers and distributors, even if it means giving up some free time. It will pay dividends in the long run. HVAC technology changes rapidly; the only way to keep ahead is to keep on learning.

* NATE (www.natex.org) is an independent certification agency that offers industry certification in several HVAC/R installation and service disciplines.

Trade Terms
Introduced in This Module

Annealed copper refrigeration (ACR) tubing: Copper tubing made especially for refrigeration and HVAC work. It is especially clean and is usually charged with dry nitrogen. The ends are sealed to prevent contamination.

Annealing: Heat treating to soften metal. Soft copper tubing is made by annealing hard copper.

Brazing: A means of joining metal using an alloy with a melting point higher than that of solder, but lower than that of the metals being joined.

Compression joint: A method of connection in which tightening of a threaded nut compresses a compression ring to seal the joint.

Flare fitting: A fitting in which one end of each tube to be joined is flared outward using a special tool. The flared tube ends mate with the threaded flare fitting and are secured to the fitting with flare nuts.

Halocarbon refrigerants: Short for halogenated hydrocarbons, a class of refrigerants that includes most of the refrigerants used in residential and small commercial air conditioning systems.

Inside diameter (ID): The distance between the inner walls of a pipe. Used as the standard measure for tubing used in heating and plumbing applications.

Insulation: A substance that retards the flow of heat.

Outside diameter (OD): The distance between the outer walls of a pipe. Used as the standard measure for ACR tubing.

Piping: A generic term used to designate thick-wall pipe that can be threaded and joined with threaded fittings.

Specifications: A document that describes the quality of the materials and work required. Specifications dictate the types of tubing, fixtures, hangers, etc., that must be used on a project.

Swaged joint: A method of creating a pipe joint in which the diameter of one of the pipes to be joined is expanded using a special tool. The other pipe then fits inside the swaged pipe.

Sweating: A method of joining pipe in which solder is applied to the joint and heated until the solder flows into the joint.

Tubing: Thin-wall pipe; generally, pipe that can be easily bent.

This module is intended to present thorough resources for task training. The following reference works are suggested for further study. These are optional materials for continued education, rather than for task training.

Cast Copper Solder-Joint Pressure Fittings, ASME B16.18, Current Edition. New York, NY: American Society of Mechanical Engineers.

Specification for Seamless Copper Water Tube, ASTM B88, Latest Edition. West Conshohocken, PA: American Society for Testing and Materials International.

Specification for Seamless Copper Pipe, ASTM B42, Latest Edition. West Conshohocken, PA: American Society for Testing and Materials International.

Wrought Copper and Copper Alloy Solder-Joint Pressure Fittings, ASME B16.22, Latest Edition. New York, NY: American Society of Mechanical Engineers.

CONTREN® LEARNING SERIES – USER UPDATE

NCCER makes every effort to keep these textbooks up-to-date and free of technical errors. We appreciate your help in this process. If you have an idea for improving this textbook, or if you find an error, a typographical mistake, or an inaccuracy in NCCER's Contren® textbooks, please write us, using this form or a photocopy. Be sure to include the exact module number, page number, a detailed description, and the correction, if applicable. Your input will be brought to the attention of the Technical Review Committee. Thank you for your assistance.

Instructors – If you found that additional materials were necessary in order to teach this module effectively, please let us know so that we may include them in the Equipment/Materials list in the Annotated Instructor's Guide.

Write: Product Development and Revision
National Center for Construction Education and Research
3600 NW 43rd St., Bldg. G, Gainesville, FL 32606

Fax: 352-334-0932

E-mail: curriculum@nccer.org

Craft _____ Module Name _____

Copyright Date _____ Module Number _____ Page Number(s) _____

Description _____

_____

_____

_____

(Optional) Correction _____

_____

_____

(Optional) Your Name and Address _____

_____

_____

Soldering and Brazing
03104-07

03104-07
Soldering and Brazing

Topics to be presented in this module include:

Overview

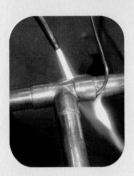

Certain types of metals can be joined by a process called soldering. This process produces a joint sealed with a low-temperature, non-magnetic solder alloy of lead and tin, or with non-lead alloys if the pipe that is being joined is used for drinking water supply. A low-temperature heat source, such as an air/acetylene torch, is used in the soldering process.

When mechanically strong, pressure-resistant joints are needed, the process used is brazing, a higher-temperature joining process using filler metals of silver or copper alloy. Because higher temperatures are needed for the brazing process, the heating equipment is typically oxygen/acetylene or oxygen/LP, each of which produces higher temperatures than soldering equipment.

Objectives

When you have completed this module, you will be able to do the following:

1. Assemble and operate the tools used for soldering.
2. Prepare tubing and fittings for soldering.
3. Identify the purposes and uses of solder and solder fluxes.
4. Solder copper tubing and fittings.
5. Assemble and operate the tools used for brazing.
6. Prepare tubing and fittings for brazing.
7. Identify the purposes and uses of filler metals and fluxes used for brazing.
8. Braze copper tubing and fittings.
9. Identify the inert gases that can be used safely to purge tubing when brazing.

Trade Terms

Alloy
Brazing
Capillary action
Flashback arrestor
Flux
Nonferrous

Oxidation
Purging
Solder
Soldering
Wetting

Required Trainee Materials

1. Paper and pencil
2. Appropriate personal protective equipment

Prerequisites

Before you begin this module, it is recommended that you successfully complete *Core Curriculum*; and *HVAC Level One*, Modules 03101-07 through 03103-07.

This course map shows all of the modules in *HVAC Level One*. The suggested training order begins at the bottom and proceeds up. Skill levels increase as you advance on the course map. The local Training Program Sponsor may adjust the training order.

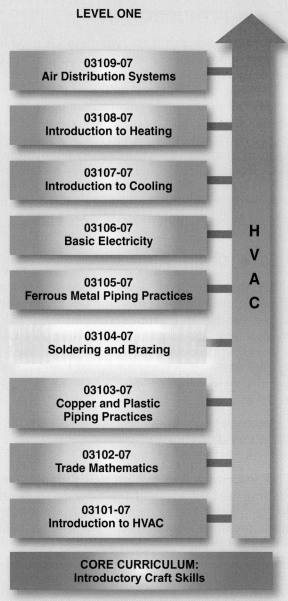

LEVEL ONE

03109-07
Air Distribution Systems

03108-07
Introduction to Heating

03107-07
Introduction to Cooling

03106-07
Basic Electricity

03105-07
Ferrous Metal Piping Practices

03104-07
Soldering and Brazing

03103-07
Copper and Plastic Piping Practices

03102-07
Trade Mathematics

03101-07
Introduction to HVAC

CORE CURRICULUM:
Introductory Craft Skills

H V A C

104CMAP.EPS

1.0.0 ◆ INTRODUCTION

Soldering and **brazing** are two methods used for joining copper tubing and fittings. Both methods fasten the metals together using a **nonferrous** filler metal that adheres to the surfaces being joined. The filler metal is distributed between the closely fitted surfaces by **capillary action**. The difference between soldering and brazing is the temperature needed to melt the filler metal in order to make the joint. Soldering uses filler metals that melt at temperatures below 800°F, usually in the 375°F to 500°F range. The filler metals used for brazing melt at temperatures above 650°F.

2.0.0 ◆ SOLDERING

Soldering, also called sweat soldering and soft soldering, is the most common method of joining copper tubing and fittings. It is of little value where strength is required and is normally used as a sealing process. Soldered joints are used in piping systems that carry liquids at temperatures of 250°F or below. Soldered joints are typically used for:

- Domestic water lines
- Sanitary drain lines
- Hydronic heating systems

Soldering involves joining two metal surfaces by using heat and a nonferrous filler metal. A nonferrous filler metal is a metal that contains no iron and is therefore nonmagnetic. The melting point of the filler metal must be lower than that of the two metals that are being joined.

Soldered joints depend on capillary action to pull and distribute the melted **solder** into the small gap between the fitting and the tubing. When the joint is filled, the solder will form a tiny bead around the joint called a fillet. Capillary action is the flow of liquid, in this case solder, into a small space between two surfaces. Capillary action is most effective when the space between the tubing and the fitting is between 0.002" and 0.005". To maintain the proper spacing, it is important to check the joint for correct alignment before soldering. *Figure 1* shows capillary action.

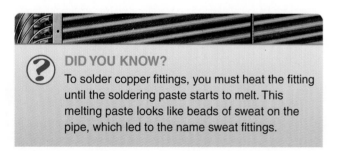

DID YOU KNOW?
To solder copper fittings, you must heat the fitting until the soldering paste starts to melt. This melting paste looks like beads of sweat on the pipe, which led to the name sweat fittings.

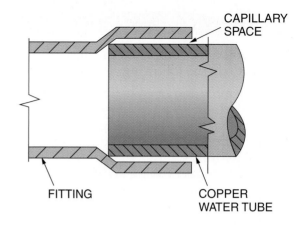

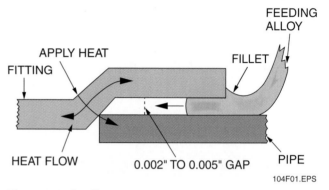

Figure 1 ◆ Capillary action.

To properly solder copper tubing and fittings, you must understand the following materials and procedures:

- Solders and soldering **fluxes**
- Preparing tubing and fittings for soldering
- Soldering joints

2.1.0 Solders and Soldering Fluxes

Solder is a nonferrous metal or metal **alloy** with a melting point below 800°F. An alloy is any substance made up of two or more metals. The Federal Safe Drinking Water Act Amendment of 1986 mandates the use of lead-free solder for drinking water supply piping. Therefore, use of soft solder composed of 50 percent tin and 50 percent lead is not permitted for joining copper pipe. The solder recommended for potable (drinking) water tubing is composed of either 95.5 percent tin, 4 percent copper, and 0.5 percent silver or 96 percent tin and 4 percent silver. Any solder containing antimony, cadmium, or lead is not used for potable water. The most common solder used on HVAC copper low-pressure refrigeration tubing is an alloy made of 95 percent tin and 5 percent antimony. The solders composed of tin-copper-silver, tin-silver, and tin-antimony are usually recommended for applications requiring greater joint strength. The tin-antimony solder melts between

430°F and 480°F and solidifies rapidly. Generally, these solders, in the form of wire, are supplied on spools for easier use.

Choosing proper flux is very important. Soldering flux performs many functions, and the wrong flux can ruin the soldered joint. Flux performs the following functions:

- It chemically cleans and protects the surfaces of the tubing and its fitting from oxidation. Oxidation occurs when the oxygen in the air combines with the recently cleaned metal. Oxidation produces tarnish or rust in metal and prevents solder from adhering.
- It allows the soldering alloy or filler metal to flow easily into the joint.
- It floats out remaining oxides ahead of the molten filler metal.
- It promotes wetting of the metals. Wetting is the process that reduces the surface tension so that the molten solder flows evenly throughout the joint.

Fluxes can be classified into three general groups: highly corrosive, less corrosive, and noncorrosive. The fluxing process must render the flux inert; that is, lacking any chemical action. If not rendered inert, the flux gradually destroys the soldered joint.

The fluxes used for joining copper tubing and copper fittings should be noncorrosive fluxes. One type is composed of water and white rosin dissolved in an organic or benzoic acid base. The best noncorrosive fluxes for joining copper pipe and fittings are compounds of mild concentrations of zinc and ammonium chloride with petroleum bases.

An oxide film begins forming on copper immediately after it has been mechanically cleaned. Therefore, it is important to apply flux immediately to all recently cleaned copper fittings and tubing. Flux should be applied to the clean metal with a brush or swab, never with your fingers. Not only is there a chance of causing infection in a cut, but there is also a chance that flux could be carried to the eyes or mouth. In addition, body contact with the cleaned fittings and tubing adds to unwanted contamination of the metal.

Flux must be stirred before each use. If a can of flux is not closed immediately after use, or if a can is not used for a considerable length of time, the chlorides separate from the petroleum base.

CAUTION

Brazing flux and soldering flux are not the same. Do not allow these fluxes to become mixed or interchanged. Carelessness can ruin work.

2.2.0 Preparing Tubing and Fittings for Solder

To prepare tubing and fittings for soldering, the tubing must be measured, cut, and reamed, and the tubing and fittings must be cleaned. It is critical to use proper cleaning techniques in order to produce a solid, leakproof joint. Use the following procedure to prepare the tubing and fittings for soldering.

Step 1 Measure the distance between the faces of the two fittings (face-to-face method). *Figure 2* shows the face-to-face method.

Step 2 Determine the cup depth engagement of each of the fittings. The cup depth engagement is the distance that the tubing penetrates the fitting. This distance can be found by measuring the fitting or by using a manufacturer's makeup chart. *Table 1* shows a manufacturer's makeup chart.

Step 3 Add the cup depth engagement of both fittings to the measurement found in Step 1 to find the length of tubing needed.

Step 4 Cut the copper tube to the correct length using a tubing cutter.

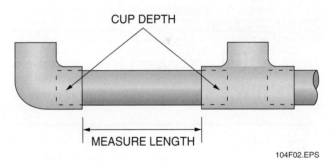

CUP DEPTH

MEASURE LENGTH

104F02.EPS

Figure 2 ◆ Face-to-face method.

Table 1	Manufacturer's Makeup Chart		
Pipe Size	**Depth of Cup**	**Pipe Size**	**Depth of Cup**
¼	⁵⁄₁₆	2	1¹¹⁄₃₂
⅜	⅜	2½	1¹⁵⁄₃₂
½	½	3	1²¹⁄₃₂
⅝	⅝	3½	1²⁹⁄₃₂
¾	¾	4	2⁵⁄₃₂
1	²⁹⁄₃₂	5	2²¹⁄₃₂
1¼	³¹⁄₃₂	6	3³⁄₃₂
1½	1³⁄₃₂	—	—

104T01.EPS

Some Characteristics of Lead-Free Solder

INSIDE TRACK

Typically, lead-free solders have a different behavior and appearance than solders with lead. Some things to be aware of are the following:

- Lead-free solders have higher melting temperatures. This can lead to using too-high temperatures.
- Lead-free solders don't wet or spread the same as lead-containing solders. Additional time is required when working with lead-free solder.
- Lead-free solder can be difficult to work with due to bridging characteristics different from lead-based solder.

Because it oxidizes rapidly, lead-free solder requires a flux that works quickly so that surfaces remain clean and free from oxidation. Most commonly used in lead-free solder is an alloy of tin, silver, and copper.

Step 5 Ream the inside of both ends of the copper tube using a reamer.

CAUTION

Take care when cleaning copper tubing and fittings to remove all of the abrasions on the copper without removing a large amount of metal. Abrasions can weaken or ruin a copper joint. Do not touch or brush away filings from the tube or fitting with your fingers because your fingers will also contaminate the freshly cleaned metal.

Step 6 Clean the tubing and fitting to a bright finish using an abrasive pad, an emery cloth, or a special copper-cleaning tool.

NOTE

Some pipe cutters are equipped with a cleaner so that copper or aluminum tubes can be cut and cleaned in one operation.

CAUTION

Do not allow more than two hours to elapse between cleaning, fluxing, and soldering the joint(s).

Step 7 Using a brush or swab, apply flux to the copper tubing and to the inside of the copper fitting socket immediately after cleaning them.

Step 8 Insert the tube into the fitting socket, and push and turn the tube into the socket until the tube touches the inside shoulder of the fitting.

Step 9 Wipe away any excess flux from the joint.

Step 10 Check the tube and fitting for proper alignment before soldering.

2.3.0 Soldering Joints

Because soldering requires relatively low heat, heating equipment that mixes acetylene, MAPP® (methylacetylene propadiene), propylene, butane, or propane gas directly with air is all that is needed. This means that you need only one tank of gas and an air/liquefied petroleum (LP) or air/acetylene torch.

Air/acetylene torches are designed to intake the correct amount of air from the atmosphere. However, air is about 80 percent nitrogen and does not support combustion. So acetylene (the fuel gas) is mixed with air to form a combustible mixture. Since the nitrogen in air does not support combustion, the flame produced by an air/acetylene torch is of a lower temperature than the flame produced by an oxygen/acetylene torch. For smaller, lighter work, acetylene is usually carried in small, easily transportable cylinders. Use the following procedure to solder a joint.

WARNING!

Always solder in a well-ventilated area because fumes from the flux can irritate your eyes, nose, throat, and lungs.

Acetylene Hazards and Costs

INSIDE TRACK

If possible, fuels other than acetylene should be used for soldering or brazing purposes. Acetylene is an unstable gas. It can become volatile and its storage tanks can explode if released at pressures exceeding 15 psi or if released from a tank at rates greater than ⅒ of the tank's contents per hour. Because of these hazards, some commercial-business insurance companies require training certification for employees who handle and use acetylene. Moreover, acetylene has a lower heat content per cubic foot than any of the other fuel gases except methane or natural gas. Butane, propane, *MAPP*®, and propylene all have twice as much heat content per cubic foot. This high-heat content allows much less fuel consumption for the same amount of heating. With the exception of *MAPP*® gas, all the other fuel gases are considerably cheaper per cubic foot than acetylene. The major advantage of a properly adjusted oxygen/acetylene torch is that higher combustion temperatures can be reached using acetylene than with any of the other gases. All other fuel gases used with oxygen produce lower temperatures and take longer to heat material for fusion welding; however, they are extensively used for brazing or cutting purposes because of their safety and lower cost.

Step 1 Obtain either an acetylene tank and related equipment or a propane bottle and torch. Obtain the required solder.

Step 2 Set up the equipment according to the manufacturer's instructions.

CAUTION

Whenever soldering or brazing with an open flame, always keep a fire extinguisher of the correct type nearby. Shield any combustible materials near the work area with a fire blanket placed a safe distance away from the material. Many fires are started accidentally by the ignition of nearby wood when using a torch.

Excess heat can damage certain components, including reversing valves and service valves. If the heat-sensitive component cannot be temporarily removed, then it should be protected by wrapping it with a wet cloth. Heat absorbing pastes can also be used to protect components from excess heat.

Step 3 Light the heating equipment according to the type of equipment you are using and the manufacturer's instructions.

WARNING!

When lighting the torch, be sure to wear gloves and goggles. Point the torch away from your body when lighting it. Use only a spark lighter to light the torch. Do not use a match, cigarette lighter, or cigarette to light the torch, because this could result in severe burns or cause the lighter to explode.

Step 4 Heat the tubing first, and then move the flame onto the fitting (*Figure 3*). Make sure not to overheat the joint.

CAUTION

The inner cone of the flame should barely touch the metal being heated. Do not direct the flame into the socket because this will burn the flux. Be sure to keep the flame moving instead of holding the flame in one place.

Step 5 Move the flame away from the joint.

Step 6 Touch the end of the solder to the area between the fitting and the tube. The solder will be drawn into the joint by capillary action. The solder can be fed upward or downward into the joint.

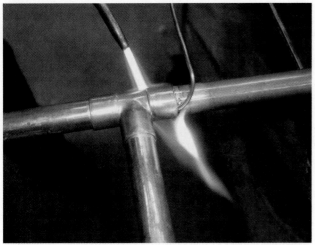

Figure 3 ◆ Heating a fitting with a torch.

104F03.EPS

NOTE

If the solder does not melt on contact with the joint, remove the solder and heat the joint again. Do not melt the solder with the flame.

DID YOU KNOW?

Two kinds of solder fittings are available with copper tubing. The first is a wrought copper fitting, which is made from copper tubing that is shaped into different types of fittings. Wrought copper fittings are generally lightweight, are smooth on the outside, and have thin walls. The second type is a cast solder fitting. This type of fitting is made using a mold. The first cast fittings had holes in the sockets to put solder in. Today, the heated copper is poured into the mold and allowed to cool. Cast copper fittings have a rough exterior and come in a wide variety of shapes. They are heavier than wrought copper fittings.

Step 7 Continue to feed the solder into the joint until a ring of solder appears around the joint, indicating that the joint is filled. On ¾" diameter tubing and smaller, the solder can be fed into the joint from one point. On larger tubing, the solder should be applied from the six o'clock position to the twelve o'clock position on the tubing. Generally, the amount of solder used is equal to the diameter of the tubing. For example, with ¾" tubing, ¾" of solder will fill the joint.

Step 8 Allow the joint to cool until the solder solidifies.

NOTE

Solder joints do not cool as quickly as brazed joints. Water can be applied to the joint to speed the cooling process.

Step 9 While the joint is still hot, wipe it clean with a soft, wet cloth to remove any flux.

3.0.0 ◆ BRAZING COPPER FITTINGS AND TUBING

Brazing, like soldering, uses nonferrous filler metals to join base metals that have a melting point above that of the filler metals. Brazing is performed above 800°F. Brazed tubing and fittings are used in:

- Low-pressure steam lines
- High-pressure refrigeration lines
- Medical gas lines
- Compressed air lines
- Vacuum lines
- Fuel lines
- Other chemical lines that need extra corrosion resistance in the piping joints

Solder Paste

Solder that has been ground into very fine particles and mixed with a liquid flux is also available. After the ends of a tube and fitting have been prepared and cleaned, a coating of the paste is applied completely around the end of the tube and a very light coating is applied to the inside of the fitting. The tube is inserted into the fitting, leaving a small bead of the paste at the edge of the fitting. The fitting is heated until the solder melts and is drawn into the joint. Additional solder is not usually required to complete the joint. Once the joint has cooled enough to solidify the solder, the joint is wiped with a wet cloth to remove the flux.

Brazing, also known as hard soldering, produces mechanically strong, pressure-resistant joints. The strength of a brazed joint results from the ability of the filler metal to adhere to the base metal. However, adhesion can occur only if the base metals are properly cleaned, the proper flux and filler metal are selected, and the clearance gap between the outside of the tubing and the inside of the fitting is 0.003" to 0.004".

NOTE

When brazing copper to brass, you must use flux because there is a higher probability of oxidation. White paste fluxes are standard for copper-to-brass brazing. A brazing rod containing 45 percent silver is normally used for this purpose.

To properly braze copper tubing and fittings, you must understand the following materials and procedures:

- Filler metals and fluxes
- Preparing tubing and fittings for brazing
- Setting up heating equipment
- Lighting an oxyacetylene torch or oxygen/ liquefied petroleum (LP) torch
- Brazing joints

3.1.0 Filler Metals and Fluxes

Filler metals used to join copper tubing are of two groups: alloys that contain 8 percent to 60 percent silver (the BAg series), and copper alloys that contain phosphorus (the BCuP series). *Table 2* lists brazing filler materials according to their American Welding Society (AWS) classification and principal elements.

WARNING!

BAg-1 and BAg-2 contain cadmium. Heating when brazing can produce highly toxic fumes. Use adequate ventilation and avoid breathing the fumes.

The two groups of filler metal differ in their melting, fluxing, and flowing characteristics. These characteristics should be considered when selecting a filler metal. When joining copper tubing, any of these filler metals can be used; however, the filler metals used most often for close tolerances are BCuP-3 and BCuP-4. BCuP-5 is used where close tolerances cannot be held, and BAg-1 is used as a general-purpose filler metal.

As with soldering, fluxes are also necessary when brazing. Brazing fluxes are applied using the same methods and rules as soldering fluxes. Brazing fluxes are more corrosive than soldering fluxes, so care must be taken never to mix a soldering flux with a brazing flux. For best results, use the flux recommended by the manufacturer of the brazing filler metals.

When copper tubing is joined to wrought copper fittings with copper-phosphorus alloys (BCuP series), flux can be omitted because the copper-phosphorus alloys are self-fluxing on copper. However, fluxes are required for joining all cast fittings.

3.2.0 Preparing Tubing and Fittings for Brazing

To prepare tubing and fittings for brazing, you must follow the same procedures as you would to prepare tubing and fittings for soldering. It is critical that proper cleaning techniques be used in

Table 2 Brazing Filler Materials

AWS Classification	Percent of Principal Element					
	Silver	Phosphorus	Zinc	Cadmium	Tin	Copper
BCuP-2	—	7–7.5	—	—	—	Balance
BCuP-3	4.75–5.25	5.75–6.25	—	—	—	Balance
BCuP-4	5.75–6.25	7–7.5	—	—	—	Balance
BCuP-5	14.5–15.5	4.75–5.25	—	—	—	Balance
BAg-1	44–46	—	14–18	23–25	—	14–16
BAg-2	34–36	—	19–23	17–19	—	25–27
BAg-5	44–46	—	23–27	—	—	29–31
BAg-7	55–57	—	15–19	—	4.5–5.5	21–23

104T02.EPS

Brazing

To achieve a sufficiently hot flame for brazing, use an oxyacetylene torch unit capable of generating temperatures of 1,400°F (760°C) or higher. Brazing requires a torch with the correct tip, regulator valves, and both oxygen and acetylene tanks. These tanks come in two sizes. The smaller B tank is commonly used. This tank has two compartments to hold the pressurized acetylene and oxygen.

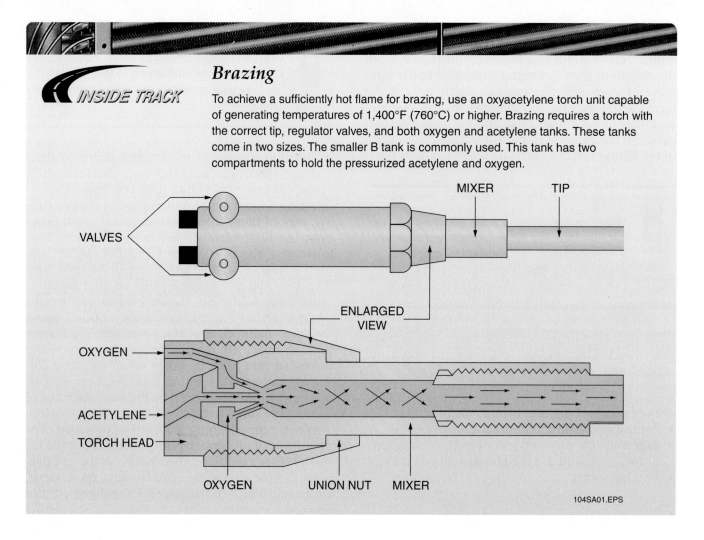

104SA01.EPS

order to produce a solid, leakproof joint. Use the following procedure to prepare the tubing and fittings for brazing.

Step 1 Measure the distance between the faces of the two fittings.

Step 2 Determine the cup depth engagement of each of the fittings.

 NOTE

The cup depths of fittings used with brazing are shorter than the cup depths of fittings used with soldering. The reason is that less penetration is needed for brazing than for soldering. This distance can be found by measuring the fitting or by using a manufacturer's makeup chart. *Figure 4* shows a manufacturer's brazing fitting makeup chart.

Step 3 Add the cup depth engagement of both fittings to the measurement found in Step 1 to find the length of tubing needed.

Step 4 Cut the copper tubing to the correct length using a tubing cutter.

Step 5 Ream the inside and outside of both ends of the copper tubing using a reamer.

 CAUTION

Take care when cleaning the tubing and fittings to remove all the abrasions on the copper without removing a large amount of metal. Abrasions can weaken or ruin a copper joint. Do not touch or brush away filings from the tube or fitting with your fingers because your fingers will also contaminate the freshly cleaned metal.

Step 6 Clean the tubing and fitting using No. 00 steel wool, an emery cloth, an abrasive pad, or a special copper-cleaning tool.

Step 7 Apply flux to the copper tubing and to the inside of the copper fitting socket immediately after cleaning them.

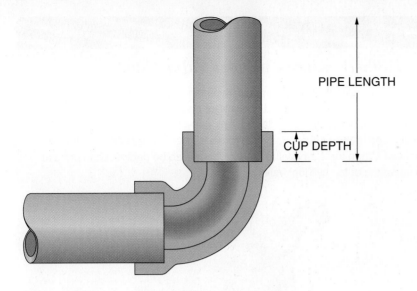

PIPE SIZE (in.)	CUP DEPTH (in.)
$^{1}/_{4}$	$^{17}/_{64}$
$^{3}/_{8}$	$^{5}/_{16}$
$^{1}/_{2}$	$^{3}/_{8}$
$^{3}/_{4}$	$^{13}/_{32}$
1	$^{7}/_{16}$
$1^{1}/_{4}$	$^{1}/_{2}$
$1^{1}/_{2}$	$^{5}/_{8}$
2	$^{21}/_{32}$
$2^{1}/_{2}$	$^{25}/_{32}$
3	$^{53}/_{64}$
$3^{1}/_{2}$	$^{7}/_{8}$
4	$^{29}/_{32}$
5	1
6	$1^{7}/_{64}$
7	$1^{7}/_{32}$
8	$1^{5}/_{16}$

104F04.EPS

Figure 4 ◆ Manufacturer's fitting makeup chart.

Step 8   Insert the tube into the fitting socket, and push and turn the tube into the socket until the tube touches the inside shoulder of the fitting.

Step 9   Wipe away any excess flux from the joint.

Step 10   Check the tube and fitting for proper alignment before brazing.

3.3.0 Setup of Brazing Heating Equipment

The brazing heating procedure differs from soldering in that different equipment is required to raise the temperature of the metals to be joined above 800°F. Actually, most brazed joints are made at temperatures between 1,200°F and 1,550°F. Because of the higher temperatures needed, oxygen/acetylene (oxyacetylene) or oxygen/LP (oxyfuel) brazing equipment is used for brazing. The flame is produced by burning a fuel gas mixed with pure oxygen. *Figure 5* shows oxyacetylene brazing equipment. Oxyfuel equipment is similar. Only the regulator for the fuel gas is different.

3.3.1 Handling Oxygen and Acetylene Cylinders

Working with oxyacetylene brazing equipment requires following basic safety precautions. Oxygen and acetylene are compressed and shipped under medium to high pressures in cylinders. Because their use is so common, technicians often get careless about handling them. Oxygen is supplied in cylinders at pressures of about 2,000 psi.

Acetylene cylinders are pressurized at about 250 psi. These cylinders should not be moved unless the protective caps are in place. Dropping a cylinder without the cap installed may result in breaking the valve off the cylinder. This allows the pressure inside to escape, propelling the cylinder like a rocket.

During use, transportation, and/or storage, oxygen and acetylene cylinders must be secured with a stout cable or chain in the upright position to prevent them from falling and injuring people or damaging equipment. When stored at the job site, oxygen and acetylene cylinders must be stored separately with at least 20' between them, or with a 5' high, ½-hour minimum fire wall separating them. Store empty cylinders away from partially full or full cylinders and make sure they are properly marked to clearly show that they are empty.

Oxygen can cause ignition even when no flame or spark is around to set it off, especially when it comes in contact with oil or grease. Never handle oxygen cylinders with oily hands or gloves. Keep grease away from the cylinders and do not use oil or grease on cylinder attachments or valves. Never use an oxygen regulator for any other gas or try to use a regulator with oxygen that has been used for other service.

A pressure-reducing regulator set for not more than 15 psig must be used with acetylene. Acetylene becomes unstable and volatile above 15 psig. The valve wrench should be left in position on open acetylene valves. This enables quick closing in an emergency. It is good practice to open the acetylene valve as little as possible, but never more than 1¼ turns.

Alternate High-Pressure Cylinder Valve Cap

High-pressure cylinders can also be equipped with a clamshell cap that can be closed to protect the cylinder valve with or without a regulator installed on the valve. This enables safe movement of the cylinder after the cylinder valve is closed. This type of cap is usually secured to the cylinder body cap threads when it is installed so that it cannot be removed. When the clamshell is closed, it can also be padlocked to prevent unauthorized operation of the cylinder valve.

CLAMSHELL OPEN TO ALLOW
CYLINDER VALVE OPERATION

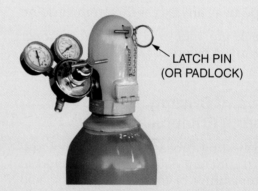

LATCH PIN
(OR PADLOCK)

CLAMSHELL CLOSED FOR MOVEMENT OR PADLOCKED
TO PREVENT OPERATION OF CYLINDER VALVE

CLAMSHELL CLOSED FOR TRANSPORT

104SA02.EPS

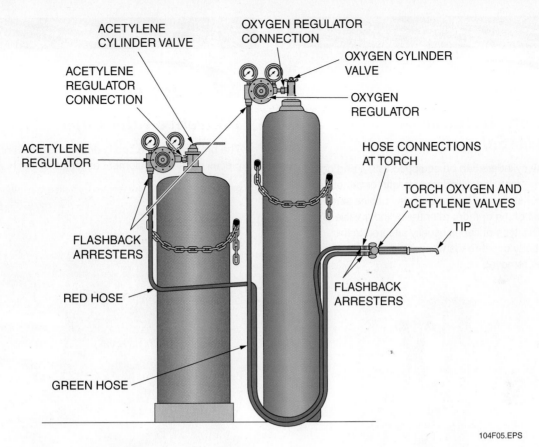

ACETYLENE
CYLINDER VALVE

OXYGEN REGULATOR
CONNECTION

ACETYLENE
REGULATOR
CONNECTION

OXYGEN CYLINDER
VALVE

OXYGEN
REGULATOR

ACETYLENE
REGULATOR

HOSE CONNECTIONS
AT TORCH

TORCH OXYGEN AND
ACETYLENE VALVES

TIP

FLASHBACK
ARRESTERS

RED HOSE

FLASHBACK
ARRESTERS

GREEN HOSE

Figure 5 ◆ Oxyacetylene or oxygen/LP gas brazing equipment.

104F05.EPS

3.3.2 Initial Setup of Oxyacetylene Equipment

Follow this procedure to set up oxyacetylene brazing equipment. The setup for oxyfuel equipment is similar.

 WARNING!
Do not handle acetylene and oxygen cylinders with oily hands or gloves. Keep grease away from the cylinders and do not use oil or grease on cylinder attachments or valves. The mixture of oil and oxygen will cause an explosion.

 WARNING!
Make sure that the protective caps are in place on the cylinders before transporting or storing the cylinders.

Step 1 Install and securely fasten the oxygen and acetylene cylinders in a bottle cart or in an upright position.

 WARNING!
Do not allow anyone to stand in front of the oxygen cylinder valve when opening because the oxygen is under high pressure (about 2,000 psig) and could cause severe injury when released.

Step 2 Install the oxygen regulator on the oxygen cylinder.
  • Remove the cylinder protective cap.
  • Open (crack) the oxygen cylinder valve just long enough to allow a small amount of oxygen to pass through the valve, then close it.
  • Turn the adjusting screw on the oxygen regulator (*Figure 6*) counterclockwise until it is loose. This will shut off the regulator output and prevent over-pressurizing of the hose and torch during hookup.

INSIDE TRACK

Alternate Acetylene Cylinder Safety Cap

Acetylene cylinders can be equipped with a ring guard cap that protects the cylinder valve with or without a regulator installed on the valve. This enables safe movement of the cylinder after the cylinder valve is closed. This type of cap is usually secured to the cylinder body cap threads when it is installed so that it cannot be removed.

104SA03.EPS

INSIDE TRACK

Protective Valve Caps

The protective valve cap is one of the most important pieces of safety equipment associated with brazing.

104SA04.EPS

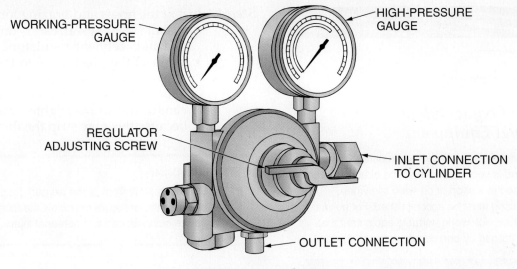

WORKING-PRESSURE GAUGE

HIGH-PRESSURE GAUGE

REGULATOR ADJUSTING SCREW

INLET CONNECTION TO CYLINDER

OUTLET CONNECTION

OXYGEN REGULATOR

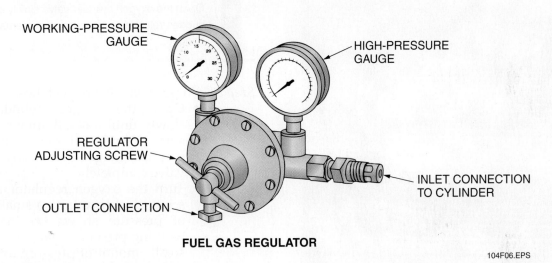

WORKING-PRESSURE GAUGE

HIGH-PRESSURE GAUGE

REGULATOR ADJUSTING SCREW

INLET CONNECTION TO CYLINDER

OUTLET CONNECTION

FUEL GAS REGULATOR

104F06.EPS

Figure 6 ◆ Oxygen and fuel gas regulators.

- Using a suitable wrench, install the oxygen regulator on the cylinder. Oxygen cylinders and regulators have right-hand threads. Tighten the nut snugly. Be careful not to overtighten the nut because this may strip the threads.

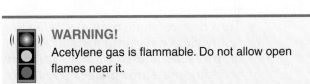

WARNING!

Acetylene gas is flammable. Do not allow open flames near it.

Step 3 Install the acetylene regulator on the acetylene cylinder.

NOTE

Acetylene is stored in the cylinder at a pressure of about 250 psig.

- Remove the cylinder protective cap.
- Open (crack) the acetylene cylinder valve, using the valve wrench, just long enough to allow a small amount of acetylene to pass through the valve, then close it.

Portable Oxyacetylene or Oxyfuel Equipment

The equipment shown in *Figure 5* is mounted on a hand truck and is very heavy. This type of equipment is typically used in a shop or on a job site when extensive brazing must be accomplished. For normal installation or service work, portable equipment that can be hand carried by one person is generally used.

104SA05.EPS

- Turn the adjusting screw on the acetylene regulator counterclockwise until it is loose. This will shut off the regulator output and prevent overpressurizing of the hose and torch during hookup.
- Using a suitable wrench, install the acetylene regulator on the cylinder. Acetylene cylinders and regulators have left-hand threads. Tighten the nut snugly. Be careful not to overtighten the nut because this may strip the threads.

Step 4 Install the hoses and brazing torch.
- Install **flashback arrestors** on the oxygen and acetylene regulators.
- Connect the green hose to the oxygen gauge and the red hose to the acetylene gauge. Tighten the hoses snugly. Be careful not to overtighten the fittings because this may strip the threads.

WARNING!
Do not stand in front of the oxygen gauge because the pressure may blow the face of the gauge outward, causing personal injury.

CAUTION
Open the oxygen cylinder valve slowly because a sudden release of pressure could damage the gauges.

Step 5 Purge (clean) the oxygen hose.
- Open the oxygen cylinder valve slowly until a small amount of pressure registers on the oxygen high-pressure gauge; then slowly open the valve completely.
- Turn the oxygen regulator adjusting screw clockwise until a small amount of pressure shows on the oxygen working-pressure gauge. Allow a small amount of pressure to build up and purge the oxygen hose, cleaning it.
- Turn the oxygen regulator adjusting screw counterclockwise until it is loose. This will shut off the regulator output.

Step 6 Purge (clean) the acetylene hose.
- Open the acetylene cylinder valve slowly until a small amount of pressure registers on the acetylene high-pressure gauge. Then open it about ½ turn.
- Turn the acetylene regulator adjusting screw clockwise until a small amount of pressure shows on the acetylene working-pressure gauge (*Figure 6*). Allow a small amount of pressure to build up and purge the acetylene hose, cleaning it.
- Turn the acetylene regulator adjusting screw counterclockwise until it is loose. This will shut off the regulator output.

Flashback Arrestors

Flashback arrestors are one-way valves that prevent a pressurized flame from traveling back up the tip and into the hoses and regulators. A flashback (back-burn) can sometimes happen if either the oxygen or the fuel gas flow rate is inadequate, or if both the oxygen and fuel gas are turned on and then ignited at the welding tip. To be effective, flashback arrestors must be installed with the flow arrow pointing toward the torch handle. Flashback is such a major concern that torch manufacturers are adding them directly to the torches.

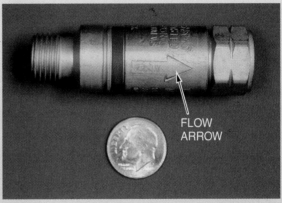

FLOW ARROW

104SA06.EPS

Step 7 Install flashback arrestors on the torch.

Step 8 Install the brazing torch on the ends of the hoses and close the valves on the torch.

> **WARNING!**
>
> Never adjust the acetylene regulator higher than 15 psig because acetylene becomes unstable and volatile at this pressure.

Step 9 Check the oxyacetylene equipment for leaks.
- Adjust the acetylene regulator adjusting screw for 10 psig on the working-pressure gauge.
- Adjust the oxygen regulator adjusting screw for 40 psig on the working-pressure gauge.

> **WARNING!**
>
> Do not use a soap with an oil base for leak testing because the mixture of oil and oxygen may cause an explosion.

- Close the oxygen and acetylene cylinder valves and check for leaks. If the working-pressure gauges remain at 10 and 40 psig, there are no leaks in the system. If the readings drop, there is a leak. Use a soap solution or commercial leak detection fluid to check the oxygen or acetylene connections for leaks.
- Open both valves on the torch to release the pressure in the hoses. Watch the working-pressure gauges until they register zero, then close the valves on the torch.
- Turn the oxygen and acetylene regulator valves counterclockwise until they are loose. This will release the spring pressure on the regulator diaphragms and completely close the regulators.

Step 10 Coil the hoses and hang them on the hose holder.

3.3.3 Lighting the Oxyacetylene Torch

After the oxyacetylene brazing equipment has been properly set up, the torch can be lit and the flame adjusted for brazing. There are three types of flames: neutral, carburizing (reducing), and oxidizing. The neutral flame burns equal amounts of oxygen and acetylene. The inner cone is bright blue in color, surrounded by a fainter blue outer flame envelope that results when the oxygen in the air combines with the superheated gases from the inner cone. A neutral flame is used for almost all fusion welding or heavy brazing applications.

A carburizing (reducing) flame has a white feather created by excess fuel. The length of the feather depends on the amount of excess fuel in the flame. The outer flame envelope is brighter than that of a neutral flame and is much lighter in color. The excess fuel in the carburizing flame produces large amounts of carbon. The carburizing flame is cooler than the neutral flame and is used for light brazing to prevent melting the base metal.

An oxidizing flame has an excess amount of oxygen. Its inner cone is shorter, with a bright blue edge and a lighter center. The cone is also more pointed than the cone of a neutral flame. The outer flame envelope is very short and often fans out at the ends. The hottest flame, it is sometimes used for brazing cast iron or other metals. *Figure 7* shows the types of flames.

Use the following procedure to light an oxyacetylene torch.

Step 1 Set up the oxyacetylene torch as discussed previously. Make sure the correct tip is installed before lighting the torch. (Refer to *Table 3* for recommended tip sizes.) Adjust regulators for pressure settings recommended by the torch manufacturer.

NOTE
Pressures are not standardized for oxyacetylene torches. Refer to the manufacturer's instructions for recommended gas pressures for the pipe size being brazed.

CARBURIZING FLAME

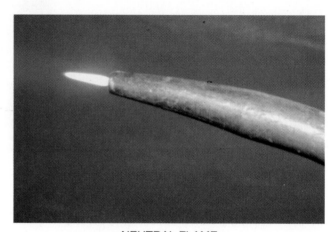
NEUTRAL FLAME

OXIDIZING FLAME

104F07.EPS

Figure 7 ◆ Types of flames.

Table 3   Tip Sizes Used for Common Pipe Sizes

Tip Size (No.)	Rod Size (Inches)	Pipe and Fitting Diameter (Inches)
4	3/32	1/4–3/8
5	1/8	1/2–3/4
6	3/16	1–1 1/4
7	1/4	1 1/2–2
8	5/16	2–2 1/2
9	3/8	3–3 1/2
10	7/16	4–6

104T03.EPS

Step 2 Adjust the torch oxygen system.

WARNING!

Do not stand in front of the oxygen gauge because the pressure may blow the face of the gauge outward, causing personal injury.

CAUTION

Open the oxygen cylinder valve slowly because a sudden release of pressure could damage the gauges.

- Open the oxygen cylinder valve slightly until pressure registers on the oxygen high-pressure gauge; then open the valve fully.
- Turn the oxygen regulator adjusting screw clockwise until pressure shows on the oxygen working-pressure gauge.
- Open the oxygen valve on the torch handle.
- Turn the oxygen regulator adjusting screw clockwise until about 20 to 25 psig registers on the oxygen working-pressure gauge.

NOTE

Always adjust the pressure with the torch valve open. When it is closed, the pressure may register higher.

- Close the oxygen valve on the torch handle.

Step 3 Adjust the torch acetylene system.
- Open the acetylene cylinder valve slightly until pressure registers on the acetylene high-pressure gauge; then open the valve about ½ turn.

NOTE

Be sure to leave the valve wrench on the acetylene cylinder valve so that the valve can be closed quickly in case of an emergency.

- Turn the acetylene regulator adjusting screw clockwise until pressure shows on the acetylene working-pressure gauge.
- Open the acetylene valve on the torch handle.
- Turn the acetylene regulator adjusting screw clockwise until about 5 psig registers on the acetylene working-pressure gauge.
- First close, then open the torch acetylene valve about ½ turn.

Torch Wrenches

Only a torch wrench, sometimes called a gang wrench, should be used to install regulators, hose connections, check valves, flashback arrestors, torches, and torch tips. The universal torch wrench shown is equipped with various size wrench cutouts for use with a variety of equipment and standard CGA components. The fittings for oxyfuel equipment are brass or bronze, and certain components are often fitted with soft, flexible, O-ring seals. The seal surfaces of the fittings or O-rings can be easily damaged by overtightening with standard wrenches. The length of a torch wrench is limited to reduce the chances of damage to fittings because of excessive torque. In some cases, manufacturers specify only hand-tightening for certain fitting connections of a torch set (tips or cutting/welding attachments). In any event, follow the manufacturer's specific instructions when connecting the components of a torch set.

104SA07.EPS

WARNING!

When lighting the torch, be sure to:

- Wear gloves and goggles.
- Hold the striker near the end of the torch tip. Do not cover the tip with the striker. Always use a striker to light the torch. Never use matches, cigarettes, or an open flame because this could result in severe burns or cause the lighter to explode. Also, make sure the torch is not pointed toward people or toward any flammable material.
- Always light the fuel gas first, then open the oxygen valve on the torch.
- Any time a flame appears from a leak in a hose, shut off the gas immediately.

NOTE

Observe the luminous cone at the tip of the nozzle and the long, greenish envelope around the flame, which is excess acetylene that represents a carburizing flame. As you continue to add oxygen, the envelope of acetylene should disappear. The inner cone will appear soft and luminous, and the torch will make a soft, even, blowing sound. This indicates a neutral flame, which is the ideal flame for welding. If too much oxygen is added, the flame will become more pointed and white in color, and the torch will make a sharp snapping or whistling sound. For brazing thin materials, a cooler, carburizing flame can be used to help prevent melting the base metal accidentally.

Step 4 Light the oxyacetylene torch.
- Hold the striker in one hand and the torch in your other hand. Strike a spark in front of the escaping acetylene gas.
- Open the acetylene valve on the torch until the flame jumps away from the tip about ¹⁄₁₆". Then close the valve until the flame just returns to the tip. This sets the proper fuel gas flow for the size tip being used.
- Open the oxygen valve on the torch slowly to add to the burning acetylene.

Step 5 Shut off the torch when finished brazing.
- Shut off the oxygen valve on the torch.
- Shut off the acetylene valve on the torch. Quickly turn off the acetylene to avoid carbon buildup.
- Shut off both the oxygen and acetylene cylinder valves completely.
- Open both valves on the torch to release the pressure in the hoses. Watch the working-pressure gauges until they register zero, then close the valves on the torch.

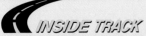

Adjusting the Fuel Gas Flame

When properly adjusted, the gas flame should return to the tip.

INITIAL FLAME ADJUSTMENT

104SA08.EPS

FINAL FLAME ADJUSTMENT

104SA09.EPS

- Turn the oxygen and acetylene regulator valves counterclockwise until loose. This will release the spring pressure on the diaphragms in the regulators.

Step 6 Coil the hoses and hang them on the hose holder.

3.4.0 Purging

Oil inside the tubing or part being brazed can vaporize when the heat of the brazing torch is applied. Oil vapor mixed with air will explode if ignited. Also, when copper is heated during brazing, it reacts with the oxygen in the air to form copper oxide. If air is allowed to flow into the tubing during brazing, copper oxide forms within the tubing. Refrigerants will later wash away the copper oxide particles, which can plug orifices, cause abrasion, and pollute the system. Other harmful chemicals will also form in the system. As a precaution, all the air must be removed from the tubing being brazed. This can be done best by purging the tubing with nitrogen.

The pressure in a nitrogen cylinder is about 2,000 psig. An accurate pressure regulator and an adjustable pressure relief valve must always be used when purging with either of these two gases. The relief valve should be adjusted to open 1 or 2 psi above the purging pressure. *Figure 8* shows a pressure regulator system. For purging, use the lowest pressure (typically 2 psi) that allows just

enough gas to keep air out of the tubing being brazed. As a rule of thumb, flow is sufficient when it can be felt with the palm of your hand. The greater the number of connections being brazed, the more important the purging procedure becomes.

 WARNING!
Never use oxygen, refrigerant, or compressed air to purge tubing. An explosion can result when oil and oxygen are mixed.

If it is necessary to braze tubing located in a confined space, make sure to attach a hose, pipe, etc., to the tubing being brazed so that the purge gas leaving the tubing is vented into the atmosphere outside. Otherwise, a hazardous atmosphere can be created within the space as the oxygen is displaced by the gas, resulting in an oxygen deficiency. OSHA regulations require that the internal atmosphere of a permit-required confined space be tested for hazardous atmosphere with a calibrated direct-reading instrument before an employee is allowed to enter the space. In addition, the atmosphere should be continuously monitored while work is in progress, in order to detect any conditions indicating the presence of a hazardous atmosphere, including an oxygen-deficient atmosphere.

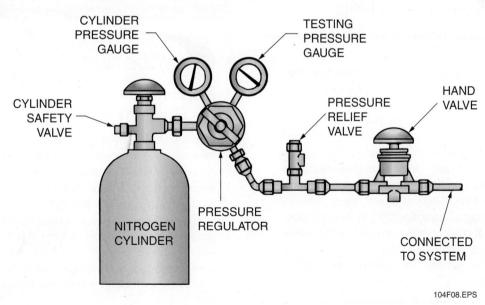

104F08.EPS

Figure 8 ◆ Nitrogen hookup.

3.5.0 Brazing Joints

Use the following procedure to braze a joint.

Step 1 Set up and light the oxyacetylene brazing equipment as described previously.
- Refer to *Table 3* for the suggested filler metal rod size for use with the size tubing being brazed.
- Adjust the torch to produce a neutral flame.

Step 2 Set up the nitrogen gas to purge the tubing following the guidelines and precautions described previously.

Step 3 Put on welding goggles with a No. 4 or 5 tint.

Step 4 Apply the heat to the tubing first. Watch the flux. It will first bubble and turn white and then melt into a clear liquid. At this time, shift the flame to the fitting and hold it there until the flux on the fitting turns clear.

Step 5 Continue to move the heat back and forth over the tubing and the fitting.

Step 6 Touch the filler metal rod to the joint. If the filler metal does not melt on contact, continue to heat and test the joint until the filler metal melts. Be careful to avoid melting the base metal.

Step 7 Hold the filler metal rod to the joint, and allow the filler metal to enter into the joint while holding the torch slightly ahead of the filler metal and directing most of the heat to the shoulder of the fitting.

Step 8 Continue to fill the joint with the filler metal until the filler metal has completely penetrated the joint. If the tubing is 2" or more in diameter, two torches can be used to evenly distribute the heat. For larger joints, small sections of the joint can be heated and brazed. Be sure to overlap the previously brazed section as you continue around the fitting. *Figure 9* shows how to work in overlapping sectors.

NOTE

Allow the fitting to receive more heat than the tubing by pausing at the fitting while continuing to move the flame back and forth. Concentrate the heat to the back of the fitting cup. For 1-inch and larger tube, it may be difficult to bring the whole joint up to temperature at one time. It will often be desirable to use an oxyfuel, multiple-orifice heating tip (rosebud) to maintain a more uniform temperature over large areas. A mild preheating of the entire fitting is recommended for larger sizes, and the use of a second torch to retain a uniform preheating of the entire fitting assembly may be necessary with the largest diameters.

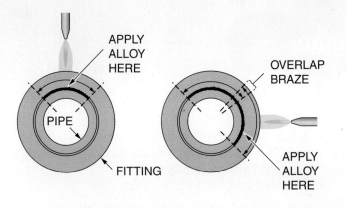

104F09.EPS

Figure 9 ◆ Working in overlapping sectors.

Step 9 After the filler metal has hardened and while it is still hot, wash the joint with warm water to clean excess, dried, or hardened brazing flux from the joint.

NOTE

If the flux is too hard to be removed with water, chip the excess flux off using a small chisel and light peen hammer, and then wash the joint with warm water. Joints clean best when they are still warm.

Step 10 Allow the joints to cool naturally.

Summary

Soldering is one of the most widely used methods for joining metals. Soldering is a procedure that fastens metals together using nonferrous metal to adhere the surfaces being joined. This filler metal is distributed between the closely fitted surfaces by capillary action. If the solder melts at a temperature above 800°F, the process is called hard soldering or brazing.

Soldering and brazing procedures require a step-by-step approach that includes:

- Cleaning the joints
- Applying flux
- Heating the parts to the correct temperature
- Bringing the nonferrous filler metal into contact with the tubing where it melts and flows into the joint by capillary action
- Cleaning and cooling the joint

If more training in soldering and brazing procedures is desired, the trainee should contact local vocational and trade schools or colleges for available courses. Also, many distributors of welding and gas supplies often have videos and other materials that they loan to interested individuals or groups.

Notes

1. The filler metals used for soldering usually melt at a temperature range of _____ .
   a. 200°F to 375°F
   b. 375°F to 500°F
   c. 750°F to 850°F
   d. 1,000°F to 1,200°F

2. The melting point of the filler metal used for soldering or brazing must be _____ the two metals that are being joined.
   a. higher than
   b. the same as
   c. lower than
   d. different from

3. Which of the following is not a characteristic of the filler metal (solder) used for soldering?
   a. Nonferrous metal
   b. Metal alloy
   c. Has a melting point below 800°F
   d. Always contains lead

4. The most commonly used solder for refrigeration tubing consists of _____ .
   a. 50 percent tin, 50 percent lead
   b. 94 percent tin, 6 percent silver
   c. 95 percent tin, 5 percent antimony
   d. 96 percent tin, 4 percent copper

5. Which of the following is not a characteristic of both soldering and brazing fluxes?
   a. Promote wetting of metals
   b. Indicate the temperature of the metal
   c. Clean and protect against oxidation
   d. Allow filler metal to flow into the joint

6. The filler metals used for brazing usually melt at a temperature range of _____ .
   a. 375°F to 500°F
   b. 750°F to 850°F
   c. 850°F to 1,200°F
   d. 1,200°F to 1,550°F

7. The oxygen cylinder used with oxyacetylene heating equipment contains oxygen stored at a pressure of about _____ .
   a. 250 psig
   b. 1,000 psig
   c. 2,000 psig
   d. 2,500 psig

8. When using oxyacetylene heating equipment for brazing, the acetylene regulator should never be adjusted to supply a pressure higher than _____ .
   a. 10 psig
   b. 15 psig
   c. 25 psig
   d. 250 psig

9. The _____ flame is used for almost all brazing applications.
   a. oxidizing
   b. carburizing
   c. feather
   d. neutral

10. To safely purge tubing while brazing, the tubing should be purged with _____ .
    a. oxygen
    b. compressed air
    c. refrigerant
    d. nitrogen

Trade Terms Quiz

1. A process that reduces the surface tension so that molten (liquid) solder flows evenly throughout the joint is called _____.

2. _____ is a method of joining metals with a nonferrous filler metal using heat below 800°F and below the melting point of the base metals being joined.

3. Metals and metal alloys that contain no iron are referred to as _____.

4. _____ is the movement of a liquid along the surface of a solid in a kind of spreading action.

5. A(n) _____ is any substance made up of two or more metals.

6. Releasing compressed gas to the atmosphere through some part or parts, such as a hose or pipeline, for the purpose of removing contaminants is called _____.

7. The process by which the oxygen in the air combines with metal to produce tarnish and rust is _____.

8. _____ is a chemical substance that prevents oxides from forming on the surface of metals as they are heated for soldering, brazing, or welding.

9. A method of joining metals with a nonferrous filler metal using heat above 800°F but below the melting point of the base metals being joined is called _____.

10. _____ is a fusible alloy used to join metals.

11. A(n) _____ is a valve that prevents the flame from traveling up the hoses.

Trade Terms

Alloy
Brazing
Capillary action

Flashback arrestor
Flux
Nonferrous

Oxidation
Purging
Solder

Soldering
Wetting

Matthew Todd, P.E.

Sales and Engineering Manager
Partner – Entek Corporation

Matt Todd followed a different course to the HVAC world than many of the people we've profiled, but like them he ended up in a satisfying and rewarding career. HVAC system design is challenging work. It takes careful analysis of the heating and cooling loads and the ability to select the right equipment with the right capacities to serve the needs of the structure.

How did you choose a career in the HVAC field?

I was hired by Carrier Air Conditioning as a sales engineer right out of college. I had friends from school who were enjoying the opportunity and my agricultural engineering studies seemed to align well with the process nature of air conditioning science. Before we started real design and sales work, Carrier put all of us through a lengthy and intensive training program, so that we could be effective as system designers.

What types of training have you been through?

In addition to my B.S. in Agricultural Engineering from Washington State University, I have had a great deal of industry training, including:

- Carrier Basic Systems Design Course in Syracuse, NY
- ASHRAE Continuing Education and Professional Development Hours courses
- ACCA Design/Build and National Conference Meetings

What kinds of work have you done in your career?

I have focused on system sales, design and installation work. I go to the job site, review the drawings, and calculate the heating and cooling loads, airflow, and indoor air quality requirements. Once I've done that preliminary work, I select the heating and cooling equipment and control system that best meet the needs of the building and satisfy the requirements established by the building owner.

Some of the important skills needed for my work include the ability to read and interpret drawings; math skills; and attention to detail. It's absolutely essential that I keep up with the state of the art in heating and cooling equipment, as well as control systems. For that reason, I spend a lot of time reading and attending any courses I can fit into my schedule.

What do you like about your job?

I like the variety. We deal with systems all the way from simple room air conditioners up to very large tonnage chilled and hot water applied systems. The best kept secret about our trade is that you have the broadest range of activities that literally touch on almost every other trade. Electrical components and circuitry are just the basics that build out into very intricate Direct Digital Control systems and computer work. Much of what we do on the service side is evaluate older systems, and come up with more energy and operationally efficient solutions for existing buildings.

The technology of our industry is constantly changing. Current technology in computer science, software, low-level signals, and wireless communications are all part of what we need to know today so that the latest advances can be factored into our work.

What factors have contributed most to your success?

Continuing education and a desire to learn. Also understanding what and where I can contribute best and doing so, while getting better at it.

What advice would you give to those new to the HVAC field?

It is a far broader industry than you might imagine. Don't get pigeon-holed into just a corner of it; try to keep an eye out for new technologies that will keep you interested and challenged – they are there, don't miss them.

Trade Terms
Introduced in This Module

Alloy: Any substance made up of two or more metals.

Brazing: A method of joining metals with a non-ferrous filler metal using heat above 800°F but below the melting point of the base metals being joined. Also known as hard soldering.

Capillary action: The movement of a liquid along the surface of a solid in a kind of spreading action.

Flashback arrestor: A valve that prevents the flame from traveling back from the tip and into the hoses.

Flux: A chemical substance that prevents oxides from forming on the surface of metals as they are heated for soldering, brazing, or welding.

Nonferrous: A group of metals and metal alloys that contain no iron.

Oxidation: The process by which the oxygen in the air combines with metal to produce tarnish and rust.

Purging: Releasing compressed gas to the atmosphere through some part or parts, such as a hose or pipeline, for the purpose of removing contaminants.

Solder: A fusible alloy used to join metals. Also known as soft soldering or sweat soldering.

Soldering: A method of joining metals with a non-ferrous filler metal using heat below 800°F and below the melting point of the base metals being joined.

Wetting: A process that reduces the surface tension so that molten (liquid) solder flows evenly throughout the joint.

Additional Resources

This module is intended to present thorough resources for task training. The following reference works are suggested for further study. These are optional materials for continued education, rather than for task training.

Brazing (VHS Video/Slides/Book), Latest Edition. Syracuse, NY: Carrier Corporation, Literature Services.

Standards and Codes, Latest Edition. New York, NY: American Society of Mechanical Engineers (ASME).

Standards and Codes, Latest Edition. Miami, FL: American Welding Society (AWS).

CONTREN® LEARNING SERIES – USER UPDATE

NCCER makes every effort to keep these textbooks up-to-date and free of technical errors. We appreciate your help in this process. If you have an idea for improving this textbook, or if you find an error, a typographical mistake, or an inaccuracy in NCCER's Contren® textbooks, please write us, using this form or a photocopy. Be sure to include the exact module number, page number, a detailed description, and the correction, if applicable. Your input will be brought to the attention of the Technical Review Committee. Thank you for your assistance.

Instructors – If you found that additional materials were necessary in order to teach this module effectively, please let us know so that we may include them in the Equipment/Materials list in the Annotated Instructor's Guide.

Write: Product Development and Revision
National Center for Construction Education and Research
3600 NW 43rd St., Bldg. G, Gainesville, FL 32606

Fax: 352-334-0932

E-mail: curriculum@nccer.org

Craft _____ Module Name _____

Copyright Date ___ Module Number ___ Page Number(s) ___

Description _____

(Optional) Correction _____

(Optional) Your Name and Address _____

Ferrous Metal Piping Practices
03105-07

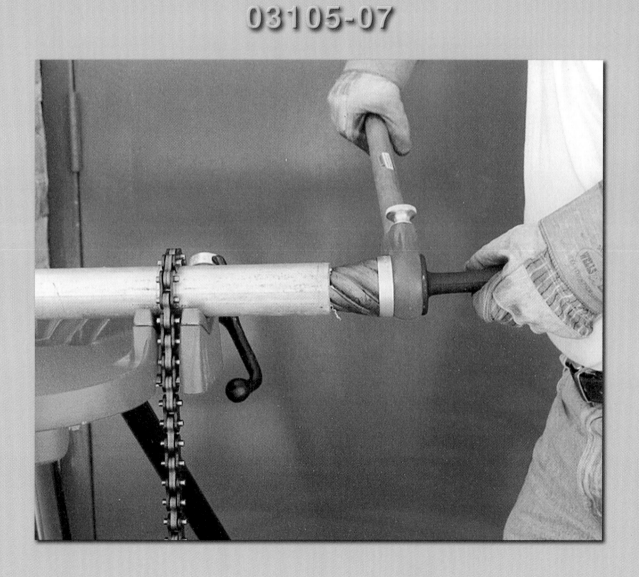

03105-07
Ferrous Metal Piping Practices

Topics to be presented in this module include:

Overview

Many large commercial systems use water to transfer heat from the indoors to the outdoors. Hydronic heating systems use hot water to deliver heat to the conditioned space. In these applications, water may be carried in galvanized steel pipe. In residential applications, black iron is used to supply natural gas to a gas-fired furnace.

Both galvanized steel and black iron pipe are joined using threaded fittings. The fittings come pre-threaded, but the pipe must be cut to size and threaded on the job. For that reason, the ability to measure, cut, and thread galvanized and black iron pipe is a necessary skill for anyone installing HVAC systems.

Objectives

When you have completed this module, you will be able to do the following:

1. Identify the types of ferrous metal pipes.
2. Measure the sizes of ferrous metal pipes.
3. Identify the common malleable iron fittings.
4. Cut, ream, and thread ferrous metal pipe.
5. Join lengths of threaded pipe together and install fittings.
6. Describe the main points to consider when installing pipe runs.
7. Describe the method used to join grooved piping.

Trade Terms

Annealing	Grooved pipe
Black iron pipe	Inside diameter
Bushing	Nipple
Cap	Nominal size
Chain vise	Outside diameter
Chain wrench	Pipe dope
Coupling	Plug
Cross	Reamer
Die	Standard yoke (pipe) vise
Elbow	Stock
Flange	Strap wrench
Galvanized pipe	Union

Required Trainee Materials

1. Paper and pencil
2. Appropriate personal protective equipment

Prerequisites

Before you begin this module, it is recommended that you successfully complete *Core Curriculum;* and *HVAC Level One,* Modules 03101-07 through 03104-07.

This course map shows all of the modules in *HVAC Level One.* The suggested training order begins at the bottom and proceeds up. Skill levels increase as you advance on the course map. The local Training Program Sponsor may adjust the training order.

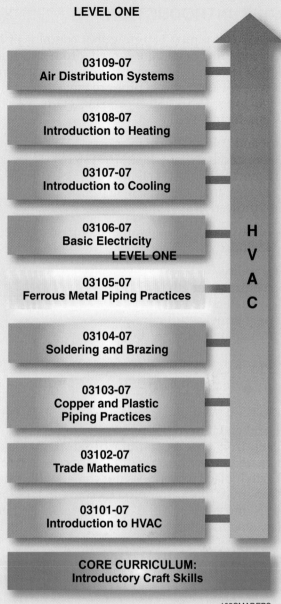

LEVEL ONE

03109-07
Air Distribution Systems

03108-07
Introduction to Heating

03107-07
Introduction to Cooling

03106-07
Basic Electricity
LEVEL ONE

03105-07
Ferrous Metal Piping Practices

03104-07
Soldering and Brazing

03103-07
Copper and Plastic
Piping Practices

03102-07
Trade Mathematics

03101-07
Introduction to HVAC

CORE CURRICULUM:
Introductory Craft Skills

H V A C

105CMAP.EPS

1.0.0 ◆ INTRODUCTION

Ferrous metal piping is piping that contains or is made from iron. Steel pipe is made from a combination of iron ore and carbon. The two most common types of carbon steel pipe are **black iron pipe** and **galvanized pipe**. Steel pipe has many uses in the field, including:

- Hot and cold water distribution
- Steam and hot water heating systems
- Gas and air piping systems
- Drainage and vent systems
- Fire protection systems

Some of the many advantages for using steel pipe include:

- Durability
- Structural strength
- Low material cost
- Low thermal conductivity
- Low expansion properties

2.0.0 ◆ STEEL PIPE

The two main types of steel pipe are black iron pipe and galvanized pipe.

Black iron pipe is manufactured exactly like galvanized pipe. The only difference between the two is that black iron pipe is not coated with zinc. Carbon gives black iron pipe its color. Black iron pipe is most often used in gas, heat, chilled water, steam, or air pressure applications. It is used where corrosion will not affect its uncoated surfaces.

Galvanized pipe is steel pipe that has been dipped in molten zinc. This protects the surfaces from abrasive or corrosive materials and gives the pipe a mottled, silvery color when new, or a dull, grayish color after time. Galvanized pipe is most often used when plumbing specifications require steel pipe.

2.1.0 Sizes and Wall Thickness

Pipe is listed in inches by its **nominal size**. For pipe sizes up to and including 12", nominal size is an approximation of the **inside diameter**. From 14" on, the nominal size reflects the **outside diameter** of the pipe. There are times when the nominal size of a pipe and its actual inside or outside diameter will differ greatly, but nominal size, not actual size, is always used to describe and select piping. *Figure 1* shows the inside and outside diameters for a 1" nominal-sized steel pipe.

There are two ways to describe the wall thickness of a pipe. The first is by schedule. As schedule numbers get larger, pipe walls get thicker and stronger. The schedule numbers used for pipe are 5, 10, 20, 30, 40, 60, 80, 100, 120, 140, and 160. It is important to remember that a schedule number only describes the wall thickness of a pipe of a given nominal size. Thus, ¾" Schedule 40 does not have the same wall thickness as 1" Schedule 40.

The second way to describe pipe wall thickness is by manufacturers' weight. There are three

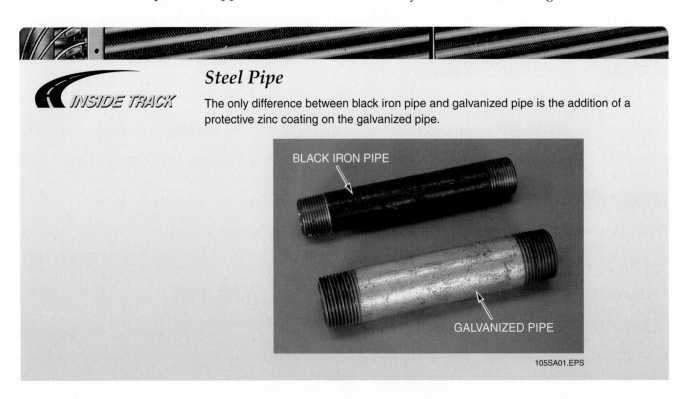

Steel Pipe

The only difference between black iron pipe and galvanized pipe is the addition of a protective zinc coating on the galvanized pipe.

INSIDE TRACK

BLACK IRON PIPE

GALVANIZED PIPE

105SA01.EPS

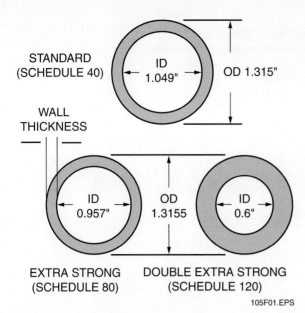

Figure 1 ◆ Ferrous pipe diameters.

classifications in common use. In ascending order of wall thickness, they are:

- STD – Standard
- XS – Extra strong
- XXS – Double extra strong

The wall thickness and the inside diameter differ with each weight. The thicker the wall, the smaller the inside diameter, but the more pressure the pipe will withstand. Standard weight will prove adequate in most piping situations; however, the two stronger weights are available when higher pressures require their use.

Because all schedules or weights for a specific pipe size have the same outside diameter, the same threading **dies** will fit all of them.

2.2.0 Threads

Steel pipe is joined either by welding or by threading the end of the pipe and using threaded fittings.

There are two American National Standard Pipe Threads, tapered pipe and straight pipe. Only tapered pipe threads are used for HVAC work because they produce leak-tight and pressure-tight connections. When tight, they also produce a mechanically rigid piping system. Tapered threads can be cut by hand with a die and **stock** or with an electric pipe threading machine.

The tapered thread used on pipe (*Figure 2*) is V-shaped with an angle of 60 degrees, very slightly rounded at the top. The taper is ⅟₁₆" per inch of length (½₂" per inch from each wall). There are about seven perfect threads and two or more imperfect threads for each joint. The actual number of perfect threads used (usable threads) depends on the size of the pipe being threaded. As shown in *Figure 2*, the first group of threads are perfect threads; they are sharp at the top and bottom. The remaining threads are non-perfect because they are not completely cut, resulting in rounded or imperfect edges. They have no sealing power. If the perfect threads are marred or broken, they also lose their sealing power.

Thread diameters refer to the nominal size of steel pipe. Threads are designated by specifying in sequence the nominal size, number of threads per inch, and the thread series symbols. For example, the thread specification ¾ – 14 NPT means:

¾ = ¾" nominal size
14 = 14 threads per inch
NPT = American National Standard Taper Pipe Thread

Taper pipe threads are engaged or made up in two phases, hand-tight engagement and wrench makeup. *Table 1* shows dimensions for hand-tight engagement as well as other NPT specifications for commonly used pipe sizes. In practice, about three turns are done by hand, followed by three or four turns with a wrench. When a pipe is threaded properly, about three threads should remain showing after the total makeup of a pipe and fitting.

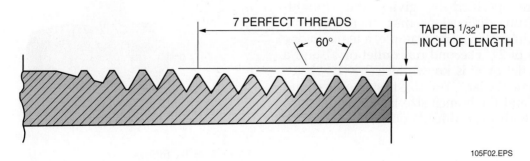

Figure 2 ◆ American National Standard taper pipe threads.

Table 1   American National Standard Taper Pipe Thread (NPT) Dimensions

Nominal Pipe Size	Threads per Inch	No. of Usable Threads	Hand-Tight Engagement	Thread Makeup	Total Thread Length
⅛	27	7	³⁄₁₆	¼	⅜
¼	18	7	¼	⅜	⁹⁄₁₆
⅜	18	7	¼	⅜	⅝
½	14	7	⁵⁄₁₆	½	¾
¾	14	8	⁵⁄₁₆	⁹⁄₁₆	¹³⁄₁₆
1	11½	8	⅜	¹¹⁄₁₆	1
1¼	11½	8	⁷⁄₁₆	¹¹⁄₁₆	1
1½	11½	8	⁷⁄₁₆	¾	1
2	11½	9	⁷⁄₁₆	¾	1¹⁄₁₆
2½	8	9	¹¹⁄₁₆	1⅛	1⁹⁄₁₆
3	8	10	¾	1³⁄₁₆	1⅝
3½	8	10	¹³⁄₁₆	1¼	1¹¹⁄₁₆
4	8	10	¹³⁄₁₆	1⁵⁄₁₆	1¾
5	8	11	¹⁵⁄₁₆	1⅜	1¹³⁄₁₆
6	8	12	¹⁵⁄₁₆	1½	1¹⁵⁄₁₆

105T01.EPS

2.3.0 Pipe Fittings

Pipe fittings for steel pipe are generally made of cast iron, malleable iron, or galvanized (zinc-coated) iron. Malleable iron is produced by prolonged **annealing** of ordinary cast iron. This process makes the iron tough. Also, it can be bent or pounded to some extent without breaking. Malleable iron fittings are typically used for gas piping. Cast-iron fittings are used for steam and hydronic system piping. Galvanized fittings are used for water piping, such as those used with exterior cooling towers and drip condensate piping systems.

2.3.1 Tees

Tees can be purchased in a great number of sizes and patterns. They are used to make a branch at a right angle to the main pipe. If all three outlets are the same size, the fitting is called a regular tee (*Figure 3*). If outlet sizes vary, the fitting is called a reducing tee.

Tees are specified by giving the straight-through (run) dimensions first, then the side-opening dimensions. For example, a tee with one run outlet of 2", a second run outlet of 1", and a branch outlet of ¾" is known as a 2 × 1 × ¾ tee (always state the large run size first, the small run size next, and the branch size last). Tees are also available with male threads on a run or branch outlet.

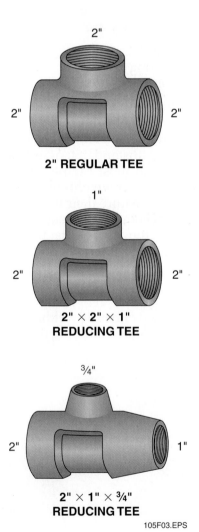

2" **REGULAR TEE**

2" × 2" × 1"
REDUCING TEE

2" × 1" × ¾"
REDUCING TEE

105F03.EPS

Figure 3 ◆ Tee fittings.

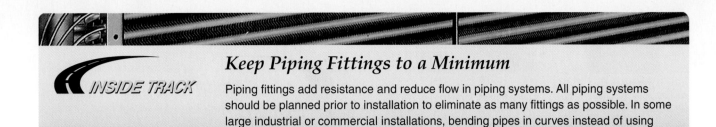

2.3.2 Elbows

Elbows, often called ells, are used to change the direction of pipe (*Figure 4*). The most common ells are the 90-degree ell; the 45-degree ell; the street ell, which has a male thread on one end; and the reducing ell, which has outlets of different sizes. Ells are also available to make 11¼-, 22½-, and 60-degree bends.

2.3.3 Unions

Unions make it possible to disassemble a threaded piping system. After disconnecting the union, the length of pipe on either end of the union may then be unscrewed. The two most common types of unions are the ground joint and the flange (*Figure 5*).

The ground joint union connects two pipes by screwing the thread and shoulder pieces onto the pipes. Then, both the shoulder and thread parts are drawn together by the collar. This union creates a gas-tight and water-tight joint.

The **flange** union also connects two separate pipes. The flanges screw to the pipes to be joined and are then pulled together with nuts and bolts. A gasket between the flanges makes this connection gas-tight and water-tight.

2.3.4 Couplings

Couplings (*Figure 6*) are short fittings with female threads in both openings. They are used to connect two lengths of pipe when making straight runs. The pipes can be of the same size or different sizes. Reducing couplings are used to join pipes of different sizes. Couplings cannot be used in place of unions because they cannot be disassembled.

2.3.5 Nipples and Crosses

Nipples (*Figure 7*) are pieces of pipe 12" or less in length and threaded on both ends. They are used to make extensions from a fitting or to join two

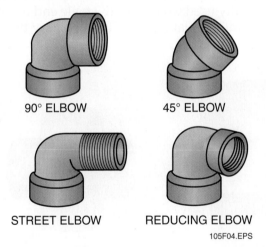

90° ELBOW   45° ELBOW

STREET ELBOW   REDUCING ELBOW

105F04.EPS

Figure 4 ◆ Elbow fittings.

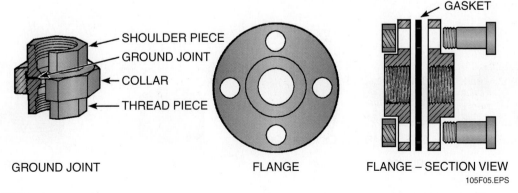

GASKET

SHOULDER PIECE
GROUND JOINT
COLLAR
THREAD PIECE

GROUND JOINT   FLANGE   FLANGE – SECTION VIEW

105F05.EPS

Figure 5 ◆ Unions.

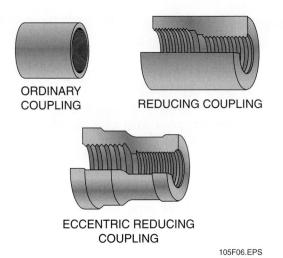

Figure 6 ◆ Couplings.

105F06.EPS

fittings. Nipples are manufactured in many sizes beginning with the close (all-thread) nipple. **Crosses** are four-way distribution devices.

2.3.6 Plugs, Caps, and Bushings

Plugs are male threaded fittings used to close openings in other fittings. There are a variety of heads (square, slotted, and hexagon) found on plugs, as shown in *Figure 8*.

Caps are fittings with a female thread. They are used for the same purpose as a plug, except that the cap fits on the male end of a pipe or nipple.

Bushings are fittings with a male thread on the outside and a female thread on the inside. They are usually used to connect the male end of a pipe to a fitting of a larger size. The ordinary bushing

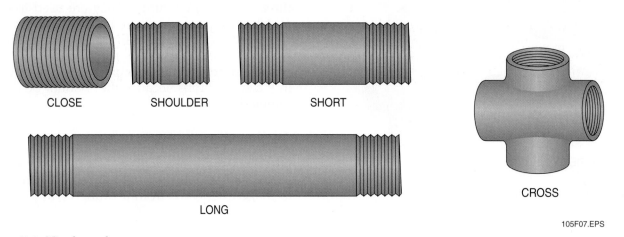

Figure 7 ◆ Nipples and crosses.

105F07.EPS

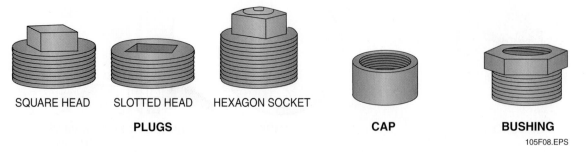

Figure 8 ◆ Plugs, caps, and bushings.

105F08.EPS

has a hexagon nut at the female end. Bushings can be used in place of a reducing fitting to accommodate a smaller pipe.

3.0.0 ◆ TOOLS AND MATERIALS

The following section describes various tools and materials used in working with ferrous metal pipe.

3.1.0 Pipe Cutters

Pipe cutters (*Figure 9*) may have from one to four cutting wheels. A single cutting wheel requires enough area so that the cutter may be passed all the way around the pipe. The more cutting wheels a cutter has, the less room it requires to cut the pipe. The pipe cutter is made of a cutting wheel, adjusting screw, and, depending on the type, either guide rollers or additional cutting wheels.

If the tube or pipe becomes mashed or flattened while being cut, or excessive force must be used when cutting, replace the cutting wheel or wheels. Lubricating oil must be applied periodically to all movable parts on the pipe cutter to ensure smooth operation. Applying cutting oil to the pipe will help when cutting ferrous pipe.

3.2.0 Reamers

After cutting a piece of pipe, a burr (rough edge) will be left on the inside of the pipe. A **reamer** (*Figure 10*) is used to remove the burr.

If the burr is left unattended, it will collect deposits and slow the flow of liquid within the pipe. Since reamers are tapered, one reamer can be used on many pipe sizes.

3.3.0 Pipe Threaders

There are two types of pipe threaders: hand threaders and power threaders. Hand threaders

(*Figure 11*) are made up of two parts, the die and the stock (handle). Dies are used to cut the threads, and the stock is the device that holds the die. The pipe die consists of the holder and cutters. Although they are mostly used manually, a hand threader could be used with a power drive or power vise, which threads automatically.

A pipe threading machine (*Figure 12*) is used when large quantities of pipe require threading. This machine rotates, threads, cuts, and reams pipe. The pipe is mounted through the machine speed chuck. After the chuck is tightened, this tool will perform one or more of these operations.

 CAUTION
Do not use the pipe threading machine to tighten fittings onto a threaded pipe, as this could cause the fitting to be over-tightened.

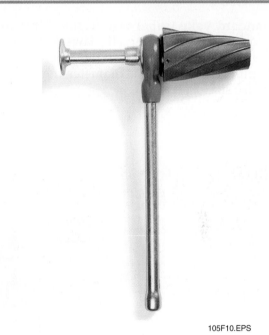

105F10.EPS

Figure 10 ◆ Pipe reamers.

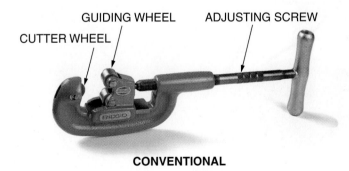

CUTTER WHEEL
GUIDING WHEEL   ADJUSTING SCREW

CONVENTIONAL

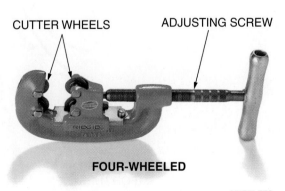

CUTTER WHEELS   ADJUSTING SCREW

FOUR-WHEELED

105F09.EPS

Figure 9 ◆ Pipe cutters.

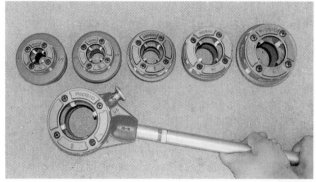

Figure 11 ◆ Hand threader.

3.4.0 Pipe Vises and Wrenches

There are many types of devices used to secure piping for various operations.

3.4.1 Vises

The **standard yoke (pipe) vise** (*Figure 13*) is the most commonly used vise. Its jaws hold pipe firmly and prevent it from turning. This vise can handle pipe from ⅛" to 3½" in diameter, depending on the vise size used.

The **chain vise** is used in the same way as the standard yoke vise; however, the chain vise can hold much larger pieces of pipe than the standard yoke vise. The chain must be kept oiled or it will become stiff. A stiff chain will make the chain vise operate poorly.

Either type of vise can be bench-mounted or mounted on a portable folding stand, as shown in *Figure 13*.

3.4.2 Wrenches

Straight and offset pipe wrenches (*Figure 14*) are used to grip and turn round stock. They have teeth that are set at an angle. This angle allows the teeth of the wrench to grip in one direction only.

The **chain wrench** is used on pipe that is over 2" in diameter. The chain must be oiled often to prevent it from becoming stiff or rusty.

The **strap wrench** is used to hold chrome-plated or other types of finished pipe. The strap wrench does not leave jaw marks or scratches on the pipe. Resin applied to the strap adds to the holding power of the wrench by reducing slippage.

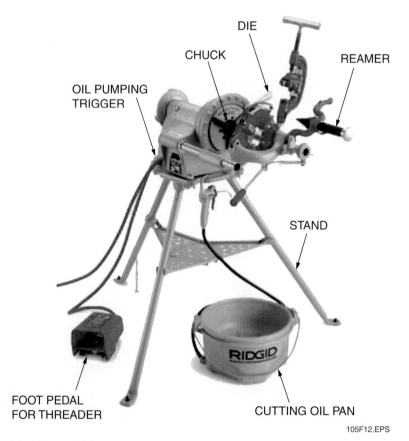

DIE

CHUCK

REAMER

OIL PUMPING
TRIGGER

STAND

FOOT PEDAL
FOR THREADER

CUTTING OIL PAN

Figure 12 ◆ Typical pipe threading machine.

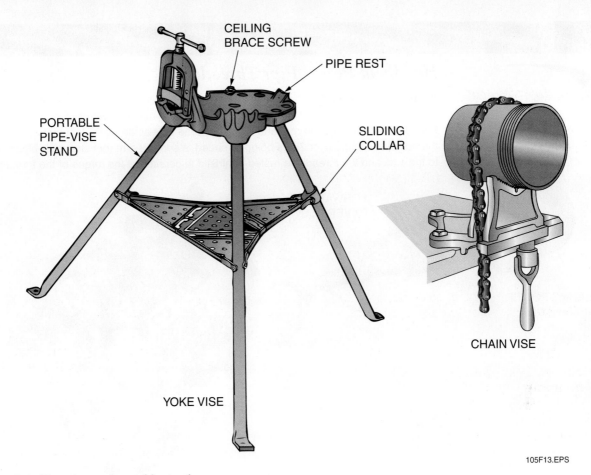

CEILING
BRACE SCREW

PIPE REST

PORTABLE
PIPE-VISE
STAND

SLIDING
COLLAR

CHAIN VISE

YOKE VISE

105F13.EPS

Figure 13 ◆ Pipe vises and portable stand.

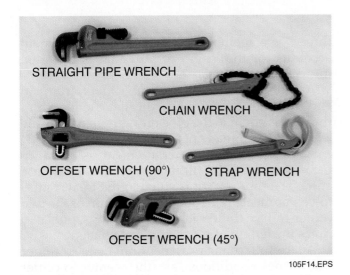

STRAIGHT PIPE WRENCH

CHAIN WRENCH

OFFSET WRENCH (90°)

STRAP WRENCH

OFFSET WRENCH (45°)

105F14.EPS

Figure 14 ◆ Pipe wrenches.

3.5.0 Pipe Dope and Tape

Joint compound, commonly called **pipe dope**, is a putty-like material applied to seal a joint and provide lubrication for assembly. The dope should be applied only to the male threads. Specific types of compound are required for various applications depending on the material to be contained in the piping. Always check to make sure the compound is compatible with the piped material. Teflon® tape made specifically as a pipe joint sealer may also be used with the proper application. No more than three wraps in the direction of the thread should be made, nor should it overlap the threaded end of the pipe. A special yellow Teflon® tape is available for use on natural gas piping.

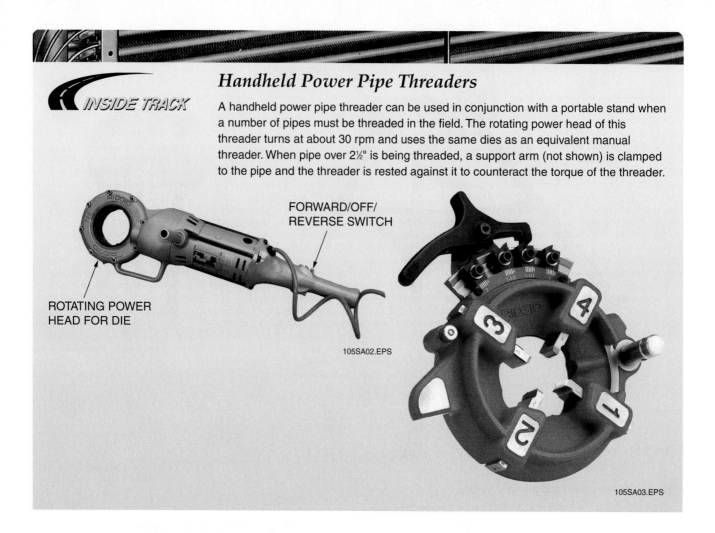

Handheld Power Pipe Threaders

A handheld power pipe threader can be used in conjunction with a portable stand when a number of pipes must be threaded in the field. The rotating power head of this threader turns at about 30 rpm and uses the same dies as an equivalent manual threader. When pipe over 2½" is being threaded, a support arm (not shown) is clamped to the pipe and the threader is rested against it to counteract the torque of the threader.

FORWARD/OFF/
REVERSE SWITCH

ROTATING POWER
HEAD FOR DIE

105SA02.EPS

105SA03.EPS

4.0.0 ◆ JOINING PROCEDURES

The following sections cover the procedures for joining threaded steel pipes.

4.1.0 Measuring

The dimensions shown on pipe drawings specify the location of center lines and/or points of center lines; they do not specify pipe lengths. This system is also used in the fabrication and installation of pipe assemblies. To determine actual pipe lengths, threaded pipe can be measured by a variety of methods: end-to-end, end-to-center, face-to-end, center-to-center, or face-to-face. *Figure 15* shows the different measuring techniques.

An end-to-end measurement is accomplished by measuring the full length of the pipe including the threads at both ends.

An end-to-center measurement is used for a piece of pipe with a fitting screwed onto one end only. The pipe length is equal to the total end-to-center measurement, minus the center-to-face dimension of the fitting, plus the length of the thread engagement. The thread engagement accounts for the length of pipe that is threaded into a fitting. See *Table 2*.

A face-to-end measurement differs from an end-to-center measurement in that the length of the pipe is equal to the pipe measurement plus the length of the thread engagement.

A center-to-center measurement is used to measure pipe with fittings screwed onto both ends. The pipe length is equal to the total center-to-center measurement, minus the sum of the two center-to-face dimensions of the fittings, plus two times the length of the thread engagement.

A face-to-face measurement can be used under the same conditions as the center-to-center method. It is figured by measuring the length of pipe plus two times the length of the thread engagement.

4.2.0 Cutting

The pipe cutter is the best tool to use when cutting steel pipe. To operate, revolve the cutter around the pipe and tighten the cutting wheel ¼ revolution

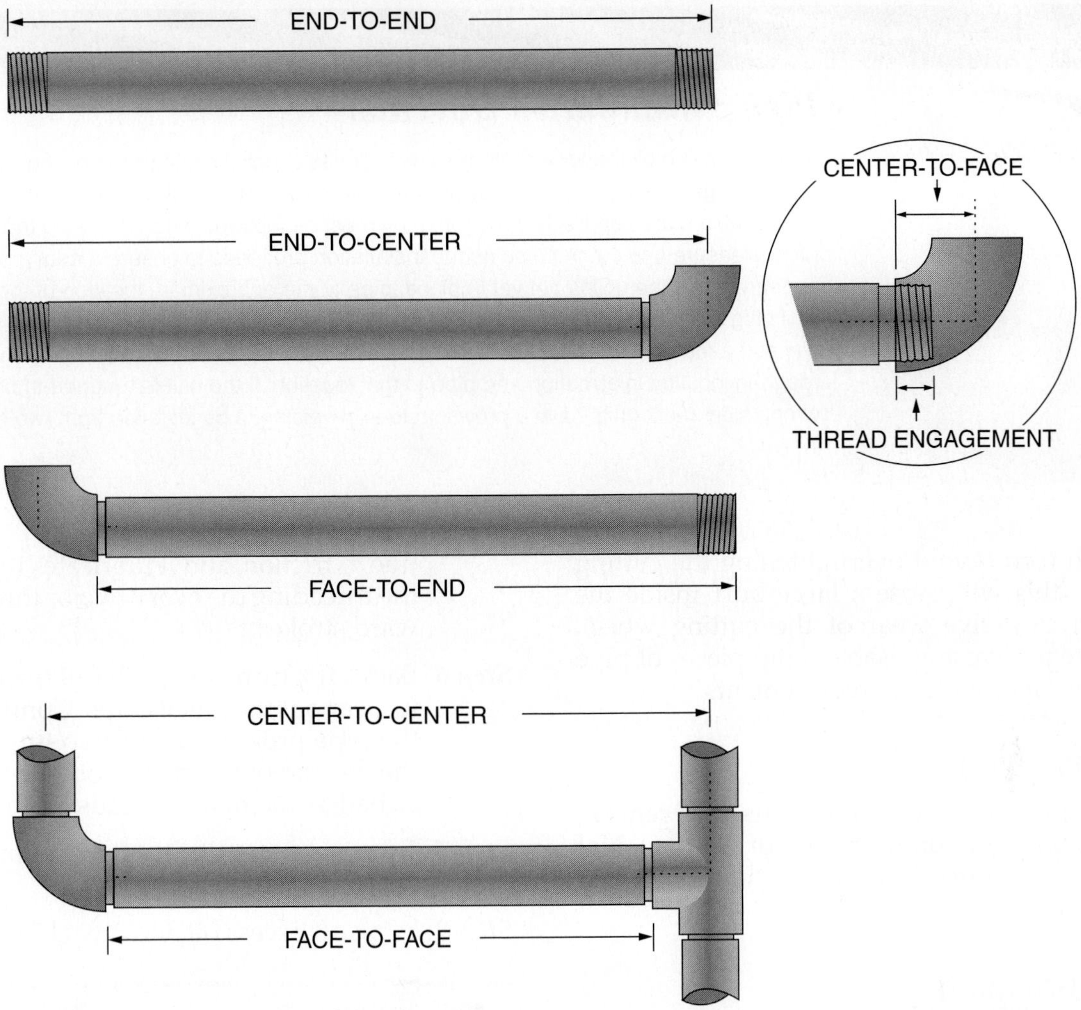

105F15.EPS

Figure 15 ◆ Pipe measuring methods.

Table 2 Pipe Engagement Allowances

Size of Pipe (In.)	Outside Diameter (In.)	Number of Threads per Inch	Total Length of Threads (In.)	Effective Length (In., Approx.)	Thread Engagement (In., Approx.)
1/4	0.54	18	5/8	3/8	3/8
3/8	0.675	18	5/8	7/16	3/8
1/2	0.84	14	13/16	9/16	1/2
3/4	1.05	14	13/16	9/16	1/2
1	1.315	11 1/2	1	11/16	9/16
1 1/4	1.66	11 1/2	1	11/16	5/8
1 1/2	1.9	11 1/2	1	3/4	5/8
2	2.375	11 1/2	1 1/16	3/4	11/16
2 1/2	2.875	8	1 9/16	1 1/8	15/16
3	3.5	8	1 5/8	1 3/16	1
4	4.5	8	1 3/4	1 5/16	1 1/16
5	5.56	8	1 13/16	1 3/8	1 3/16
6	6.625	8	1 15/16	1 1/2	1 1/4

105T02.EPS

Piping Installation Direction

INSIDE TRACK

When installing piping for a system such as a gas supply line, start the piping at the source (a gas meter) and work toward the unit being installed. If piping drawings and specifications are supplied for the job, they must be followed when installing the piping runs. Measure and cut the pipe as the installation proceeds to ensure a neat appearance. If the unit is not yet in place, pipe to the approximate location of the unit and plug or cap the pipe. The final piping can be accomplished when the unit is in place. If a material specification sheet exists for the job, check the sheet for the unit rough-in location information and pipe to that location. If the unit is roughed-in on the wrong side, most units have a provision to allow piping to be installed from two or more sides.

with each turn. Avoid overtightening the cutting wheel, as this will cause a large burr inside the pipe and excessive wear of the cutting wheel. Make sure to save any usable scrap pieces of pipe for nipples or other fit-up requirements.

4.3.0 Reaming

Once the pipe is cut, a reamer is used to remove the burrs that form on the inside of the pipe. Not removing the burrs can cause blockage and restrict liquid flow.

4.4.0 Threading

Threads may be cut by hand with a hand threader and vise or an electric pipe threading machine. The following two subsections provide general guidelines for threading pipe.

4.4.1 *Using a Hand Threader and Vise*

To cut threads using a hand threader and vise, proceed as follows:

Step 1 Select the correct size die for the pipe being threaded.

Step 2 Inspect the die to see if the cutters are free of nicks and wear.

Step 3 Lock the pipe securely in a vise.

Step 4 Slide the die over the end of the pipe, guide end first.

Step 5 Push the die against the pipe with the heel of one hand. Take three or four short, slow, clockwise turns. Be careful to keep the die pressed firmly against the pipe. When enough thread is cut to keep the die firmly on the pipe, apply some thread-cutting oil. This oil prevents the pipe from overheating

due to friction, and it lubricates the die. Oil the threading die every two or three downward strokes.

Step 6 Back off ¼ turn after each full turn forward to clear out the metal chips. Continue until the pipe projects one or two threads from the die end of the stock. Too few threads is as bad as too many threads.

Step 7 To remove the die, rotate it counterclockwise.

Step 8 Wipe off excess oil and any chips.

WARNING!

Use a rag, not your bare hands, when wiping the pipe. The chips are sharp and could cause cuts.

4.4.2 *Using a Bench or Tripod Threading Machine*

Each threading machine is slightly different. Become familiar with the manufacturer's operating procedures before attempting to operate any threading machine. Also, become thoroughly familiar with the maintenance and safety instructions for the machine. A poorly maintained machine is a safety hazard. To thread using a power threading machine:

Step 1 Select the pipe stock, install the correct size die, and inspect it for nicks.

WARNING!

Do not wear loose-fitting clothing, jewelry, or loose-fitting gloves when using any pipe threading machine. Tie back long hair.

Step 2 Mount the pipe stock into the chuck (*Figure 16*). Long stock must have additional support.

Step 3 Check the pipe and die alignment.

Step 4 With the power switch in the forward position and the machine running, move the tool carriage and start the die on the end of the pipe (*Figure 17*). Apply cutting oil during the threading operation.

Step 5 Cut threads until two threads appear at the other end of the die (*Figure 18*). Stop threading.

Step 6 Reverse the machine and back off the die until it is clear of the pipe.

Step 7 Remove the pipe from the machine chuck. Be careful not to mar the threads.

Step 8 Using a rag, wipe the pipe clean of oil and metal chips.

4.5.0 Assembling

Apply pipe dope to the male threads before assembling a pipe connection. Do not apply the compound to the threads of the fitting. An applicable paste compound or Teflon® tape may be used.

When using Teflon® tape, apply the tape in a clockwise direction from the end of the pipe, the same direction as the fitting turns. Be sure to check the local codes to see if the use of tape is permitted. Apply two to three layers, starting two to three threads from the end of the pipe.

Start the fitting onto the threaded pipe by hand. Turn the fitting clockwise. Finish tightening the fitting using pipe wrenches (*Figure 19*).

4.6.0 Installing Steel Piping

The various types of hangers and supports used to install steel piping are the same as previously described for non-ferrous metal piping systems.

The main thing to remember is that the method used to install a piping system is generally defined by the builder's specifications. Air conditioning and heating installations planned by architects and engineering consultants include plans and specifications, which completely describe the proposed installation. After the job has been awarded to a successful bidder, the engineering drawings are supplemented by the working drawings of the installation contractor. After the working drawings and submittals are approved by the engineer and architect, the work proceeds according to specifications.

SPEED CHUCK

PIPE CLAMPED IN SPEED CHUCK

105F16.EPS

Figure 16 ◆ Pipe mounted in a threading machine.

MOVABLE TOOL CARRIAGE

105F17.EPS

Figure 17 ◆ Starting the die on the end of the pipe.

105F18.EPS

Figure 18 ◆ Pipe threads at completion.

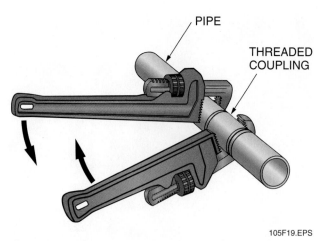

PIPE

THREADED
COUPLING

105F19.EPS

Figure 19 ◆ Tightening pipe using two pipe wrenches.

Each section of the specifications is numbered for ready reference and describes one portion of the work. Specifications are based upon code or ordinances and must be followed precisely. A specification for pipe hangers, for example, may read as follows: "All piping shall be supported with hangers spaced not more than 10" apart (on center). Hangers shall be the malleable iron, split-ring type and shall be as manufactured by XYZ Hangers, Inc. or other approved vendor."

Although piping drawings should be available for every piping system that is installed, this is not always the case. Therefore, the HVAC installer often has to select the best route and install the piping. This is known as field-fabricating or field-routing a pipe run. In this case, the HVAC installer has to determine the placement of the pipe hangers based on the size, weight, and type of pipe being run. In order for a hanger system to do its job, it must support the pipe at regular intervals.

Evenly spacing the hangers prevents any individual hanger from being overloaded. *Table 3* lists the recommended maximum hanger spacing intervals for carbon steel pipe.

Table 3 only serves as a guide. Always check the job site specifications when determining pipe hanger spacing, and if the pipe sags, add more hangers. The most important thing to remember is that the pipe must run straight without sagging.

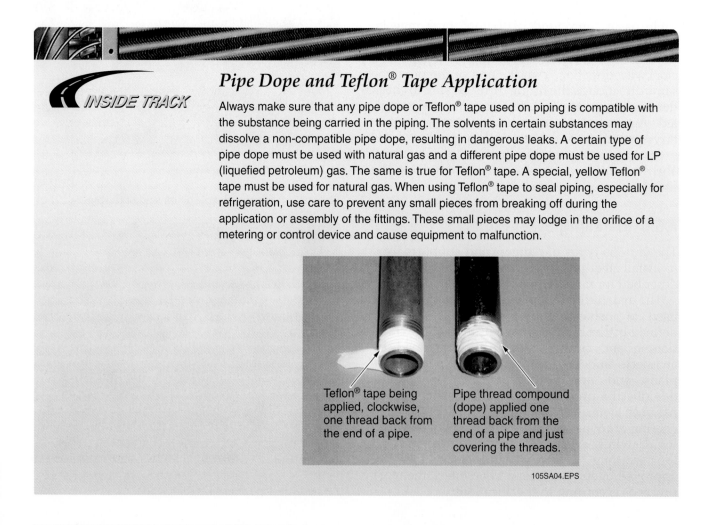

Pipe Dope and Teflon® Tape Application

INSIDE TRACK

Always make sure that any pipe dope or Teflon® tape used on piping is compatible with the substance being carried in the piping. The solvents in certain substances may dissolve a non-compatible pipe dope, resulting in dangerous leaks. A certain type of pipe dope must be used with natural gas and a different pipe dope must be used for LP (liquefied petroleum) gas. The same is true for Teflon® tape. A special, yellow Teflon® tape must be used for natural gas. When using Teflon® tape to seal piping, especially for refrigeration, use care to prevent any small pieces from breaking off during the application or assembly of the fittings. These small pieces may lodge in the orifice of a metering or control device and cause equipment to malfunction.

Teflon® tape being applied, clockwise, one thread back from the end of a pipe.

Pipe thread compound (dope) applied one thread back from the end of a pipe and just covering the threads.

105SA04.EPS

Table 3 Recommended Maximum Hanger Spacing Intervals for Carbon Steel Pipe

Pipe Size	Rod Diameter	Maximum Spacing
Up to 1¼"	⅜"	8'
1½" and 2"	⅜"	10'
2½" to 3½"	½"	12'
4" and 5"	⅝"	15'
6"	¾"	17'
8" to 12"	⅞"	22'

105T03.EPS

Some other points to consider when installing steel piping systems are:

- To the extent possible, keep pipe runs straight and avoid excessive use of elbows, tees, and other fittings.
- Do not use black iron or lead pipe in a water supply system.
- Use piping, fittings, valves, etc., that will provide trouble-free service during the expected life of the piping system.
- Make provisions in hydronic and steam piping systems to reduce noise and avoid damage from water hammer and vibration.
- Install valves in the piping system so selected portions of the piping can be isolated for repairs.
- Install gas piping systems in accordance with the latest editions of the *National Fuel Gas Code* (*ANSI Z223.1*), local codes, and the related equipment manufacturer's installation instructions. Similarly, install oil piping systems in accordance with the National Board of Fire Underwriters and local regulations.
- In gas supply piping systems, use a pipe compound resistant to the action of liquefied petroleum gases on all threaded pipe connections.

5.0.0 ◆ GROOVED PIPE

Grooved pipe is so called because grooves instead of welded, flanged, or threaded joints are used for coupling. Each joint in a grooved piping system serves as a union, allowing easy access to any part of the piping system for cleaning or servicing. Grooved piping systems have a wide range of applications and can be used with a wide variety of piping materials. The most common uses of grooved piping systems are in oil fields, industrial facilities, mining, municipal systems, and fire protection systems. The following types of piping can be joined by grooved couplings:

- Carbon steel
- Stainless steel
- Aluminum
- PVC plastic
- High-density polyethylene
- Ductile iron

A standard grooved piping coupling consists of a rubber gasket and two housing halves that are bolted together. The housing halves are tightened together until they touch, so no special torquing of the housing bolts is required.

The grooved piping system offers varied mechanical benefits, including the option of rigid or flexible couplings. Rigid and flexible couplings can be incorporated as needed into any system to take full advantage of the characteristics of each. Rigid couplings create a rigid joint useful for risers, mechanical rooms, and other areas where positive clamping with no flexibility within the joints is desired. Flexible couplings provide allowance for controlled pipe movement that occurs with expansion, contraction, and deflection. Flexible couplings may eliminate the need for expansion joints, cold springing, or expansion loops, and will provide a virtually stress-free piping system. *Figure 20* shows examples of rigid and flexible grooved pipe couplings.

5.1.0 Preparing Pipe Ends

Grooved pipe can be delivered to the job site pre-cut to length and grooved, or it can be cut and grooved on the job. Use of the grooved piping method is based on the proper preparation of a groove in the pipe end to receive the coupling housing key. The groove serves as a recess in the pipe with enough depth to secure the coupling housing, yet at the same time leaves enough wall thickness for a full pressure rating. Groove preparation varies with different pipe materials and wall thicknesses. Many types of tools are available to properly groove pipe in the shop or in the field. The two methods of forming a groove in pipe are roll grooving and cut grooving. *Figure 21* shows pipe grooving equipment, while *Figure 22* shows details of pipe grooves.

The dimensions indicated by the letters in *Figure 22* must comply with engineering specifications at your job site. The A dimension is the distance from the pipe end to the groove and provides the gasket seating area. This area must be free from indentations, projections, or roll marks to provide a leakproof sealing seat for the gasket. The B dimension is the groove width; it controls expansion and angular deflection based

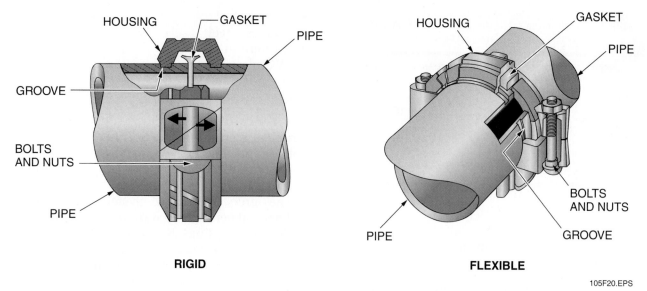

RIGID

FLEXIBLE

105F20.EPS

Figure 20 ◆ Rigid and flexible grooved pipe couplings.

PORTABLE FIELD PIPE GROOVER

SHOP PIPE GROOVER

105F21.EPS

Figure 21 ◆ Pipe grooving equipment.

on its distance from the pipe end and its width in relation to the width of the coupling housing key. The C dimension is the proper diameter tolerance and is concentric with the outside diameter of the pipe. The D dimension must be changed if necessary to keep the C dimension within the stated tolerances. The F dimension is used with the standard roll only and gives the maximum allowable pipe end flare. The T dimension is the lightest grade or minimum thickness of pipe suitable for roll or cut grooving. The R dimension is the

radius necessary at the bottom of the groove to eliminate a point of stress concentration for cast-iron and PVC plastic pipe.

5.1.1 Roll Grooving

Power roll-grooving machines are used to roll grooves at the ends of pipe to prepare the piping for groove-type fittings and couplings. Power grooving machines are available to groove 2" to 16" standard and lightweight steel pipe,

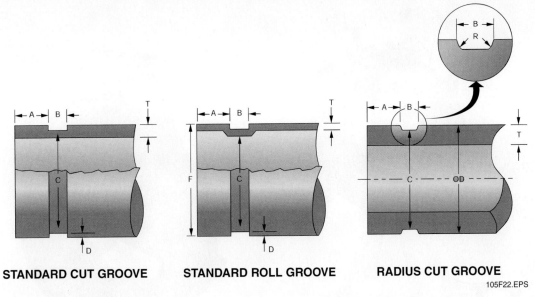

STANDARD CUT GROOVE   **STANDARD ROLL GROOVE**   **RADIUS CUT GROOVE**

105F22.EPS

Figure 22 ◆ Details of pipe grooves.

aluminum pipe, stainless steel pipe, and PVC plastic pipe.

Roll grooving removes no metal from the pipe but forms a groove by displacing the metal instead. Since the groove is cold-formed, it has rounded edges that reduce the pipe movement after the joint is made up.

5.1.2 Cut Grooving

Cut grooving differs from roll grooving in that a groove is cut into the pipe. Cut grooving is basically intended for standard weight or heavier pipe. The cut removes less than one-half of the pipe wall, which is less depth than thread cuts.

Cut-grooving machines (*Figure 23*) are designed to be driven around a stationary pipe. This creates a groove that is of uniform depth and is concentric with the outside diameter of the pipe.

5.2.0 Selecting Gaskets

There are many types of synthetic rubber gaskets available to provide the option of selecting grooved piping products for the widest range of applications. In order to provide maximum life for the service intended, the proper gasket selection and specification is required.

Several factors must be considered in determining the best gasket to use for a specific service. The foremost consideration is the temperature of the product flowing through the pipe. Temperatures beyond the recommended limits decrease gasket life. Also, the concentration of the product, duration of service, and continuity of service must be considered because there is a direct relationship

between temperature, continuity of service, and gasket life. It should also be noted that there are services for which gaskets are not recommended. Always refer to the manufacturer's recommendations for selecting and installing gaskets.

5.3.0 Installing Grooved Pipe Couplings

The procedure used to join grooved pipe is as follows (see *Figure 24*):

Step 1   Make sure that the gasket is suitable for its intended service. Some manufacturers color code their gaskets. Apply a thin coat of lubricant to the gasket lips and the outside of the gasket.

Step 2   Check the pipe ends. To ensure a leakproof seal, they must be free from indentations, projections, or roll marks.

Step 3   Install the gasket over the pipe end. Be sure the gasket lip does not hang over the pipe end.

Step 4   Align and bring the two pipe ends together. Slide the gasket into position and center it between the grooves on each pipe. Be sure that no part of the gasket extends into the groove on either pipe.

Step 5   Assemble the housing segments loosely, leaving one nut and bolt off to allow the housing to swing over the joint.

Step 6   To install the housing, swing it over the gasket and into position in the grooves on both pipes.

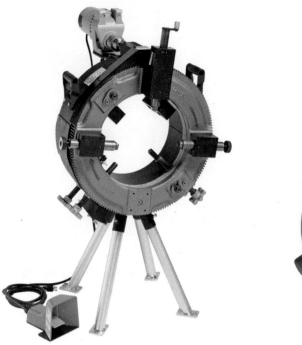

(A) POWER CUT GROOVER

(B) MANUAL CUT GROOVER

105F23.EPS

Figure 23 ◆ Cut-grooving machines.

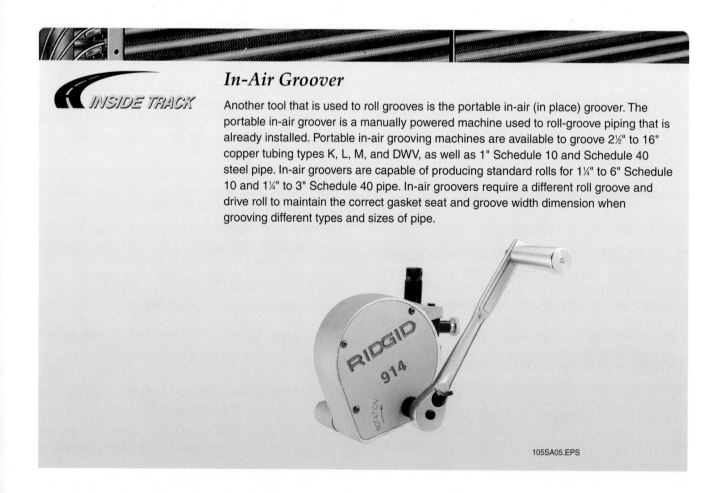

In-Air Groover

Another tool that is used to roll grooves is the portable in-air (in place) groover. The portable in-air groover is a manually powered machine used to roll-groove piping that is already installed. Portable in-air grooving machines are available to groove 2½" to 16" copper tubing types K, L, M, and DWV, as well as 1" Schedule 10 and Schedule 40 steel pipe. In-air groovers are capable of producing standard rolls for 1¼" to 6" Schedule 10 and 1¼" to 3" Schedule 40 pipe. In-air groovers require a different roll groove and drive roll to maintain the correct gasket seat and groove width dimension when grooving different types and sizes of pipe.

105SA05.EPS

CHECK PIPE ENDS

LUBRICATE GASKET

INSTALL GASKET

JOIN PIPE ENDS AND CENTER
GASKET BETWEEN GROOVES

INSTALL HOUSINGS

INSTALL BOLTS AND NUTS

TIGHTEN NUTS

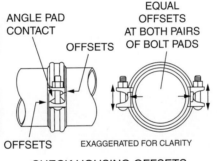

CHECK HOUSING OFFSETS

HOUSING OFFSETS/COMPLETED
JOINT

Always refer to instructions supplied with the product for complete information regarding pipe preparation, installation, product inspection, and safety requirements.

105F24.EPS

Figure 24 ◆ Joining grooved pipe.

Step 7 Insert the remaining bolt and nut. Be sure that the bolt track head engages into the recess in the housing.

Step 8 Tighten the nuts alternately and equally to maintain metal-to-metal contact at the angle bolt pads.

6.0.0 ◆ FLANGED PIPE

In some cases, larger pipes and valves are joined with flange fittings (*Figure 25*). Flange fittings are joined with gaskets and bolts. As previously discussed under grooved fittings, the proper selection of gaskets is important in establishing and maintaining a tight seal. Flanged fittings commonly use a ⅛" gasket. Flange pipe systems use the same types of fittings as other systems, including tees, elbows, crosses, and reducers.

A torque wrench is used to tighten flange bolts. The bolts must be tightened in a specific pattern, as shown in *Figure 26*. Rather than tightening the bolts to the full torque the first time around, torque is applied in increments to prevent the joint from becoming distorted.

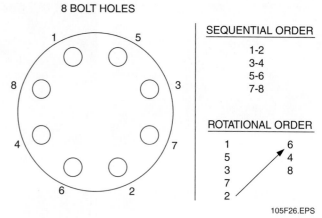

8 BOLT HOLES

SEQUENTIAL ORDER

1-2
3-4
5-6
7-8

ROTATIONAL ORDER

1   6
5   4
3   8
7
2

105F26.EPS

Figure 26 ◆ Bolt-tightening pattern.

105F25.EPS

Figure 25 ◆ Flanged valve.

Summary

Carbon steel pipe is used in many HVAC applications. Black steel pipe is used most often for gas and air-pressure applications. Galvanized steel pipe is often used for hydronic heating and cooling systems. It is electroplated with zinc to protect the pipe from abrasive or corrosive materials.

Steel pipe is categorized by both its size (inside or outside diameter) and strength. Schedule 40 is a standard-weight pipe, Schedule 80 is an extrastrong weight, and Schedule 120 is a double extrastrong weight.

Steel pipe can be threaded or grooved. Threads used on carbon steel pipe are standardized. This means that thread sizes on all manufactured pipe products are the same. Threaded pipe can be measured in several ways, depending on whether there are fittings on one or two ends. Incorrect measuring will mean your pipe may be cut too long or too short for the installation. A variety of tools are available for working with and cutting carbon steel pipe. Familiarity with tool use and maintenance will ensure that your workmanship is of the best quality.

All pipe requires adequate support during installation to keep it from sagging, bending, and losing proper alignment. Codes and building specifications indicate which hangers, fasteners, and connectors are appropriate for different applications. Seismically active areas may require additional supports.

Notes

1. The inside diameter of Schedule 80 pipe is
   _____ .
   a. greater than that of Schedule 40
   b. less than that of Schedule 40
   c. less than that of Schedule 120
   d. the same as that of Schedule 40 and 120

2. Of the following types of pipe, which is the
   strongest?
   a. Schedule 5
   b. Schedule 20
   c. Schedule 40
   d. Schedule 120

3. The standard number of threads per inch
   for a pipe whose nominal size is ½" is _____.
   a. 11½ threads per inch
   b. 14 threads per inch
   c. 17 threads per inch
   d. 18 threads per inch

4. A tee fitting described as 2 × 1½ × 1 has a
   _____ .
   a. large run size of 2", small run size of 1½",
      and branch size of 1"
   b. large run size of 1", small run size of 1½",
      and branch size of 2"
   c. large run size of 1½", small run size of 2",
      and branch size of 1"
   d. large run size of 2", small run size of 1",
      and branch size of 1½"

5. Pipe nipples are used to _____ .
   a. close openings in other fittings
   b. connect two lengths of pipe when mak-
      ing straight runs
   c. change the direction of the pipe run
   d. make extensions from a fitting or to join
      two fittings

6. A measurement of the full length of pipe
   including the threads describes a(n) _____
   measurement.
   a. end-to-center
   b. end-to-end
   c. face-to-end
   d. face-to-face

7. The threading die should be oiled _____
   when threading pipe.
   a. after the threading of each piece of pipe
   b. every 10 downward strokes
   c. every two or three downward strokes
   d. once a day

8. Before using a threading machine to thread
   pipe, you should always _____ .
   a. spray the pipe dies with cutting oil
   b. become familiar with the operating and
      safety instructions for the machine
   c. wipe the unthreaded end of the pipe
      with cutting oil
   d. apply pipe dope to the unthreaded end
      of the pipe

9. Generally, 2" steel pipe should be supported with metal hangers at _____ intervals.
   a. 4'
   b. 6'
   c. 8'
   d. 10'

10. Which of the following joining methods is
    used with grooved piping systems?
    a. Threading
    b. Flanging
    c. Welding
    d. Gasket and two housing halves bolted
       together

Trade Terms Quiz

1. A pipe fitting with female threads on both ends is known as a(n) _____.

2. A(n) _____ has external threads and is used to close the opening in a fitting.

3. The tool insert used to cut external threads in a pipe is a(n) _____.

4. _____ pipe is carbon steel pipe that has been coated with zinc.

5. The pipe fitting with male threads on the outside and female threads on the inside is called a(n) _____.

6. A(n) _____ is a short length of pipe with male threads on both ends.

7. When you specify the approximate dimensions of a pipe, you are referring to its _____ size.

8. The tool used to hold a die when cutting external threads on a pipe is called a(n) _____.

9. A(n) _____ is a fitting used to join two lengths of pipe.

10. A(n) _____ is mated to a fitting with gaskets and couplings.

11. A female threaded pipe fitting used to close off the end of a pipe is called a(n) _____.

12. A device used to clamp pipe and other round objects is a(n) _____.

13. The type of pipe that gets its coloring from the carbon in the steel is _____ pipe.

14. The pipe fitting with four female-threaded openings is called a(n) _____.

15. An angled pipe fitting with two open ends is a(n) _____.

16. A measurement across the opening of a pipe is called the _____.

17. The tool used to remove burrs from the end of a pipe is the _____.

18. A(n) _____ is a device with one movable jaw that is used to secure pipe while working on it.

19. A special type of wrench used to turn large pipe is a(n) _____.

20. The putty-like substance used to seal threaded pipe joints is called _____.

21. The measurement across a pipe that includes the wall thickness is known as the _____.

22. A process in which a material is heated then cooled to give it strength is called _____.

23. A tool that uses a nylon web to grip a pipe is called a(n) _____.

24. A(n) _____ is a flat plate attached to a pipe or fitting and used to join pipe and fittings.

Trade Terms

Annealing	Coupling	Grooved pipe	Plug
Black iron pipe	Cross	Inside diameter	Reamer
Bushing	Die	Nipple	Standard yoke (pipe) vise
Cap	Elbow	Nominal size	Stock
Chain vise	Flange	Outside diameter	Strap wrench
Chain wrench	Galvanized pipe	Pipe dope	Union

Dennis Lazard, Sr.

Electrician and Air Conditioning Supervisor
Agrifos Fertilizer

Dennis Lazard, Sr., is the air conditioning supervisor for Agrifos Fertilizer (formerly Mobile Mining & Mineral). A resident of Baytown, Texas, Dennis took his first job in the construction field as an electrician in 1974. While working as an electrician, he assisted air conditioning technicians, which spurred his interest in HVAC.

Tell us about your background in HVAC.
In 1989 I began my career with Mobile Mining & Mineral in Pasadena, Texas. Initially, I worked as an electrician, but later I became the air conditioning shop supervisor.

I began my formal air conditioning education in 1992 at Lee College in Baytown, Texas, where I earned a certificate in air conditioning and refrigeration. In 1995, I became the air conditioning supervisor at Agrifos Fertilizer, and obtained my air conditioning contractor's license in 1996. In 1999, I earned an associate's degree in applied science in air conditioning. While pursuing my degree, I worked as a tutor and mentor to the air conditioning students at Lee College.

Tell us about your position as Agrifos Fertilizer's air conditioning supervisor.
I am responsible for all of the heating, ventilating, and air conditioning equipment at the plant. It is exciting to see HVAC needs within the plant satisfied, and I like to see that people appreciate my skill.

What qualities must an individual have to succeed in the HVAC trade?
Succeeding in HVAC takes a lot of hard work and dedication. Outside encouragement and tutoring were important to me.

What advice do you have for new trainees?
The HVAC field holds many opportunities for someone who wants to own and run a business. There are many technical and supervisory positions in the HVAC trade, too. Of all the jobs in the construction industry, you should try HVAC first because there is a wide variety of work involved in the field. There is always something new to learn, which keeps the trade stimulating. Knowing HVAC also makes you an asset to yourself, your family, and your friends. Stay with it, and I think you'll be satisfied.

Trade Terms
Introduced in This Module

Annealing: A process in which a material is heated, then cooled to strengthen it.

Black iron pipe: Carbon steel pipe that gets its black coloring from the carbon in the steel.

Bushing: A pipe fitting with male threads on the outside and female threads on the inside. It is most often used to connect the male end of a pipe to a fitting of a larger size.

Cap: A female pipe fitting that is closed at one end. It is used to close off the end of a piece of pipe.

Chain vise: A device used to clamp pipe and other round metal objects. It has one stationary metal jaw and a chain that fits over the pipe and is clamped to secure the pipe.

Chain wrench: An adjustable tool for holding and turning large pipe up to 4" in diameter. A flexible chain replaces the usual wrench jaws.

Coupling: A pipe fitting containing female threads on both ends. Couplings are used to join two pipes in a straight run or to join a pipe and fixture.

Cross: A pipe fitting with four female openings at right angles to one another.

Die: A tool insert used to cut external threads by hand or machine.

Elbow: An angled pipe fitting having two openings. It is used to change the direction of a run of pipe.

Flange: A flat plate attached to a pipe or fitting and used as a means of attaching pipe, fittings, or valves to the piping system.

Galvanized pipe: Carbon steel pipe that has been coated with zinc to prevent rust.

Grooved pipe: A piping method for connecting piping systems. The use of grooved piping eliminates the need for threading, flanging, or welding when making connections. Connections are made with gaskets and couplings installed using a wrench and lubricant.

Inside diameter: The measurement made across the inside width (internal opening) of a pipe.

Nipple: A short length of pipe that is used to join fittings. It is usually less than 12" long and has male threads on both ends.

Nominal size: The approximate dimension(s) by which standard material is identified.

Outside diameter: The measurement made across the outside width of a pipe, including the wall thickness plus the internal opening.

Pipe dope: A putty-like pipe joint material used for sealing threaded pipe joints.

Plug: A pipe fitting with external threads and head that is used for closing the opening in another fitting.

Reamer: A tool used to remove the burr from the inside of a pipe that has been cut with a pipe cutter.

Standard yoke (pipe) vise: A holding device used to hold pipe and other round objects. It has one movable jaw that is adjusted with a threaded rod.

Stock: A tool used to hold and turn dies when cutting external threads.

Strap wrench: A tool for gripping pipe. The strap is made of nylon web.

Union: A pipe fitting used to join two lengths of pipe. It permits disconnecting the two pieces of pipe without cutting.

This module is intended to present thorough resources for task training. The following reference work is suggested for further study. This is optional material for continued education, rather than for task training.

Pipefitter's Handbook, Latest Edition. New York, NY: Industrial Press, Inc.

CONTREN® LEARNING SERIES – USER UPDATE

NCCER makes every effort to keep these textbooks up-to-date and free of technical errors. We appreciate your help in this process. If you have an idea for improving this textbook, or if you find an error, a typographical mistake, or an inaccuracy in NCCER's Contren® textbooks, please write us, using this form or a photocopy. Be sure to include the exact module number, page number, a detailed description, and the correction, if applicable. Your input will be brought to the attention of the Technical Review Committee. Thank you for your assistance.

Instructors – If you found that additional materials were necessary in order to teach this module effectively, please let us know so that we may include them in the Equipment/Materials list in the Annotated Instructor's Guide.

Write: Product Development and Revision
National Center for Construction Education and Research
3600 NW 43rd St., Bldg. G, Gainesville, FL 32606

Fax: 352-334-0932

E-mail: curriculum@nccer.org

Craft _____ Module Name _____

Copyright Date _____ Module Number _____ Page Number(s) _____

Description _____

_____

_____

_____

(Optional) Correction _____

_____

_____

(Optional) Your Name and Address _____

_____

_____

Basic Electricity
03106-07

03106-07
Basic Electricity

Topics to be presented in this module include:

Overview

Air conditioning, heating, and refrigeration systems use electricity to obtain operating power for their compressors, fan motors, and other load devices. In addition, electrical switching devices, sensors, and indicators are used in the control circuits of these systems. Some of these control systems are quite complex, containing thermostats, as well as numerous switching devices. Many systems use microprocessor controls, and some systems used in commercial building are managed from a central computer.

Most of the problems an HVAC technician encounters on a service call involve the electrical system. Therefore, it is essential that the technician understand electrical circuits and be able to interpret complex circuit diagrams. Without these skills, diagnosing system problems would be pure guesswork.

Objectives

When you have completed this module, you will be able to do the following:

1. State how electrical power is distributed.
2. Describe how voltage, current, resistance, and power are related.
3. Use Ohm's law to calculate the current, voltage, and resistance in a circuit.
4. Use the power formula to calculate how much power is consumed by a circuit.
5. Describe the differences between series and parallel circuits and calculate loads in each.
6. Describe the purpose and operation of the various electrical components used in HVAC equipment.
7. State and demonstrate the safety precautions that must be followed when working on electrical equipment.
8. Make voltage, current, and resistance measurements using electrical test equipment.
9. Read and interpret common electrical symbols.

Trade Terms

Alternating current (AC)
Ammeter
Ampere (amp)
Analog meter
Clamp-on ammeter
Conductor
Contactor
Continuity
Current
Digital meter
Direct current (DC)
Electromagnet
HACR (heating, air conditioning, and refrigeration) circuit breaker
Induction
In-line ammeter
Insulator
Ladder diagram
Line duty
Load
Multimeter
Ohm
Pilot duty
Power
Pressurestat
Rectifier
Relay
Resistance
Short circuit
Slow-blow fuse
Solenoid
Solid state
Starter
Transformer
Volt
Voltage
Watts

Required Trainee Materials

1. Paper and pencil
2. Appropriate personal protective equipment

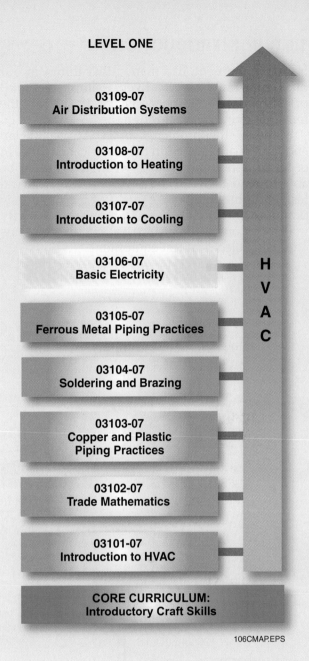

LEVEL ONE

03109-07
Air Distribution Systems

03108-07
Introduction to Heating

03107-07
Introduction to Cooling

03106-07
Basic Electricity

03105-07
Ferrous Metal Piping Practices

03104-07
Soldering and Brazing

03103-07
Copper and Plastic Piping Practices

03102-07
Trade Mathematics

03101-07
Introduction to HVAC

CORE CURRICULUM:
Introductory Craft Skills

H V A C

106CMAP.EPS

Prerequisites

Before you begin this module, it is recommended that you successfully complete *Core Curriculum*; and *HVAC Level One*, Modules 03101-07 through 03105-07.

This course map shows all of the modules in *HVAC Level One*. The suggested training order begins at the bottom and proceeds up. Skill levels increase as you advance on the course map. The local Training Program Sponsor may adjust the training order.

1.0.0 ◆ INTRODUCTION

HVAC equipment, like many other things in our lives, runs on electricity. Even a simple gas-fired or oil-fired furnace needs electricity to run the fans that circulate air through the building.

Many of the problems an HVAC technician encounters are located in the electrical circuits. In order to determine what is wrong, you must be able to read a circuit diagram and use electrical test equipment to make measurements at key points in the circuit. Your training in electricity will focus on reading electrical circuit diagrams and using electrical test equipment.

2.0.0 ◆ ELECTRICITY

It is important for an HVAC technician to understand the basic principles and terms involved in electrical **power** generation and distribution.

2.1.0 Electrical Power Generation and Distribution

Electricity comes from electrical generating plants (*Figure 1*) operated by utilities like your local power company. Steam from coal-burning or nuclear power plants is used to power huge generators called turbines, which generate electricity. There are also hydroelectric power plants where water flowing over dams is used to drive turbines.

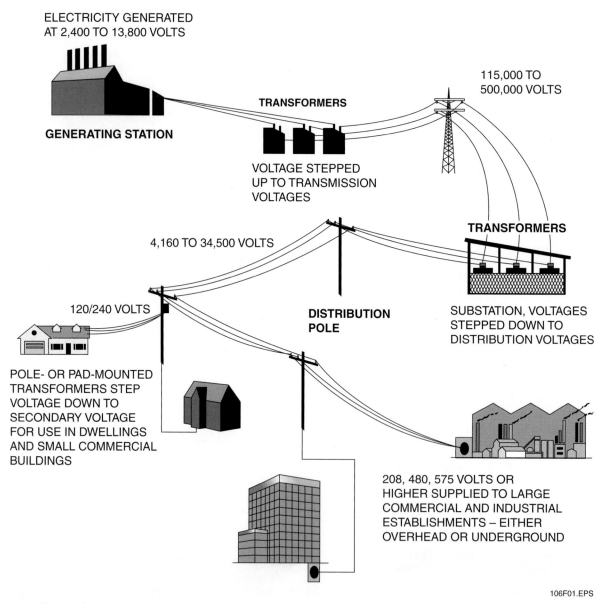

ELECTRICITY GENERATED
AT 2,400 TO 13,800 VOLTS

GENERATING STATION

TRANSFORMERS

VOLTAGE STEPPED
UP TO TRANSMISSION
VOLTAGES

115,000 TO
500,000 VOLTS

TRANSFORMERS

4,160 TO 34,500 VOLTS

SUBSTATION, VOLTAGES
STEPPED DOWN TO
DISTRIBUTION VOLTAGES

DISTRIBUTION
POLE

120/240 VOLTS

POLE- OR PAD-MOUNTED
TRANSFORMERS STEP
VOLTAGE DOWN TO
SECONDARY VOLTAGE
FOR USE IN DWELLINGS
AND SMALL COMMERCIAL
BUILDINGS

208, 480, 575 VOLTS OR
HIGHER SUPPLIED TO LARGE
COMMERCIAL AND INDUSTRIAL
ESTABLISHMENTS – EITHER
OVERHEAD OR UNDERGROUND

106F01.EPS

Figure 1 ◆ Electrical power distribution.

The electrical power that travels through long-distance transmission lines may be as high as 500,000 **volts** (V). High voltages are used to reduce line losses. Devices known as **transformers** are used to step the **voltage** down to lower levels as it reaches electrical substations and eventually our homes, offices, and factories. The voltage we receive at home is usually about 240V. At the wall outlet where we plug in small appliances such as televisions and toasters, the voltage is about 120V (*Figure 2*). Electric stoves, clothes dryers, water heaters, and central air conditioning systems usually require the full 240V. Commercial buildings and factories may receive anywhere from 208V to 575V. This depends on the amount of power their machines consume.

2.2.0 Current, Voltage, and Resistance

An electrical **current** is caused by the movement of electrons from one point to another. Electrons are the negatively charged particles that exist in all matter. When a difference in the number of electrons exists between two points, electrons will flow from the negative point (the one with more electrons) to the positive point (the one with fewer electrons). The difference in electrical potential between the two points is called voltage.

Lightning is a good example of natural electron flow. A storm cloud has a negative charge with respect to the earth. Lightning occurs when the difference in potential between the cloud and the earth becomes so great that the air between them conducts a current.

In the common 12V car battery, a chemical reaction causes one of the poles to be negative with respect to the other. If you connect a light bulb or other electrical device between the negative (−) and positive (+) poles of the battery, electrons will flow from the negative pole to the positive pole (*Figure 3*).

In order to harness the potential energy of the battery, a **resistance**, such as a lightbulb, is connected between the two battery poles. Resistance

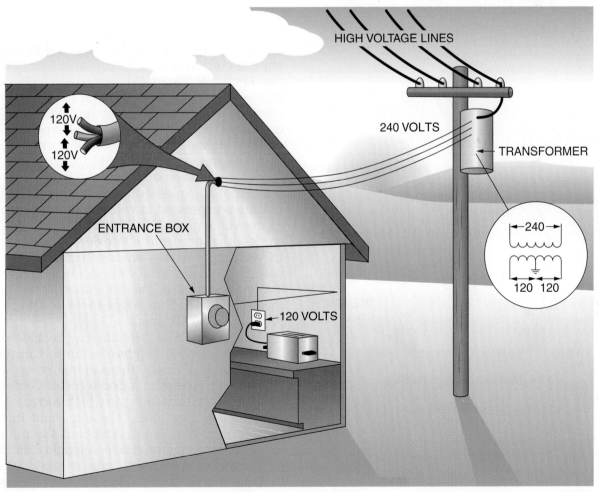

Figure 2 ◆ Internal power distribution.

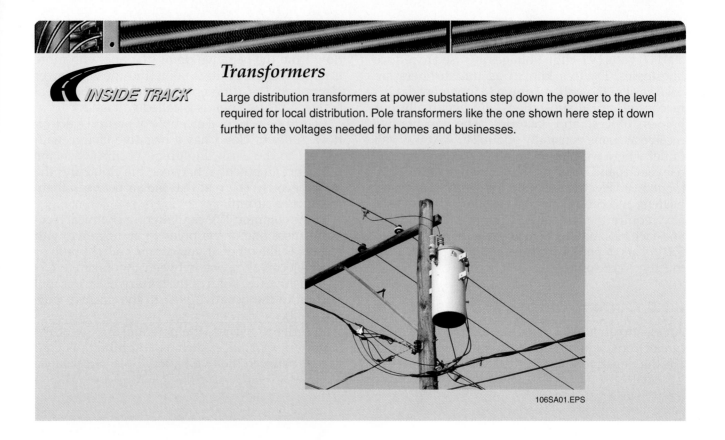

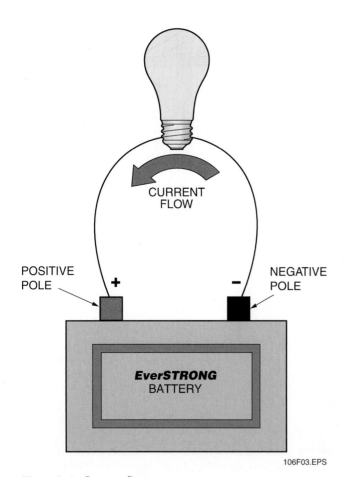

Figure 3 ◆ Current flow.

is the property that causes a material to resist or impede current flow. The resistance in a circuit is also known as a **load**. When the switch closes, the moving electrons flow through the bulb, which converts the electrical energy into light. The bulb consumes the electrical energy in the process. The amount of energy consumed is known as power, and is expressed in **watts** (W). A 60W light bulb consumes 60 watts of power.

Current is expressed in **amperes** or amps (A), voltage is expressed in volts (V), and resistance is expressed in **ohms** (Ω). The AC voltages you will work with range from 24V to 600V. The occasional DC voltages you encounter will generally be low voltages in the 5V to 15V range. Occasionally, you may encounter a very high-voltage circuit, such as the ignitor circuit for a gas furnace, which may produce a 10,000V spark.

Currents in HVAC circuits may be as low as a few microamps (millionths of an amp). Currents in the milliamp, or mA (thousandths of an amp), range are more common. Some circuits, especially motor circuits, may carry very high currents in the range of 100A or more. The important thing to remember is that even currents in the milliamp range can be dangerous under the right conditions.

Resistances are usually large values expressed in thousands of ohms (kilohms or K) or millions of ohms (megohms or M). In some cases, a resistance value may be only a few ohms. For example, the

 *INSIDE TRACK*

Lightning

A lightning strike can carry a current of 20,000 amps (A) and reach temperatures of 54,000°F. No wonder people in the path of, or near, a lightning strike are often killed.

106SA02.EPS

 *INSIDE TRACK*

Voltage

Eighteenth-century physicist Alessandro Volta theorized that when certain objects and chemicals come into contact with each other, they produce an electric current. Believing that electricity came from contact between metals only, Volta coined the term metallic electricity. To demonstrate his theory, Volta placed two discs, one of silver and the other of zinc, into a weak acidic solution. When he linked the discs together with wire, electricity flowed through the wire. Thus, Volta introduced the world to the battery, also known as the Voltaic pile. Now Volta needed a term to measure the strength of the electric push or the flowing charge; the volt is that measure. Voltage is known by other names as well, including electromotive force (EMF) and potential difference. Voltage can be viewed as the potential to do work.

thermistors (heat-sensitive resistors) used as temperature-sensing devices and as an aid in motor startup fall into this category.

3.0.0 ◆ AC AND DC VOLTAGE

The kind of electricity produced by a battery is known as **direct current (DC)**. Automobiles, portable stereos, calculators, and flashlights are good examples of devices that use DC voltage. Note that all of them are battery-powered. The electricity supplied by your local utility is **alternating current (AC)**. Almost all HVAC devices use AC. Modern units use DC motors or electronic circuit boards that require DC voltage.

Rather than using batteries, which need to be replaced or recharged, such units contain special circuits called **rectifiers** that convert AC to DC. The device that allows you to plug your calculator or other portable device into a wall socket is also a rectifier.

4.0.0 ◆ ELECTRICAL CIRCUIT CHARACTERISTICS

Voltage, current, resistance, and power are closely related. If you know any two of them, you can determine the other two. For example, if you know how much voltage is available and how much power the load consumes, you can figure out how much current the circuit will draw. For example, suppose you want to plug a 1,600W electric heater into a household circuit that is protected by a 15A circuit breaker. How do you determine if you can safely add the heater without overloading the circuit and tripping the circuit breaker?

The wrong way would be to plug in the heater and see if the circuit breaker trips. A better (and less dangerous) way is to calculate how much current is used (drawn) by each appliance in the circuit, including the heater. This can be done using the equations discussed in the following sections.

4.1.0 Ohm's Law

Ohm's law is a formula for calculating voltage, current, or resistance. These values are expressed as E, I, and R, respectively. *Figure 4* shows the three variations of Ohm's law. The pyramid provides an easy way to remember the formula. If you know two of the values, cover the unknown value with your finger to see how to solve the problem. For example, if you know the voltage and resistance, covering the I shows that you divide E by R.

NOTE

Ohm's law applies only to resistive circuits. Because of the nature of alternating current, circuits containing motors, relay coils, and other inductive devices do not act the same as pure resistances. This subject is covered in a later module.

For example, in a 120V circuit containing a 60Ω load, the current flow is 2A.

$$I = \frac{E}{R} = \frac{120V}{60\Omega} = 2A$$

4.2.0 Electrical Power

Most load devices are rated by the power they consume, rather than the resistance they offer. The electric heater we discussed earlier in the module has a rating of 1,600W. Ohm's law will not help you determine how much current the heater will draw, because two variables are missing (resistance and current). In such instances, you must use the power formula: power (P) = voltage (E) × current (I). In this case, you would use I = P ÷ E because current is the unknown value. (See *Figure 5*.) In a 120V circuit, the 1,600W heater would draw 13.3A (1,600W ÷ 120V). The 15A circuit breaker would accommodate the heater, but just barely. If other appliances in the circuit are operating at the same time as the heater, the circuit breaker will probably trip.

Electric motors may be rated in equivalent horsepower, rather than watts. A large swimming pool pump, for example, might contain a 2-horsepower (hp) electric motor. There is a simple conversion from horsepower to watts: 1hp = 746W. Therefore, a 2hp motor would consume 1,492W. In a 12V circuit, it would draw more than 12A:

$$I = \frac{P}{E} = \frac{1,492W}{120V} = 12.43A$$

	LETTER SYMBOL	UNIT OF MEASUREMENT
CURRENT	I	AMPERES (A)
RESISTANCE	R	OHMS (Ω)
VOLTAGE	E	VOLTS (V)

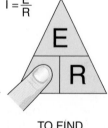

$E = I \times R$

TO FIND VOLTAGE

$I = \frac{E}{R}$

TO FIND CURRENT

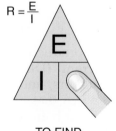

$R = \frac{E}{I}$

TO FIND RESISTANCE

106F04.EPS

Figure 4 ◆ Ohm's law.

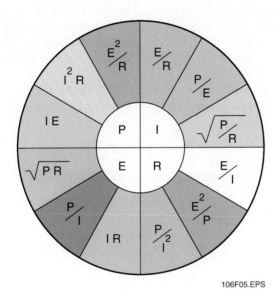

106F05.EPS

Figure 5 ◆ Power formula.

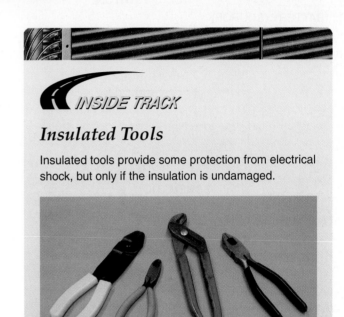
5.0.0 ◆ ELECTRICAL CIRCUITS

A basic electrical circuit is shown in *Figure 6*. An electrical circuit is a closed loop that contains a voltage source, a load, and **conductors** to carry current. It usually contains a switching device to turn the current on and off.

A conductor is a material that readily carries an electric current. Most metals are good conductors. Copper and aluminum are the most common conductors. Gold and silver are better conductors, but are too expensive, except in tiny electronic circuits. The minerals found in water make it an excellent conductor. That is why it is dangerous to work with electricity or use electrical appliances in a wet or damp environment.

An **insulator** is the opposite of a conductor. It inhibits the flow of electricity. Rubber is a good insulator. Tools used in electrical trades are often insulated with rubber to prevent electrical shock.

5.1.0 Series Circuits

A series circuit provides only one path for current flow. The total resistance of the circuit is equal to the sum of the individual resistances. The 12V series circuit in *Figure 7* has two 30Ω loads. The total resistance is therefore 60Ω. The amount of current flowing in the circuit is 0.2A.

$$I = \frac{E}{R} = \frac{12V}{60\Omega} = 0.2A$$

If there were five 30Ω loads, the total resistance would be 150Ω. The current flow is the same through all the loads. The voltage measured across any load (voltage drop) depends on the resistance of that load. The sum of the voltage drops equals the total voltage applied to the circuit. Circuits containing loads in series are uncommon in HVAC work. An important trait of a series circuit is that if the circuit is open at any point, no current will flow. For example, if you have five lightbulbs connected in series and one of them blows, all five lights will go off.

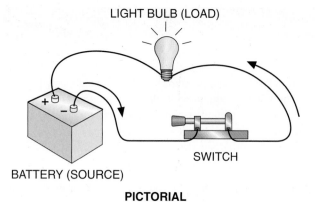

LIGHT BULB (LOAD)

BATTERY (SOURCE)

PICTORIAL

LIGHT BULB

BATTERY   SWITCH

SCHEMATIC

106F06.EPS

Figure 6 ◆ Electrical circuit.

5.2.0 Parallel Circuits

In a parallel circuit, each load is connected directly to the voltage source; therefore, the voltage drop is the same through all loads. The source sees the circuit as two or more individual circuits containing one load each. In the parallel circuit in *Figure 7*, the source sees three circuits, each containing a 30Ω load. The current flow through any load is determined by the resistance of that load. Thus the total current drawn by the circuit is the sum of the individual currents. The total resistance of a parallel circuit is calculated differently from that of a series circuit. In a parallel circuit, the total resistance is less than the smallest of the individual resistances.

For example, each of the 30Ω loads draws 0.4A at 12V; therefore, the total current is 1.2A:

$$I = \frac{E}{R} = \frac{12V}{30\Omega}$$

0.4A per circuit × three circuits = 1.2A

Now, Ohm's law can be used again to calculate the total resistance:

$$R = \frac{E}{I} = \frac{12V}{1.2A} = 10\Omega$$

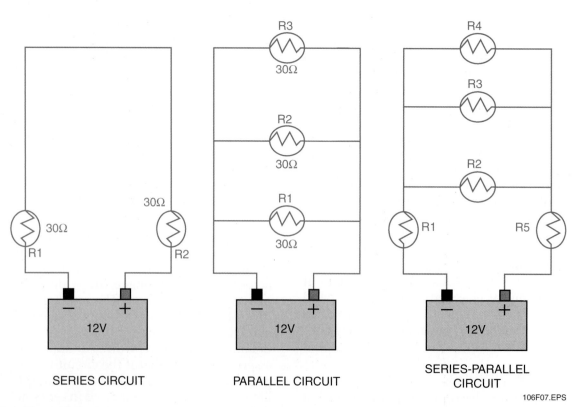

SERIES CIRCUIT   PARALLEL CIRCUIT   SERIES-PARALLEL CIRCUIT

106F07.EPS

Figure 7 ◆ Types of circuits.

Is It a Series Circuit?

When the term series circuit is used, it refers to the way the loads are connected. The same is true for parallel and series-parallel circuits. You will rarely, if ever, find loads in an HVAC system connected in series, or in a series-parallel arrangement. The simple circuit shown here illustrates this point. At first glance, you might think it is a series-parallel circuit. On closer examination, you can see that there are only two loads, and they are connected in parallel. Therefore, it is a parallel circuit. The control devices are wired in series with the loads, but only the loads are considered in determining the type of circuit.

24VAC

COOLING THERMOSTAT   DEFROST RELAY

INDOOR FAN RELAY

HIGH-PRESSURE SWITCH   LOW-PRESSURE SWITCH

COMP. CONTACTOR

106SA04.EPS

This one was simple because all the resistances were the same value. The process is the same when the resistances are different, but the current calculation has to be done for each load. The individual currents are added to get the total current.

Unlike series circuits, parallel circuits continue working even if one circuit opens. Household circuits are wired in parallel. Almost all the HVAC load circuits you encounter will be parallel circuits.

Either of the following formulas can be used to convert parallel resistances to a single resistance value. The first one works only when there are two resistances in parallel. The second is used when there are three or more.

$$\text{Total resistance} = \frac{R1 \times R2}{R1 + R2}$$

$$\text{Total resistance} = \frac{1}{\dfrac{1}{R1} + \dfrac{1}{R2} + \dfrac{1}{R3}}$$

Example:

1. The total resistance of the parallel circuit below is 6Ω.

10Ω 15Ω

$$\frac{R1 \times R2}{R1 + R2} = \frac{10 \times 15 = 150}{10 + 15 = 25} = 6$$

106SA05.EPS

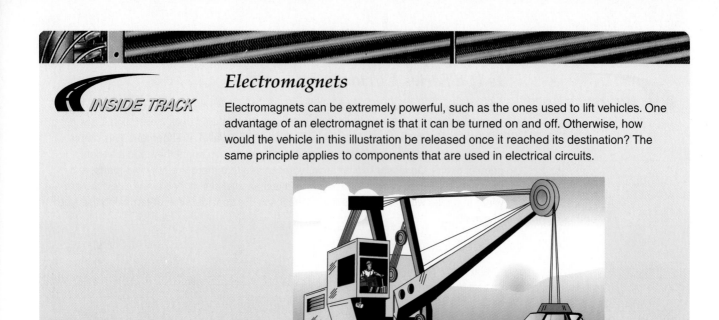
2. The total resistance of the parallel circuit below is 4.76Ω.

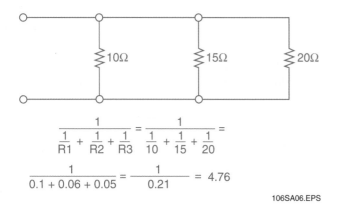

$$\frac{1}{\dfrac{1}{R1} + \dfrac{1}{R2} + \dfrac{1}{R3}} = \frac{1}{\dfrac{1}{10} + \dfrac{1}{15} + \dfrac{1}{20}} =$$

$$\frac{1}{0.1 + 0.06 + 0.05} = \frac{1}{0.21} = 4.76$$

106SA06.EPS

5.3.0 Series-Parallel Circuits

Electronic circuits often contain a hybrid arrangement known as a series-parallel circuit or combination circuit (*Figure 7*). It is unlikely, however, that you will ever have to determine the electrical characteristics of one of these circuits. If it becomes necessary, the parallel loads must be converted to their equivalent series resistance. Then the load resistances are added to determine total circuit resistance.

6.0.0 ◆ MAGNETISM

The operation of many electrical components relies on the power of magnetism. Motors, **relays**, transformers, and **solenoids** are examples. Magnetized iron generates a magnetic field consisting of magnetic lines of force, also known as magnetic flux lines (*Figure 8*). Magnetic objects within the field will be attracted or repelled by the magnetic field. The more powerful the magnet, the more powerful the magnetic field around it. Each magnet has a north pole and a south pole. Opposing poles attract each other; like poles repel each other.

Electricity also produces magnetism. Current flowing through a conductor produces a small magnetic field around the conductor. If the conductor is coiled around an iron bar, the result is an **electromagnet** (*Figure 9*) that attracts and repels other magnetic objects just like an iron magnet. This is the basis on which electric motors and other components operate.

PERMANENT MAGNET

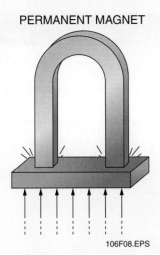

106F08.EPS

Figure 8 ◆ Magnetism.

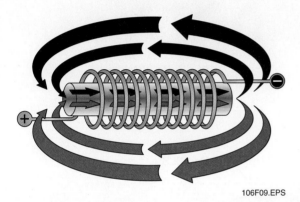

106F09.EPS

Figure 9 ◆ Electromagnet.

7.0.0 ◆ ELECTRICAL COMPONENTS

Circuit diagrams, often known as wiring diagrams or wiring schematics, use symbols to represent the electrical components and connections. In this section, you will begin to learn about electrical components and how they are shown on wiring schematics. Electrical components generally fall into two categories: load devices and control devices. Some devices (relays, for example) contain both an energy-consuming element (load) and a control element (switch contacts). These are treated as control devices. The symbols commonly used on wiring schematics are shown in the *Appendix*.

7.1.0 Load Devices

Any device that consumes electrical energy and does work in the process is a load device. Loads convert electrical energy into some other form of energy such as light, heat, or mechanical energy. They have resistance and consume power. An electric motor is a load; so is the burner on an electric stove. As electrical current flows through the burner element, it is converted into heat.

7.1.1 Motors

Electric motors convert electrical energy into mechanical energy. They are used primarily to drive compressors and fans. Electric motors are the most common loads found in HVAC systems. *Figure 10* shows some of the ways that motors are depicted on schematic wiring diagrams. Because it is likely that three or more motors will appear on an HVAC circuit diagram, a letter code such as IFM (indoor fan motor) is used to identify the function of a particular motor. The same is true for other components.

7.1.2 Electric Heaters

Electric heaters are also called resistance heaters because they are made of high-resistance material that consumes a large amount of electricity, converting it into heat. They are somewhat like the burners on an electric stove. The diagram symbol for an electric heater is usually the same as that for a resistor (*Figure 11*).

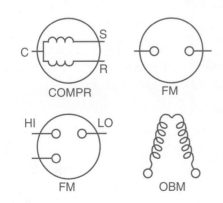

LEGEND

COMPR = COMPRESSOR
FM = FAN MOTOR
FM = FAN MOTOR, MULTIPLE SPEED
OBM = OIL BURNER MOTOR

106F10.EPS

Figure 10 ◆ Schematic symbols for electric motors.

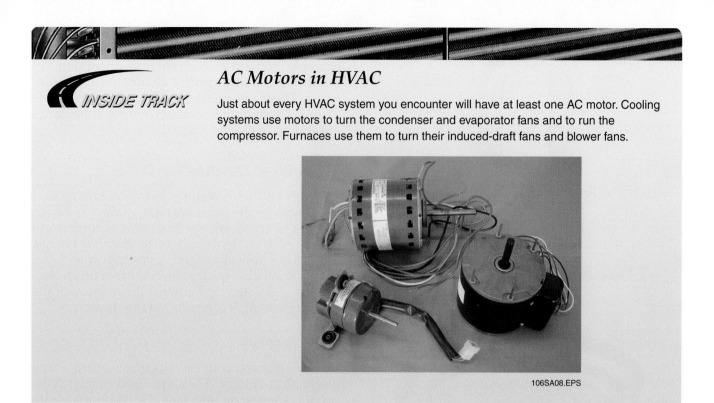
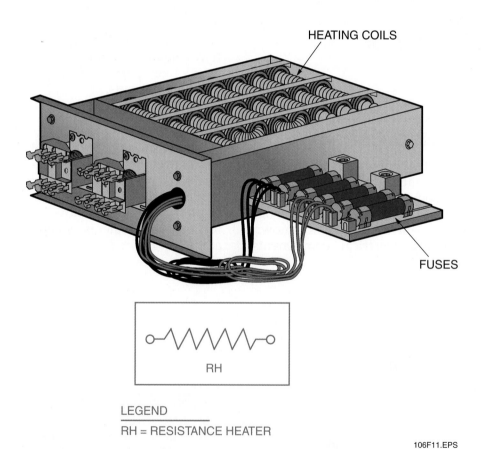

HEATING COILS

FUSES

LEGEND

RH = RESISTANCE HEATER

106F11.EPS

Figure 11 ◆ Resistance heater.

7.1.3 Lights

In an HVAC system, a lightbulb (*Figure 12*) is classified as a secondary load. It is used to signal an operator about the status of the equipment. It is usually represented as a circle with the lens color indicated.

7.2.0 Control Devices

There are many types of control devices. Their primary function is to turn loads on and off.

7.2.1 Switches

Switches stop and start the flow of current to other control devices or loads. Electrical switches are classified according to the force used to operate them (i.e., manual, magnetic, temperature, light, moisture, and pressure).

The simplest type of switch is one that makes (closes) or breaks (opens) a single electrical circuit. More complicated switches control several circuits. The switching action is described by the number of poles (number of electrical circuits to the switch) and the number of throws (number of circuits fed by the switch). *Figure 13* shows some of the common switch arrangements, including:

1. Single-pole, single-throw (SPST)
2. Double-pole, single-throw (DPST)
3. Double-pole, double-throw (DPDT)

An example of a manual switch is the main power disconnect switch (*Figure 14*). It is usually mounted on the unit. It allows the technician to remove power from the system for safety purposes. As shown by the schematic diagram in *Figure 14*, some disconnects are fused so that power is removed at the unit in case of an electrical overload. The *National Electrical Code*® and many local building codes require that a power disconnect device be located within sight of an air conditioning unit. Some disconnects, like the one shown in *Figure 14*, have a handle on the side to turn the power on and off. Others have a plug that is removed to turn the power off.

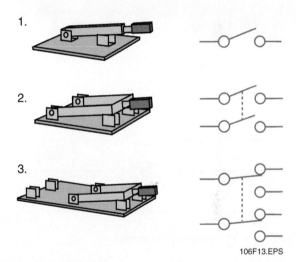

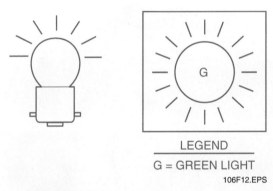

LEGEND

G = GREEN LIGHT

106F12.EPS

Figure 12 ◆ Light.

106F13.EPS

Figure 13 ◆ Switches.

Electric Heaters in Air Conditioning Systems

INSIDE TRACK

Air conditioning systems and heat pumps often contain electric resistance heaters. Local codes may even require that a heat pump have enough electric heat capacity to support up to 100% of the heating demand in the event of a compressor failure. An air conditioning system with 15,000W (15 kilowatts [kW]) of supplementary electric heaters would draw about 65A in a 230V system. Many homes do not have that much extra capacity in their electrical service. In older homes, for example, the entire electrical service may be only 100A.

It is not uncommon when replacing a furnace with a heat pump to find it necessary to expand the electrical service. This can represent a significant cost to the building owner. The sales engineer who specifies the system must take such factors into account when pricing the job.

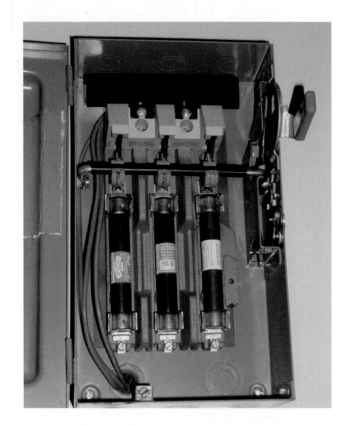

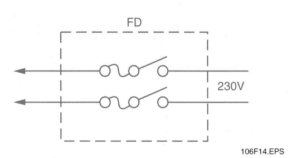

Figure 14 ◆ Fused power disconnect.

CIRCUIT MAKES ON TEMPERATURE DROP FOR HEATING

CIRCUIT MAKES ON TEMPERATURE RISE FOR COOLING

106F15.EPS

Figure 15 ◆ Thermostatic switch.

A thermostatic switch (*Figure 15*) opens and closes in response to changes in temperature. The thermostat shown on the left closes on a temperature drop. It would thus be used to control a heating system. A cooling thermostat would close on a temperature rise. Thermostats are the primary control devices in heating and cooling systems. Other temperature-controlled switches provide protection from current or heat overloads. They are commonly used in compressor circuits. Heat-sensing switches are also used as limit switches to shut off furnaces in case of a malfunction.

Pressure-sensing switches, also known as **pressurestats**, are used in a variety of ways (*Figure 16*). One common use is to shut off an air

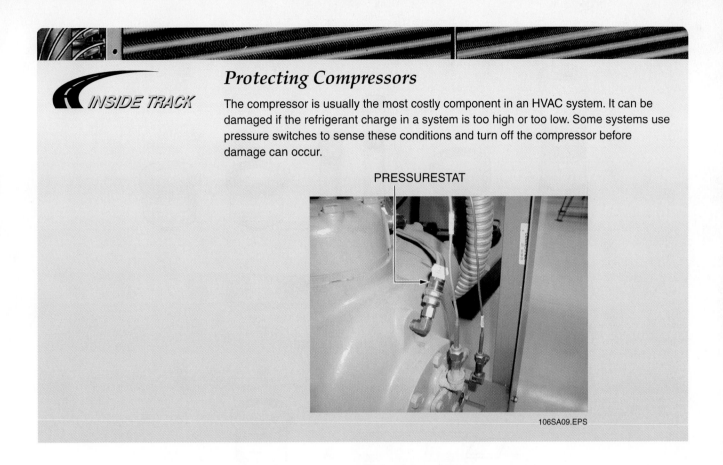
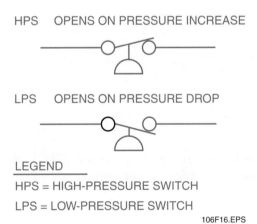

HPS   OPENS ON PRESSURE INCREASE

LPS   OPENS ON PRESSURE DROP

LEGEND
HPS = HIGH-PRESSURE SWITCH
LPS = LOW-PRESSURE SWITCH

106F16.EPS

Figure 16 ◆ Pressure switches.

conditioning compressor if the system pressure becomes too high or too low because of a problem in the system.

Light-operated switches are sometimes used to sense flame. If the flame in an oil furnace goes out, for example, the light-sensing switch will shut off the flow of oil to the combustion chamber.

Moisture-operated switches are used in humidifiers and dehumidifiers. A strand of hair or nylon, which expands and contracts with changes in moisture, can be used to activate the unit.

7.2.2 *Fuses and Circuit Breakers*

Fuses and circuit breakers protect components and wiring against damage from current surges or **short circuits**. By definition, a short circuit occurs when current flow bypasses the load. An example would be when conductors touch due to defective insulation. Because there is no load, current flow is uninhibited and could burn up the wiring, causing a fire. Many home fires are caused by short circuits.

See *Figure 17*. A fuse contains a metal strip that melts when the current exceeds the rated capacity of the fuse (e.g., 20A). Because the strip is in series with the circuit, the current flow is cut off when the strip opens. The fuse will blow more quickly in the presence of a short circuit than it will in the presence of a current overload. Once a fuse has blown, it must be replaced.

Compressor motors require a large startup current. The initial current surge, known as locked rotor amps (LRA), may be six times that of its normal running current, known as full load amps (FLA). The LRA on many compressors is more than 100A. Special delayed-opening fuses (**slow-blow fuses**) are used in air conditioning equipment. These fuses will not blow unless the current surge is sustained.

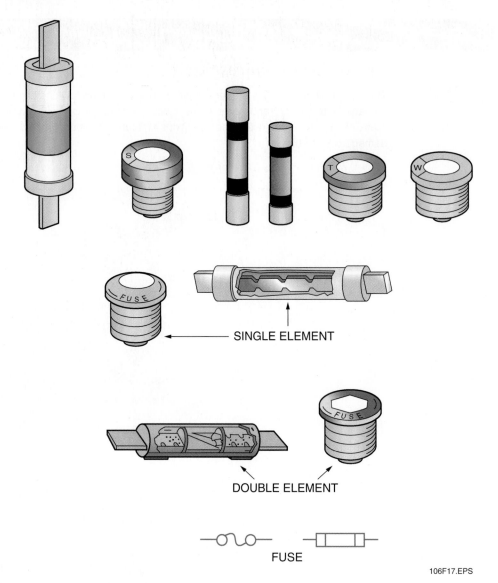

SINGLE ELEMENT

DOUBLE ELEMENT

FUSE

106F17.EPS

Figure 17 ◆ Fuses.

Circuit breakers (*Figure 18*) serve the same purpose as fuses. Most modern construction uses circuit breakers. The big advantage of circuit breakers is that they operate like switches and can be reset when they trip. If a circuit breaker trips more than once, however, it is a sign of a circuit problem. Special **HACR (heating, air conditioning, and refrigeration) circuit breakers** are used for air conditioning equipment for the same reason that slow-blow fuses are used.

Thermal trip circuit breakers contain a spring-loaded metal element that opens when a high current causes it to overheat. Magnetic trip breakers contain a small coil of wire that magnetically opens the contacts when the current is excessive.

A problem known as a ground fault occurs when a live conductor touches another conducting substance, such as the metal frame of a unit.

Ground fault circuit interrupter (GFCI) devices are installed to detect and prevent such problems.

7.2.3 Solenoids

A solenoid (*Figure 19*) is an electromagnet. When a current flows through the solenoid, the magnetism produced by the current is used to attract or repel a nearby object. Solenoids are used for a variety of purposes, such as opening and closing valves and switching devices. A GV next to the solenoid symbol indicates that this solenoid is used to control a gas valve, which feeds natural or LP gas to a gas-fired furnace.

7.2.4 Relays, Contactors, and Starters

These are magnetically controlled switching devices consisting of a coil (solenoid) and one or

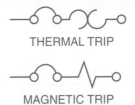

THERMAL TRIP

MAGNETIC TRIP

106F18.EPS

Figure 18 ◆ Circuit breakers.

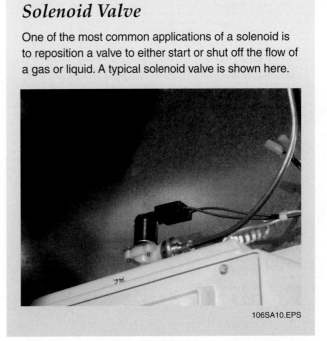
more sets of movable contacts that open and close in response to current flow in the coil. In other words, they are electrically operated switching devices.

Figure 20 shows different types of relays. When a current flows through the relay coil, a magnetic field is created. This field causes the contacts to change position. Some of the contacts close, providing a path for current; these are called normally open or N.O. contacts because they are

SOLENOID

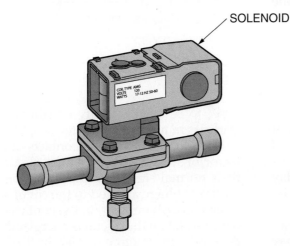

SOL

LEGEND

SOL = SOLENOID (RELAY COIL)

106F19.EPS

Figure 19 ◆ Solenoid.

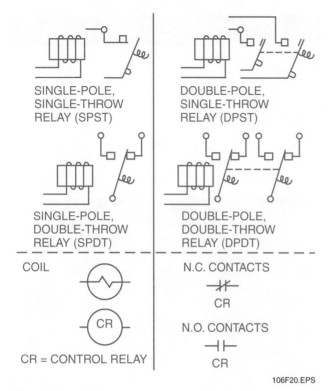

SINGLE-POLE, SINGLE-THROW RELAY (SPST)

DOUBLE-POLE, SINGLE-THROW RELAY (DPST)

SINGLE-POLE, DOUBLE-THROW RELAY (SPDT)

DOUBLE-POLE, DOUBLE-THROW RELAY (DPDT)

COIL

N.C. CONTACTS
CR

CR

N.O. CONTACTS
CR

CR = CONTROL RELAY

106F20.EPS

Figure 20 ◆ Relays.

open when the relay is de-energized. Other contacts open; these are called normally closed or N.C. contacts because they are closed when the relay is de-energized. Normally closed contacts have a slash through them. The coil and contacts of a relay will be shown at different locations on a diagram because each set of contacts controls a different circuit. You can find them because the coil and contacts will have the same name; in this case, CR for control relay. The same pole-throw descriptions used for switches also apply to relays and contactors.

Figure 21 shows an example of how a relay might be used in an electrical control circuit. The thermostatic switch on the left closes when the temperature in the space exceeds a selected value (setpoint). The closed thermostat completes a current path to the relay coil. The coil produces a magnetic field that changes the position of the relay contacts. The normally open contacts close, completing a path to the fan motor. The normally closed contacts open, causing the FAN OFF light to go out. When the thermostat opens, the current stops and the contacts return to their original positions.

A **contactor** is basically a heavy-duty relay designed to carry large currents. They are often used to start and stop compressors.

Starters are used to start large motors. They usually contain overload protection devices and may include an on-off switch. The *National Electrical*

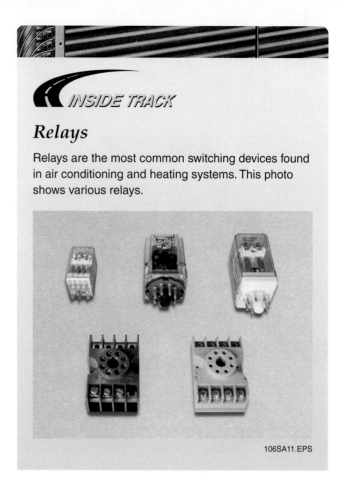
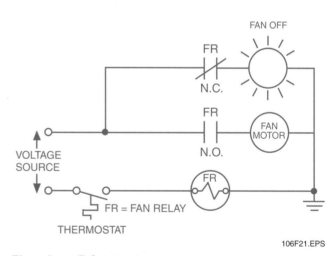

FAN OFF

FR
N.C.

FR
N.O.

FAN MOTOR

VOLTAGE SOURCE

FR

FR = FAN RELAY

THERMOSTAT

106F21.EPS

Figure 21 ◆ Relay circuit.

Code® (NEC®) will specify whether a contactor or starter is required for a motor.

Keep in mind that the contacts of a relay, contactor, or switch will appear in a circuit diagram in their normal state (i.e., the condition they are in with no voltage applied to the circuit). The reader must mentally reposition the contacts to see what happens when the circuit is energized. One way to do that is to imagine a slash in the N.O. contact and no slash in the N.C. contact.

Time delay relays use a thermal element or digital control to delay the energizing or de-energizing of a relay. Delays may range from a few seconds to a few hours. A typical use in HVAC is to keep a furnace fan running for 30 to 60 seconds after the burners have shut off. The fan extracts residual heat from the heat exchangers. This improves heating efficiency.

7.2.5 Transformers

A transformer (*Figure 22*) is used to raise or lower a voltage. The transformer usually consists of two or more coils of wire wound around a common iron core. When a current passes through the primary coil (winding) of the transformer, the resulting magnetic field cuts through the other coil (secondary winding), creating a current in that coil. This process is known as **induction**. Depending on the number of turns of wire in each secondary winding, the voltage induced in the secondary will be stepped down (less than) or stepped up (greater than) from that in the primary. The transformer shown in *Figure 22* is a step-down transformer with two secondary windings. Each secondary winding acts as the power source for a circuit.

Transformers are used extensively in power distribution systems. In HVAC systems, a transformer is commonly used to step down the 120V source to 24V to operate the control circuits. It is common to find 24V control circuits in HVAC

equipment because low-voltage circuits are less dangerous. They are also less expensive to build because low-voltage circuits can use lighter-gauge wire and lighter-duty components.

7.2.6 Overload Protection Devices

Overload devices stop the flow of current when safe current or temperature limits are exceeded (see *Figure 23*). Most compressor motor circuits will have one or more of these devices. Thermal overload devices are often embedded in the windings of the motor. Magnetic overloads operate much the same as relays.

Protective devices are placed into two categories. **Line duty** devices are directly in line with the voltage source. In **pilot duty** devices, a small coil of wire senses current and acts as a magnet to open contacts in the motor control circuit when the current is too high.

Some overload devices reset automatically when the cause of the overload has been removed. Others must be manually reset, usually by pressing a button on the device. The overload device embedded in the motor windings is an automatic reset device because it is not accessible.

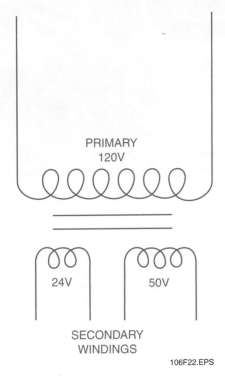

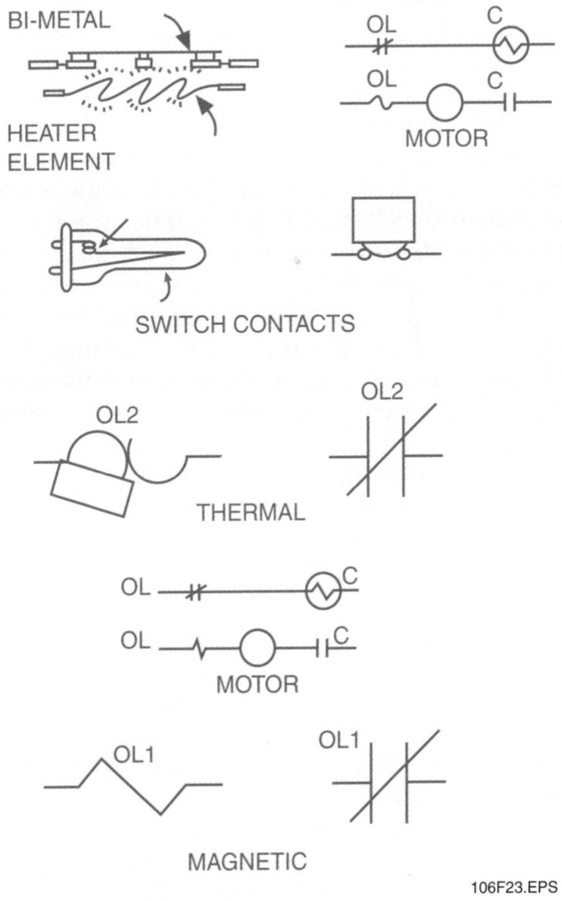

PRIMARY
120V

24V   50V

SECONDARY
WINDINGS

106F22.EPS

Figure 22 ◆ Transformer.

BI-METAL

HEATER
ELEMENT

OL   C
OL   C
MOTOR

SWITCH CONTACTS

OL2

OL2

THERMAL

OL   C
OL   C
MOTOR

OL1

OL1

MAGNETIC

106F23.EPS

Figure 23 ◆ Overload devices.

8.0.0 ◆ ELECTRICAL SAFETY

Working with electrical devices always involves some degree of danger. You must be constantly alert to the possibility of electric shock and take the necessary precautions to protect yourself and others.

8.1.0 The Effect of Current

The amount of current that passes through the human body determines the outcome of an electrical shock. The higher the voltage, the greater the current, and the greater the chance for a fatal shock.

 WARNING!

High voltage, defined as 600V or more, is almost ten times as likely to kill as low voltage. On the job, you spend most of your time working on or near lower voltages. However, lower voltages can also kill. For example, portable, electrically operated hand tools can cause severe injuries or death if the frame or case of the tool becomes energized. Many electrical accidents can be prevented by following proper safety practices, including the use of GFCIs and proper grounding.

Electrical current flows along the path of least resistance to return to its source. If you come in contact with a live conductor, you become a load. *Figure 24* shows how much resistance the human body presents under various circumstances and how this converts to amps or milliamps when the voltage is 110V. Note that the potential for shock increases dramatically if the skin is damp. A cut will also reduce your resistance. Currents of less than 1A can severely injure and even kill a person (*Table 1*).

This information has been included in this training module to help you gain respect for the environment where you work and to stress how important safe working habits really are.

INSIDE TRACK

Transformers

There are many kinds of transformers. The pole-mounted transformer pictured earlier in the module is used to step down the high voltage on the local utility lines to a level that can be used by lighting, equipment, and appliances in a building. It is large and very heavy. Other transformers, such as those found in power substations, can be nearly as big as a house. The transformers used in air conditioning equipment are usually small enough to hold in the palm of your hand.

Nearly all residential and light commercial HVAC equipment contains at least one step-down transformer. To protect the transformer, many control circuits contain a fuse or circuit breaker in the transformer secondary circuit. If you are troubleshooting an HVAC system and have power to the transformer primary, but control circuit voltage (usually 24V) is not present, check the fuse or circuit breaker before replacing the transformer.

CONTROL
TRANSFORMER

106SA12.EPS

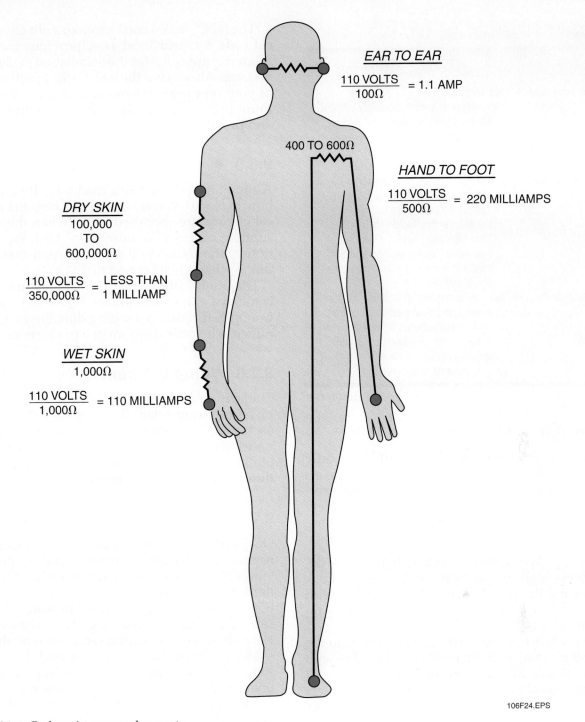

Figure 24 ◆ Body resistances and currents.

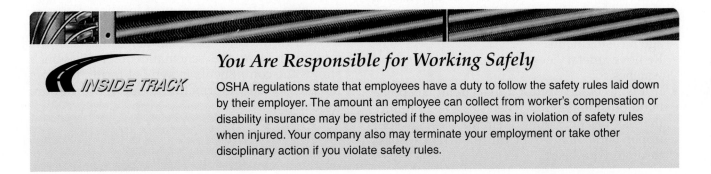

You Are Responsible for Working Safely

OSHA regulations state that employees have a duty to follow the safety rules laid down by their employer. The amount an employee can collect from worker's compensation or disability insurance may be restricted if the employee was in violation of safety rules when injured. Your company also may terminate your employment or take other disciplinary action if you violate safety rules.

Table 1 Current Effects on the Human Body

Current Value	Typical Effects
Less than 1mA	No sensation.
1mA to 20mA	Sensation of shock, possibly painful. May lose some muscular control between 10mA and 20mA.
20mA to 50mA	Painful shock, severe muscular contractions, breathing difficulties.
50mA to 200mA	Up to 100mA, same symptoms as above, only more severe. Between 100mA and 200mA ventricular fibrillation may occur. This typically results in almost immediate death unless special medical equipment and treatment are available.
Over 200mA	Severe burns and muscular contractions. The chest muscles contract and stop the heart for the duration of the shock, followed by death unless special medical equipment and treatment are available.

106T01.EPS

8.2.0 Safety Practices

HVAC technicians work with potentially deadly levels of electricity every day. They can do so because good safety practices have become second nature to them. Here are some general safety practices to follow whenever you are working with electricity.

- Unless it is essential to work with the power on, always shut off electricity at the source. Lock and tag the power switch in accordance with company or site procedures and OSHA requirements.
- Use a voltmeter to verify that the power to the unit is actually off. Remember that even though the power may be switched off, there is still potential at the input side of the shutoff switch.
- Use protective equipment such as rubber gloves.
- Use insulated tools.
- Short components to ground before touching de-energized wires.
- Do not kneel on the ground when making voltage measurements.
- Remove metal jewelry such as rings and watches.
- If it is necessary to test a live circuit, keep one hand outside the unit when possible to reduce the risk of completing a circuit through your upper body that might cause current to pass through your heart.

The *NEC*®, when used together with the electrical code for your local area, provides the minimum requirements for the installation of electrical systems. Always use the latest edition of the *NEC*® as your on-the-job reference. It specifies the minimum provisions necessary for protecting people and property from electrical hazards.

9.0.0 ◆ CIRCUIT DIAGRAMS

A circuit diagram is like a road map. If you know how to read it, you can determine how the electrical circuits are supposed to act when the unit is running properly. You can then check the operation of the circuit to find the portion that is not doing what it is supposed to do.

Diagrams supplied by manufacturers will come in a variety of formats. *Figure 25* is just one example. It contains a wiring diagram and a simplified schematic diagram of a gas furnace.

9.1.0 Wiring Diagram

The wiring diagram shows how the wiring is physically connected. It also shows the color of each wire. The wiring diagram is helpful if it is necessary to rewire a unit. When you are troubleshooting, it will help you find a physical location at which to make a measurement.

9.2.0 Simplified Schematic Diagram

On this diagram, the wire color and physical connection information are removed and the pictorial views of the components are replaced with standard electrical diagram symbols. This simplification makes the diagram easier to read. Also, the simplified schematic is arranged to make it easy to trace circuits. **Ladder diagrams**, such as the one shown in *Figure 26*, may be provided. The power source is shown as the uprights, and each load line and its related control devices is represented as a rung of the ladder. The legend on the diagram explains the component abbreviations used on the diagram.

The diagram provided by the manufacturer may also contain a component location diagram. It shows where the electrical components are located in the unit. This will help you make the transition from the schematic diagram, which shows the components in symbol form, but does not provide any information about where they are located.

WIRING DIAGRAM

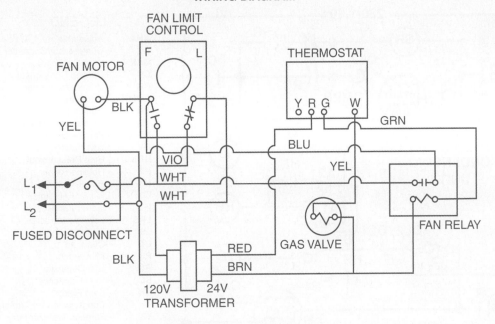

SIMPLIFIED SCHEMATIC

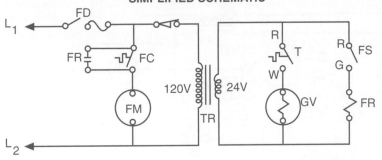

106F25.EPS

Figure 25 ◆ Circuit diagram.

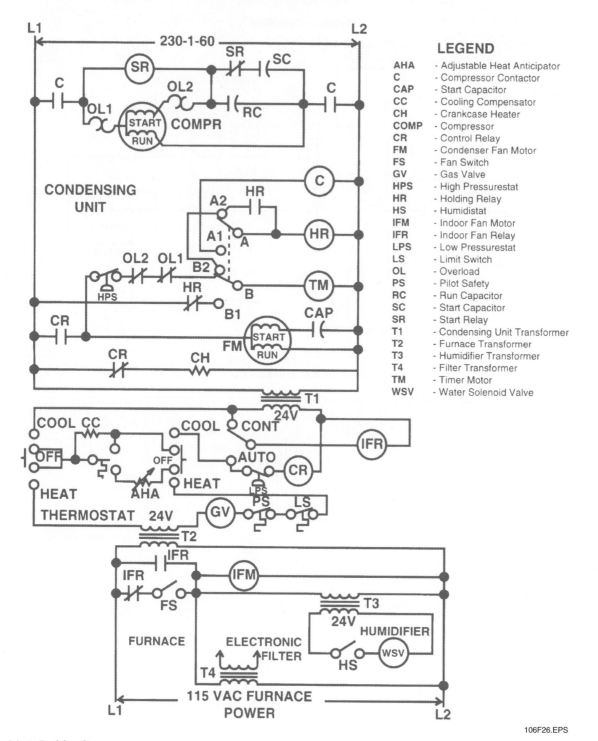

LEGEND

AHA	- Adjustable Heat Anticipator
C	- Compressor Contactor
CAP	- Start Capacitor
CC	- Cooling Compensator
CH	- Crankcase Heater
COMP	- Compressor
CR	- Control Relay
FM	- Condenser Fan Motor
FS	- Fan Switch
GV	- Gas Valve
HPS	- High Pressurestat
HR	- Holding Relay
HS	- Humidistat
IFM	- Indoor Fan Motor
IFR	- Indoor Fan Relay
LPS	- Low Pressurestat
LS	- Limit Switch
OL	- Overload
PS	- Pilot Safety
RC	- Run Capacitor
SC	- Start Capacitor
SR	- Start Relay
T1	- Condensing Unit Transformer
T2	- Furnace Transformer
T3	- Humidifier Transformer
T4	- Filter Transformer
TM	- Timer Motor
WSV	- Water Solenoid Valve

106F26.EPS

Figure 26 ◆ Ladder diagram.

Tracing a Circuit

Each switch or set of relay contacts in a circuit represents a condition that must be met before the compressor can operate. If the thermostat is calling for cooling and the compressor isn't running, there is a good chance that one of the conditions is not being met.

When you are troubleshooting the circuit with a voltmeter, start by placing your meter probes across the entire circuit, as shown in the illustration here. This verifies that voltage is applied to the circuit. Then, move the hot probe to the next component in the series chain and read the voltage. Keep doing this until the meter registers no voltage. The last component you jumped (or its related wiring) will be the defective component. This technique is known as hopscotch troubleshooting.

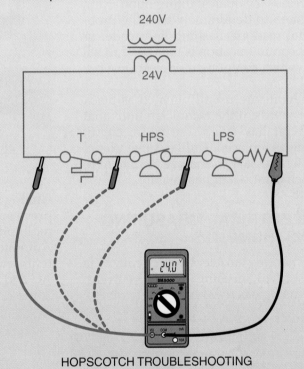

HOPSCOTCH TROUBLESHOOTING

106SA13.EPS

10.0.0 ◆ ELECTRONIC CONTROLS

Electronic circuits use **solid state** timing, switching, and sensing devices to control loads and protective circuits. Electronic circuits operate at much lower voltage and current levels than electromechanical devices such as relays and solenoids. In addition, they are more reliable because they have no moving parts. Most electronic circuits consist of microminiature components mounted on printed circuit boards (*Figure 27*). The components are mounted to the top of the board. The copper runs found on the underside of the board serve the same purpose as wiring. Circuits that can be used in a variety of products are often packaged in sealed modules.

106F27.EPS

Figure 27 ◆ Printed circuit board.

Microcomputers allow building occupants or managers to custom-tailor the operation of an HVAC system to their needs (*Figure 28*). Programmable thermostats, for example, allow the occupant to program a different operating temperature for different times of day. Microcomputers are able to receive a large amount of information and make decisions based on the information and the instructions in their programs. One of the most important features of these circuits is that they offer many more capabilities and more precise control than systems controlled by conventional circuits. Microprocessor-controlled systems are often self-diagnosing. When there is a problem, the microprocessor evaluates information from around the system to determine where the problem is. A digital readout, flashing light code, or other means of communication is then used to tell the technician which component or electronic circuit to check.

Electronic circuits are usually treated as black boxes. In other words, if the technician finds that the control circuit has failed, the entire circuit board or module is replaced. Unlike conventional circuits, it is not necessary to analyze the circuit to figure out which component has failed.

11.0.0 ◆ ELECTRICAL MEASURING INSTRUMENTS

When troubleshooting an electrical circuit, it is usually necessary to measure voltage, current, and resistance using electrical meters (*Figure 29*).

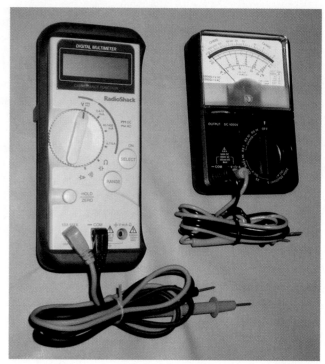

106F29.EPS

Figure 29 ◆ Analog and digital meters.

Analog meters, which require the reader to interpret a scale, are still around, but have been largely replaced by direct-reading **digital meters**. **Ammeters** are used to measure alternating current; **multimeters** are commonly used to measure AC and DC voltage, resistance, and direct current. Some can read AC current in the milliamp range.

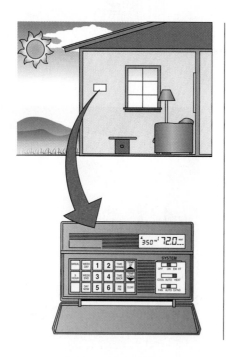

106F28.EPS

Figure 28 ◆ Electronic control of HVAC systems.

11.1.0 Ammeter

Ammeters are often used to check motor circuits. A **clamp-on ammeter** (*Figure 30*) is normally used to measure AC. The jaws of the ammeter are placed around the wire conductor. Current flowing through the wire creates a magnetic field, which induces a proportional current in the ammeter jaws. This current is read by the meter movement and appears as a direct readout or, on an analog meter, as a deflection of the meter needle.

In-line ammeters (*Figure 31*) are less common. This type of meter must be connected in series with the circuit, which means that the circuit must be opened.

Aside from following good safety practices, there are a few things to remember when measuring current:

- If the ammeter jaws are dirty or misaligned, the meter will not read correctly.
- When using an analog meter, always start at a high range and work down to avoid damaging the meter.
- Do not clamp the meter jaws around two different conductors at the same time, or an inaccurate reading will result.

11.2.0 Multimeters

Multimeters have a selector on the meter that allows the user to select AC or DC voltage (voltmeter), resistance (ohmmeter), or current (ammeter). See *Figure 32*. On an analog meter, the range of values to be read must also be selected.

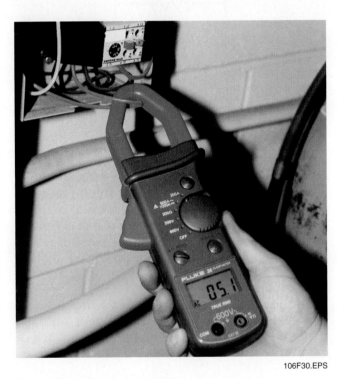

106F30.EPS

Figure 30 ◆ Clamp-on ammeter.

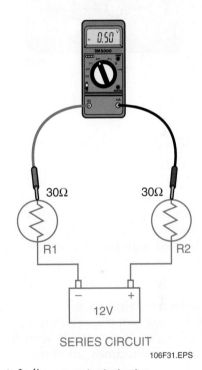

SERIES CIRCUIT

106F31.EPS

Figure 31 ◆ In-line ammeter test setup.

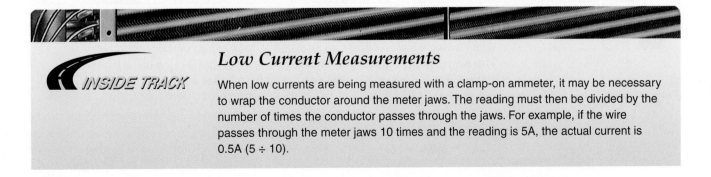

Figure 32 ◆ Digital multimeter.

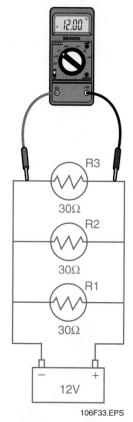

106F33.EPS

Figure 33 ◆ Voltmeter connection.

11.2.1 Voltage Measurements

A voltmeter must be connected in parallel with (across) the component or circuit to be tested (*Figure 33*). If a circuit function is not operating, the voltmeter can be used to determine if the correct voltage is available to the circuit. Voltage must be checked with power applied.

11.2.2 Resistance Measurements

An ohmmeter contains an internal battery that acts as a voltage source. Therefore, resistance measurements are always made with the system power shut off. Sometimes, an ohmmeter is used

to measure resistance in a load; motor windings are a good example. More often, an ohmmeter is used to check **continuity** in a circuit. A wire or closed switch offers negligible resistance. With the ohmmeter connected as in *Figure 34*, and the three switches closed, the current produced by the ohmmeter battery will flow unopposed, and the meter will show zero resistance. The circuit has continuity; i.e., it is continuous. If a switch is open, however, there is no path for current and the meter will see infinite resistance; that is, a lack of continuity.

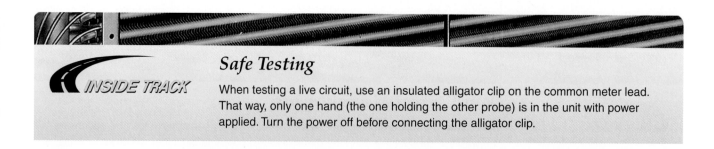

Safe Testing

When testing a live circuit, use an insulated alligator clip on the common meter lead. That way, only one hand (the one holding the other probe) is in the unit with power applied. Turn the power off before connecting the alligator clip.

Using Ohm's Law

Suppose the resistive load in a circuit is expected to be 50Ω. You want to know if the resistance value of the load is correct, but the battery in your multimeter is dead, so you can't measure resistance. You measure the applied voltage and find that it is 230V. Using an ammeter, you find that the load is drawing 4.5A. What is the resistance of the load?

The allowable tolerance of the resistance value for this load is ±5%. Is the resistance of this load within this range?

Electrical Safety

Every year in the U.S., there are approximately 700 deaths resulting from electrical accidents. Electrical accidents are the third leading cause of death in the workplace.

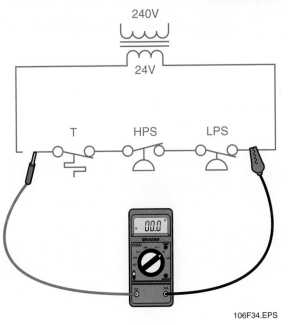

106F34.EPS

Figure 34 ◆ Ohmmeter connection for continuity testing.

CAUTION

Never measure resistance in a live (energized) circuit. The voltage present in a live circuit may permanently damage the meter. Higher quality meters are equipped with a fuse or circuit breaker to protect the meter if this occurs. Also, a circuit in parallel with the circuit you are testing can provide a current path that will make it appear as though the circuit you are testing is good, even when it is not. The way to avoid this problem is to isolate the device you are testing from the rest of the circuit by disconnecting it, as shown in *Figure 34*.

Summary

All HVAC systems contain electrical circuits. In order to install and service HVAC systems, the HVAC technician must know how electrical components work, how to read circuit diagrams, and how to use electrical test equipment. When working with electricity, knowing and following proper safety practices is extremely important. Failure to develop good safety habits and follow safety rules can result in injury or death to yourself or your co-workers.

Notes

1. A transformer is used to _____ .
   a. transform electrical energy into another form of energy
   b. raise or lower a voltage
   c. consume power
   d. increase the resistance of a circuit

2. The difference in potential between the two poles of a car battery is caused by _____ .
   a. the current flowing between the two poles
   b. the circuit resistance
   c. the amount of power consumed by the load
   d. a chemical reaction

3. Current is measured in _____ .
   a. milliwatts per square meter
   b. volts
   c. amps
   d. miles per hour

4. A(n) _____ converts AC voltage to DC voltage.
   a. voltmeter
   b. inverter
   c. rectifier
   d. transformer

5. How much current is drawn by a 120V circuit containing a 1,500W load?
   a. 125mA
   b. 0.08A
   c. 12.5A
   d. 180,000A

6. How much power is consumed by a 120V parallel circuit that contains a 100Ω load and an 80Ω load and draws 2.7A?
   a. 44W
   b. 80W
   c. 120W
   d. 324W

7. How much current will be drawn by a 120V series circuit containing a 100Ω load and a 50Ω load?
   a. 0.8W
   b. 0.8A
   c. 1.25A
   d. 8A

8. A DPDT switch can control _____ circuit(s).
   a. 1
   b. 2
   c. 3
   d. 4

9. Fuses are rated on the basis of _____ .
   a. how fast they blow
   b. how much current they can withstand before they blow
   c. full load amps (FLA)
   d. locked rotor amps (LRA)

10. It is okay to work in a unit with the power on, as long as you wear gloves.
   a. True
   b. False

Solve the following problems using the applicable formula.

11. If a series circuit with a 12Ω resistance draws 10A, the source voltage is _____ V.

12. If a 120V circuit draws 3A, the resistance of the circuit is _____ Ω.

13. The total resistance of the circuit below is _____ Ω.

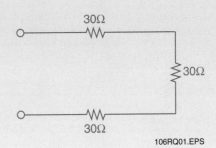

106RQ01.EPS

14. If the circuit in Question 13 has a 120V supply, the current draw will be _____ A.

15. The current through each resistor in the circuit of Question 13 is _____ A.

16. The total resistance of the parallel circuit below is _____ Ω.

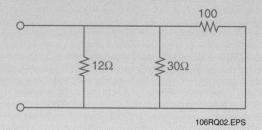

100

12Ω   30Ω

106RQ02.EPS

17. If the circuit in Question 16 draws 3A, the source voltage must be _____ V.

18. A resistance heater that produces 5kW of heat in a 240V circuit will draw _____ A.

19. In the circuit described in Question 18, the resistance of the heating elements is _____ Ω.

20. In a 120V circuit containing two parallel resistances of equal value, the voltage drop across each resistance is _____ V.

Trade Terms Quiz

1. An unbroken circuit path has _____.

2. The amount of electrical energy used by an electrical load device is called _____.

3. The electrical control device with a built-in trip delay that is designed specifically for HVAC circuits is a(n) _____.

4. The type of electrical diagram in which the loads are arranged between vertical lines is known as a(n) _____.

5. A material that readily carries electricity is a(n) _____.

6. The magnetic control device used for switching high-voltage circuits is the _____.

7. The motor control device that opens the motor control circuit, which then opens the motor circuit, is known as a(n) _____ device.

8. Resistance in a DC circuit is measured in _____.

9. A meter that uses a needle to indicate a value is a(n) _____.

10. An electromagnetic coil used to control a mechanical device is called a(n) _____.

11. Electrical power is measured in _____.

12. If the electricity changes direction on a cyclic basis, it is known as _____.

13. When a current is generated in a conductor by placing it in a magnetic field, the process is known as _____.

14. The type of ammeter that is placed in series with the load is the _____.

15. A device that converts AC voltage into DC voltage is the _____.

16. A magnetic switching device used to control heavy-duty motors is the _____.

17. A device with two or more coils of wire wrapped around a common core is a(n) _____.

18. Electromotive force is another name for _____.

19. When a conductor bypasses a load, resulting in a large current surge, it is called a(n) _____.

20. The flow of electrons in a circuit is known as _____.

21. When a protective device is in series with the load, it is known as a(n) _____ device.

22. The pressure-sensitive switch used to protect a compressor is called a(n) _____.

23. The _____ is the unit of measure for current flow in an electrical circuit.

24. A(n) _____ is a test meter able to read voltage, current, and resistance.

25. The device used to sense current using jaws that are clamped around a conductor is the _____.

26. A measuring device that provides a direct numerical reading is the _____.

27. The electrical device that converts energy from one form to another is the _____.

28. A fuse with a time delay is known as a(n) _____.

29. A battery is a common source of
_____ .

30. The opposition to the flow of electricity is
called _____ .

31. When a current flows through a coil of wire
wrapped around an iron core, a(n)
_____ is created.

32. A magnetically operated switching device
used in low-current circuits is the
_____ .

Trade Terms

Alternating current (AC)
Ammeter
Ampere (amp)
Analog meter
Clamp-on ammeter
Conductor
Contactor
Continuity
Current
Digital meter

Direct current (DC)
Electromagnet
HACR (heating, air
conditioning, and
refrigeration) circuit
breaker
Induction
In-line ammeter
Insulator
Ladder diagram

Line duty
Load
Multimeter
Ohm
Pilot duty
Power
Pressurestat
Rectifier
Relay
Resistance

Short circuit
Slow-blow fuse
Solenoid
Solid state
Starter
Transformer
Volt
Voltage
Watts

Joe Moravek

HVAC Instructor, Author
Hunton Trane, Houston, TX

Joe Moravek came to the HVAC trade in a roundabout way, but once in it, he adapted quickly. He went from student to college instructor and eventually authored an HVAC textbook.

How did you choose a career in the HVAC field?
The HVAC industry chose me. In the mid '70s I worked for a Community Act Program in their weatherization program. My job was to conduct energy audits using a 12-page DOE audit manual. It was a load calculation program that evaluated the energy-saving priorities of a home. Anyway, after doing many of these I became interested in the interconnection between energy use and HVAC. My next step was to learn more about the function and operation of an HVAC system. Once I learned how they worked, my interest evolved into how to troubleshoot and repair these systems.

What types of training have you been through?
I have formal training in how to be an educator. Additionally, I have taken college courses in HVAC. I never finished the HVAC program because the college hired me to teach their courses. I have attended many weeks of continuing education throughout my career.

What kinds of work have you done in your career?
I was a ground radio mechanic in the Air Force. I have worked in energy conservation programs with three different agencies. I have been an HVAC mechanic and a mechanical inspector with the City of Houston. Next, I was the lead HVAC instructor at Lee College for 14 years. Currently I am the training coordinator at Hunton Trane. In addition to coordinating training, I teach classes to our dealers and service techs. Mixed in with all of these experiences I write air conditioning reviews and related books.

What do you like about your job?
I like teaching and learning. I enjoy learning new ideas from my students as well as attending advancement continuing education classes. I like to develop classes and critique fellow trainers' classes. There is always something new to learn in the HVAC trade.

What factors have contributed most to your success?
Motivation and finding a job I like. When I had a job I did not like I would wait for the right opportunity to switch. In order to learn the technical side of the trade I took a significant pay cut to become an experienced technician. It was worth it since I am happy in my work. Being satisfied with your work is more important than the money you make. I feel I am lucky to have found a job I like.

What advice would you give to those new to the HVAC field?
It is a challenging field, but a very interesting field. It is always changing. There is always something new to learn. You will never learn it all. You need to decide what type of HVAC work you would like to do. Set your goals and strive to obtain them. There are many opportunities in the HVAC trade.

Trade Terms
Introduced in This Module

Alternating current (AC): An electrical current that changes direction on a cyclical basis.

Ammeter: A test instrument used to measure current flow.

Ampere (amp): The unit of measurement for current flow. The magnitude is determined by the number of electrons passing a point at a given time.

Analog meter: A meter that uses a needle to indicate a value on a scale.

Clamp-on ammeter: A current meter in which jaws placed around a conductor sense the magnitude of the current flow through the conductor.

Conductor: A material that readily conducts electricity; also, the wire that connects components in an electrical circuit.

Contactor: A control device consisting of a coil and one or more sets of contacts used as a switching device in high-voltage circuits.

Continuity: A continuous current path. Absence of continuity indicates an open circuit.

Current: The rate at which electrons flow in a circuit, measured in amperes.

Digital meter: A meter that provides a direct numerical reading of the value measured.

Direct current (DC): An electric current that flows in one direction. A battery is a common source of DC voltage.

Electromagnet: A coil of wire wrapped around a soft iron core. When a current flows through the coil, magnetism is created.

HACR (heating, air conditioning, and refrigeration) circuit breaker: A circuit breaker with a built-in trip delay used specifically in HVAC circuits because of the power surge that occurs with compressor startup.

Induction: To generate a current in a conductor by placing it in a magnetic field and moving the conductor or magnetic field.

In-line ammeter: A current-reading meter that is connected in series with the circuit under test.

Insulator: A device that inhibits the flow of current. Opposite of conductor.

Ladder diagram: A simplified schematic diagram in which the load lines are arranged like the rungs of a ladder between vertical lines representing the voltage source.

Line duty: A protective device connected in series with the supply voltage.

Load: A device that converts electrical energy into another form of energy (heat, mechanical motion, light, etc.). Motors are the most common loads in HVAC systems.

Multimeter: A test instrument capable of reading voltage, current, and resistance.

Ohm: The unit of measurement for electrical resistance.

Pilot duty: A protective device that opens the motor control circuit, which then shuts off the motor.

Power: The amount of energy (measured in watts) consumed by an electrical load. Power = voltage × current.

Pressurestat: A pressure-sensitive switch used to protect compressors.

Rectifier: A device that converts AC voltage to DC voltage.

Relay: A magnetically operated device consisting of a coil and one or more sets of contacts. Used in low-current circuits.

Resistance: The opposition to the flow of electrons (i.e., load).

Short circuit: A situation in which a conductor bypasses the load, causing a very high current flow.

Slow-blow fuse: A fuse with a built-in time delay.

Solenoid: An electromagnetic coil used to control a mechanical device such as a valve or relay contacts.

Solid state: Having to do with semiconductors.

Starter: A magnetic switching device used to control heavy-duty motors.

Transformer: Two or more coils of wire wrapped around a common core. Used to raise and lower voltages.

Volt: The unit of measurement for voltage.

Voltage: A measure (in volts) of the electrical potential for current flow; also known as electromotive force (EMF).

Watts: The unit of measure for power consumed by a load.

Schematic Symbols

SWITCHES

DISCONNECT	MAGNETIC CIRCUIT BREAKER	THERMAL CIRCUIT BREAKER	LIMIT			MAINTAINED POSITION
			SPRING RETURN			
			NORMALLY OPEN	NORMALLY CLOSED	NEUTRAL	
			HELD CLOSED	HELD OPEN		

Liquid Level		Vacuum & Pressure		Temp. Actuated		Air or Water Flow	
LOW	HIGH	LOW	HIGH	NORMALLY OPEN (1)	NORMALLY CLOSED (2)	NORMALLY OPEN (1)	NORMALLY CLOSED (2)

CONDUCTORS		FUSES	COILS			
NOT CONNECTED	CONNECTED	OR	RELAYS, TIMERS, ETC.	OVERLOAD THERMAL	SOLENOID	TRANSFORMER

PUSHBUTTONS

SINGLE CIRCUIT		DOUBLE CIRCUIT	MUSHROOM CIRCUIT	MAINTAINED CONTACT	
NORMALLY OPEN	NORMALLY CLOSED				

TIMER CONTACTS CONTACT ACTION IS RETARDED WHEN COIL IS:				GENERAL CONTACTS STARTERS, RELAYS, ETC.		
ENERGIZED		DE-ENERGIZED		OVERLOAD THERMAL	NORMALLY OPEN	NORMALLY CLOSED
NORMALLY OPEN	NORMALLY CLOSED	NORMALLY OPEN	NORMALLY CLOSED			

(1) **Make on rise**
(2) **Make on fall**

106A01.EPS

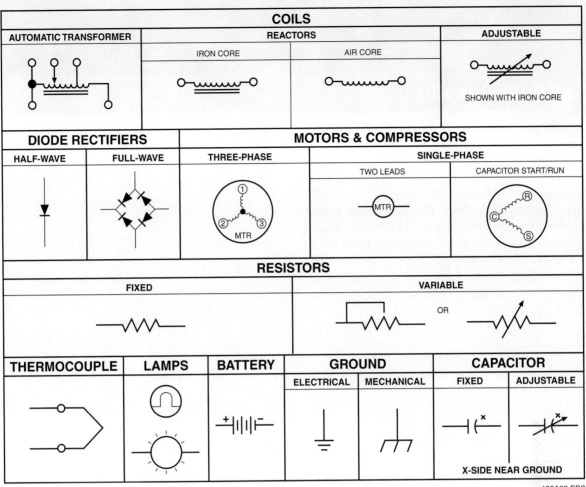

106A02.EPS

This module is intended to present thorough resources for task training. The following reference works are suggested for further study. These are optional materials for continuing education rather than for task training.

Vest Pocket Guide to the National Electrical Code®, Latest Edition. Quincy, MA: National Fire Protection Association.

NCCER makes every effort to keep these textbooks up-to-date and free of technical errors. We appreciate your help in this process. If you have an idea for improving this textbook, or if you find an error, a typographical mistake, or an inaccuracy in NCCER's Contren® textbooks, please write us, using this form or a photocopy. Be sure to include the exact module number, page number, a detailed description, and the correction, if applicable. Your input will be brought to the attention of the Technical Review Committee. Thank you for your assistance.

Instructors – If you found that additional materials were necessary in order to teach this module effectively, please let us know so that we may include them in the Equipment/Materials list in the Annotated Instructor's Guide.

Write: Product Development and Revision
National Center for Construction Education and Research
3600 NW 43rd St., Bldg. G, Gainesville, FL 32606

Fax: 352-334-0932

E-mail: curriculum@nccer.org

Craft _____ Module Name _____

Copyright Date _____ Module Number _____ Page Number(s) _____

Description _____

_____

_____

_____

(Optional) Correction _____

_____

_____

(Optional) Your Name and Address _____

_____

_____

Introduction to Cooling
03107-07

03107-07
Introduction to Cooling

Topics to be presented in this module include:

Overview

Cooling is accomplished by removing heat, not by generating cold air. In an air conditioner, a heat exchanger—the evaporator—in the conditioned space is used to transfer heat to a chemical refrigerant flowing through the heat exchanger. The refrigerant is then routed to another heat exchanger—the condenser—where the heat that was removed from the conditioned space is transferred to the outdoor air. This process is made possible by changing the temperature and pressure of the chemical refrigerant as it flows through the system.

Different systems may use different methods of accomplishing this process, but they all work on the same principle. Therefore, once you understand temperature-pressure relationships and how they affect the refrigeration cycle, you will understand the operating principles of all refrigeration systems.

Objectives

When you have completed this module, you will be able to do the following:

1. Explain how heat transfer occurs in a cooling system, demonstrating an understanding of the terms and concepts used in the refrigeration cycle.
2. Calculate the temperature and pressure relationships at key points in the refrigeration cycle.
3. Under supervision, use temperature- and pressure-measuring instruments to make readings at key points in the refrigeration cycle.
4. Identify commonly used refrigerants and demonstrate the procedures for handling these refrigerants.
5. Identify the major components of a cooling system and explain how each type works.
6. Identify the major accessories available for cooling systems and explain how each works.
7. Identify the control devices used in cooling systems and explain how each works.
8. State the correct methods to be used when piping a refrigeration system.

Trade Terms

Absolute pressure
Atmospheric pressure
British thermal unit (Btu)
Cold
Conduction
Conductor
Convection
Enthalpy
Floodback
Fluorocarbons
Gauge pressure
Halocarbons
Halogens
Heat
Heat content
Hydrocarbons
Insulators
Latent heat

Latent heat of condensation
Latent heat of fusion
Latent heat of vaporization
Pressure
Radiation
Refrigerant
Refrigeration
Sensible heat
Slug
Specific heat
Subcooling
Superheat
Thermistor
Thermocouple
Ton of refrigeration
Total heat

Required Trainee Materials

1. Paper and pencil
2. Appropriate personal protective equipment

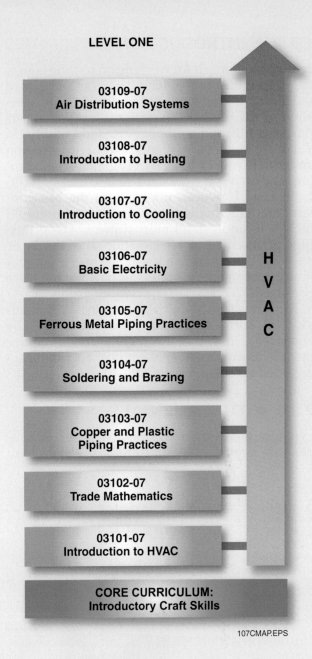

LEVEL ONE

03109-07
Air Distribution Systems

03108-07
Introduction to Heating

03107-07
Introduction to Cooling

03106-07
Basic Electricity

03105-07
Ferrous Metal Piping Practices

03104-07
Soldering and Brazing

03103-07
Copper and Plastic Piping Practices

03102-07
Trade Mathematics

03101-07
Introduction to HVAC

CORE CURRICULUM:
Introductory Craft Skills

H V A C

107CMAP.EPS

Prerequisites

Before you begin this module, it is recommended that you successfully complete *Core Curriculum*; and *HVAC Level One*, Modules 03101-07 through 03106-07.

This course map shows all of the modules in *HVAC Level One*. The suggested training order begins at the bottom and proceeds up. Skill levels increase as you advance on the course map. The local Training Program Sponsor may adjust the training order.

1.0.0 ◆ INTRODUCTION

To prepare yourself for working on cooling equipment, you must understand the basic concepts of **refrigeration**. You must also understand the operation of the mechanical refrigeration system: its purpose, function, components, and conditions. There are many types of mechanical refrigeration systems. If you try to learn refrigeration by learning how each one works, it will be a long and difficult task. However, if you learn the basics of refrigeration provided in this module, you should be able to understand most systems. The principles of mechanical refrigeration and the basic parts used in a system are the same, no matter how big or small the system or how the parts are packaged.

From this study, your ability to install and service all sorts of cooling equipment will be enhanced. This is true whether the cooling equipment is used for personal comfort air conditioning, food preservation, or industrial processes.

2.0.0 ◆ FUNDAMENTALS

In nature, there is nothing from which **heat** or temperature is totally absent. **Cold** is a relative term for temperature. It means an object has less heat energy (making it colder) than another object.

Refrigeration is the transfer of heat from a place or object where it is not wanted. Simply defined, refrigeration is cooling by the removal of heat. Air conditioners and refrigerators do not pump cold into a space. They take heat out of the space or object to be cooled and move it outside (*Figure 1*). A chemical fluid known as **refrigerant** circulates through a refrigeration or air conditioning system. The refrigerant absorbs heat from the refrigerated space, then carries it to a location outside the space. You will learn more about refrigerants later in this module.

2.1.0 Heat

To understand refrigeration you must understand heat. Like light, electricity, and magnetism, heat is a form of energy. Heat can be measured and controlled. Like other forms of energy, it can do work. Its ability to do work depends on two characteristics: temperature and **heat content** (quantity).

2.1.1 Temperature

Temperature compares the degree of hotness or coldness of any object or substance. The intensity of heat is measured in degrees (°) with a thermometer. Two temperature scales are normally

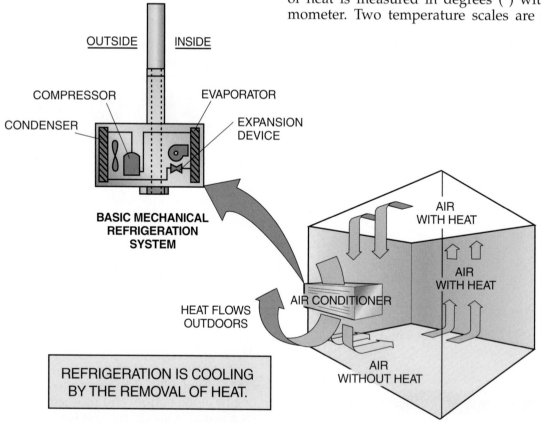

Figure 1 ◆ Refrigeration—transfer of heat.

used for measuring temperature. The one you will use most often is the Fahrenheit scale (*Figure 2*). The other scale commonly used worldwide, and for scientific work in the U.S., is the Celsius or Centigrade scale.

On the Fahrenheit scale (abbreviated °F), water boils at 212°F and freezes at 32°F. The distance between the two points is divided into 180 equal parts. On the Celsius scale (abbreviated °C), water boils at 100°C and freezes at 0°C. The distance between the two points is divided into 100 equal parts. The absolute zero point shown on the temperature scale in *Figure 2* is the theoretical point where all molecular motion stops, resulting in zero heat content.

The following formulas are used in temperature conversions:

$$°C = \frac{5}{9}\,(°F - 32°)$$
$$°F = (\frac{9}{5} \times °C) + 32°$$

2.1.2 Heat Content

Heat content, or the quantity of heat, is the amount of heat energy contained in a substance. Heat content is measured by the **British thermal unit (Btu)** in the English system. In the metric system, the joule is the measure of heat content. One Btu is the amount of heat needed to raise the temperature of one pound of water one degree Fahrenheit. (One pound of water is equal to about a pint of water.) For example, by heating 10 pounds of water from 40°F to 50°F (difference of 10°F), 100 Btus of heat are added to the water. The opposite is also true. If the 10 pounds of water had been cooled 10°F, 100 Btus would have been removed.

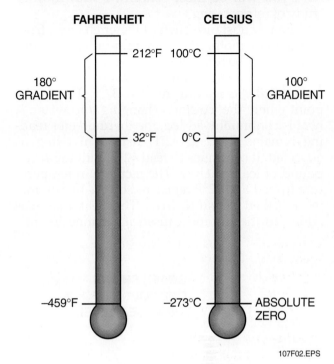

TEMPERATURE IS THE MEASURE OF THE INTENSITY OF HEAT IN A SUBSTANCE.

107F02.EPS

Figure 2 ◆ Fahrenheit and Celsius temperature scales.

2.1.3 Sensible and Latent Heat

Depending on the heat content and temperature, materials or substances can exist in three states: solid, liquid, and gas.

Using water as an example, the three states are ice (solid), water (liquid), and steam or vapor (gas).

Pressure Affects Boiling Point

INSIDE TRACK

As altitude increases, atmospheric pressure decreases. Because temperature and pressure are directly related, a liquid will boil at lower temperatures as the altitude increases. In a high-altitude location like Denver, Colorado, it takes longer to cook most foods than it does in locations closer to sea level. The opposite condition occurs in a pressure cooker. The high pressure inside the cooker causes boiling to occur at a higher temperature, so the contents cook faster than normal.

Chemical refrigerants are designed to boil at relatively low temperatures, so heat passing over the evaporator at normal room temperature will cause the refrigerant in the evaporator to boil. In an R-22 system, for example, the lowside pressure is about 69 psig, which corresponds to a refrigerant temperature of 40°F. Because of the chemical composition of the refrigerant, the heat absorbed from the conditioned space will cause the refrigerant to boil.

In a refrigeration system, the refrigerant exists in either a liquid or gas state, depending on its heat content. For this reason, these two states are the ones you will be most concerned with in your study of refrigeration.

When a substance such as water changes from one state to another, something peculiar happens (*Figure 3*). If heat is added to one pound of ice at 0°F, a thermometer would show a rise in temperature until the reading reaches 32°F. This is the point where the ice starts changing into water. If heat is continually added, the thermometer reading remains fixed at 32°F instead of rising as expected. It continues to read 32°F until the entire pound of ice is melted. The increase in temperature from 0°F to 32°F registered by the thermometer is called sensible heat. The heat that was added to the ice and caused its change in state

from a solid to a liquid, but that did not register on the thermometer, is called latent heat.

- Sensible heat is heat that can be sensed by a thermometer or by touch.
- Latent heat is heat energy that is absorbed or rejected when a substance is changing state (solid to liquid, liquid to gas, or vice versa) without a change in the measured temperature.
- Sensible heat plus latent heat equals total heat (enthalpy).

If we continue to add heat to the pound of water after all the ice has melted, the thermometer will once again show an increase in temperature until the temperature reaches 212°F. At this point, the water starts boiling and changes state from water into steam or water vapor. As even more heat is added, the thermometer reading remains

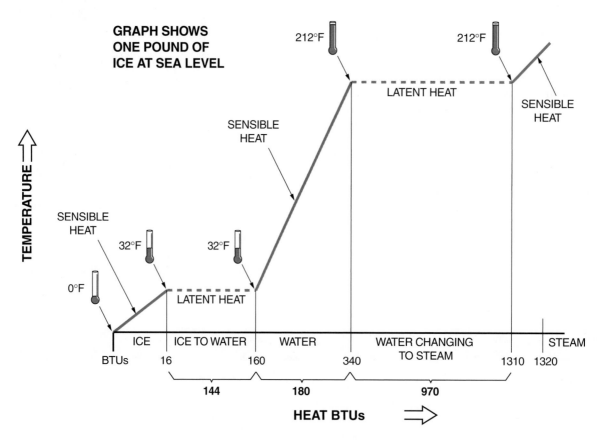

Figure 3 ◆ Changing states of water.

at 212°F until all the water has turned into steam. If we continue to add heat after all the water has been converted to steam, the thermometer will once again register sensible heat called **superheat**. Superheat is the measurable heat added to the vapor once a liquid has reached its boiling point and has been completely changed into a vapor.

As shown in *Figure 3*, it takes a great deal more heat to cause a change in state than is needed for a degree change in temperature. It required 144 Btus of latent heat to melt the pound of ice before the temperature began to rise. This is 144 times as much heat as is needed to raise the temperature of water one degree. The change from water to steam required an even greater amount of latent heat. It took 970 Btus to change the water to steam. It is important to remember that none of this latent heat registered on the thermometer.

The latent heat added or removed in changing to and from the solid, liquid, and vapor states has several special names, as shown in *Figure 4*.

- **Latent heat of fusion** – The heat gained or lost in changing to or from a solid state (ice to water or water to ice).
- **Latent heat of vaporization** – The heat gained in changing from a liquid to a vapor (water to steam). The temperature at which this occurs is known as the boiling point. In refrigeration work, the boiling point is also known as the saturation temperature.
- **Latent heat of condensation** – The heat given up or removed from a vapor in changing back to a liquid state (steam to water).

As mentioned previously, the two states you are most concerned with in learning about refrigeration are the liquid and vapor states. Another term used in refrigeration work concerning changes in the state of matter is **subcooling**. Subcooling is the reverse of superheat. It is the temperature of a liquid when it is cooled below its condensing temperature. For example,

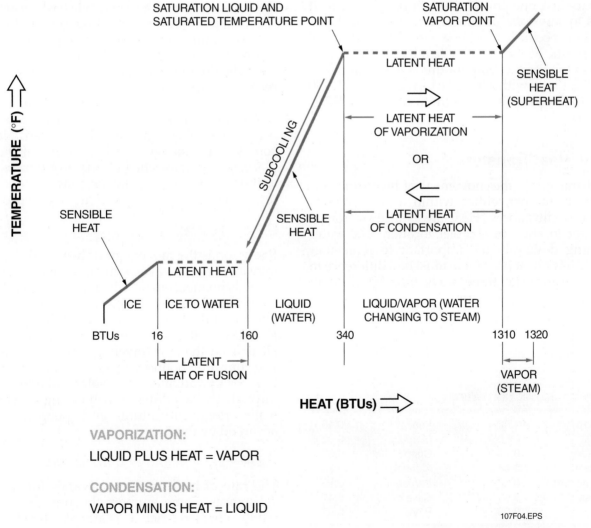

Figure 4 ◆ Change of state terminology.

the condensing temperature of water is 212°F. If the water is subcooled 10°, the temperature is cooled to 202°F.

2.1.4 Specific Heat Capacity

Specific heat is the amount of heat required to raise the temperature of one pound of a substance one degree Fahrenheit. In the previous examples, we saw that water in the liquid state has a specific heat of one Btu per pound of water, per each degree Fahrenheit of temperature change (one Btu/lb/°F). The specific heat of each substance is different. *Table 1* shows the specific heat values for some common substances.

Specific heats vary from substance to substance, but also from one state of a substance to another state. As shown in *Table 1*, liquid water has a specific heat of 1.00. Ice, which is solid water, has a specific heat of 0.50. Substances with lower specific heat numbers are the most easily heated. Those with higher numbers require more heat. The result is that it takes twice as many Btus of heat to raise one pound of water one degree as it does to raise one pound of ice one degree. The effect of specific heat is shown in *Figure 3*. Because the specific heat of ice is 0.50, it took only 16 Btus of heat to raise the temperature of the ice from 0°F to 32°F. Water with a specific heat of 1.00 took 180 Btus to raise the temperature from 32°F to 212°F, or 180°.

2.2.0 Heat Transfer

Heat transfer is the movement of heat from one place to another, either within a substance or between substances. Heat always flows from a warmer location to a cooler location, like water running downhill. It's important to remember that to have heat flow there must be a difference in temperature. The three ways (see *Figure 5*) to move or transfer heat are:

- Conduction
- Convection
- Radiation

Table 1	Specific Heat Values		
Water	1.00	Iron	0.10
Ice	0.50	Mercury	0.03
Air (dry)	0.24	Copper	0.09
Steam	0.48	Alcohol	0.60
Aluminum	0.22	Kerosene	0.50
Brass	0.09	Olive oil	0.47
Lead	0.03	Pine	0.67

107T01.EPS

2.2.1 Conduction

Conduction is a means of heat transfer in which heat is moved from molecule to molecule within a substance. When these molecules are heated, they move about, colliding with one another. These collisions continue in a direction toward the cooler part of the material, causing the movement of heat in the same direction. For example, when copper tubing is heated by a torch, the molecules in the tubing nearest the torch get heated first and begin to move and collide with the nearby molecules in the tubing. These molecules then collide with other molecules, causing the tubing to become heated. The result is that the heat is carried by conduction from the heated end toward the cold end.

2.2.2 Convection

Convection is the transfer of heat by the flow of liquid or gas caused by a temperature differential. As shown in *Figure 5*, air near the fireplace is heated by conduction and becomes warmer than the air in the rest of the room. Since warm air rises, the heated air moves toward the ceiling. In doing so, it gives up heat as it goes upward, and then settles back down to the floor as it cools. The cooler air at the floor level moves toward the fireplace to replace the rising warm air. It too will warm, rise, give off heat, and settle back down. This circulation of air is accomplished via convection. Convection can be either natural or forced. Natural convection is shown by the example of the fireplace. Forced convection uses fans or pumps, such as those found in home heating and air conditioning systems, to speed up the circulation process.

2.2.3 Radiation

Radiation is the movement of heat in the form of invisible rays or waves, similar to light. Like light, it needs no medium on which to travel. Radiation takes place free of convection. It travels in straight lines from the heat source to the point where it is absorbed without heating the space in between. Heat from the sun traveling through space and warming our homes is a good example of heat transfer by radiation. The solar radiation comes through the windows in a building, strikes the walls, floors, furniture, and people, and is absorbed by them.

2.2.4 Conductors and Insulators

The rate of heat conduction varies for different substances. Some support the transfer of heat, while others restrict it. Materials in which the transfer of heat by conduction occurs easily are

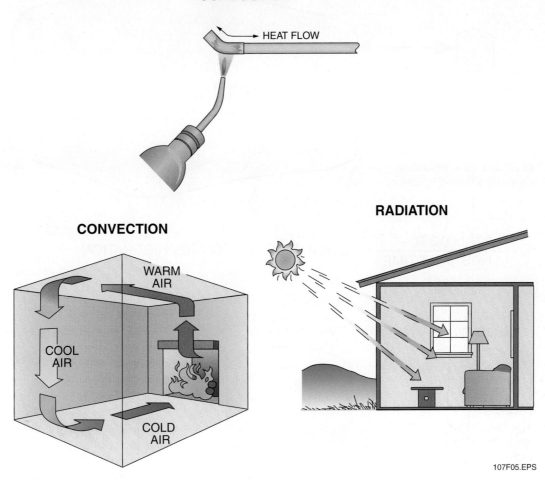

CONDUCTION

HEAT FLOW

CONVECTION

WARM AIR

COOL AIR

COLD AIR

RADIATION

107F05.EPS

Figure 5 ◆ Heat movement.

called **conductors**. **Insulators** are materials that resist heat transfer by conduction. Cork, fiberglass, and polyurethane foam are examples of insulators. Most metals are good conductors of heat. Copper and aluminum are used in refrigeration systems because of their good heat conduction ability.

2.2.5 Rate of Heat Transfer

The rate of heat transfer describes how fast heat can be added to or removed from an object or between objects. It usually is expressed in two ways. One way is in Btus per hour or Btuh. Another way is by the ton, which is a much larger unit of measure. The ton is commonly used in refrigeration work to describe the heat or cooling load for a space, or the capacity of a piece of equipment or system.

One **ton of refrigeration** is defined as 12,000 Btus per hour or 12,000 Btuh (*Figure 6*). The ton is based on the amount of heat required to melt one ton of ice in a 24-hour period. As you learned earlier, one pound of ice at 32°F absorbs 144 Btus of

heat while melting. Assuming that it takes one hour to melt, the rate of heat transfer is 144 Btuh. Since a ton of ice contains 2,000 pounds, a total of 288,000 Btus per day, or 12,000 Btus per hour (288,000 ÷ 24) is the amount of heat required to melt the ice.

2.3.0 Pressure

Pressure is defined as force per unit area. This is normally expressed in pounds per square inch. For pressures below atmospheric pressure, the pressure is measured in inches of mercury. Depending on the state of a substance, pressure may be exerted in one direction, several directions, or all directions (*Figure 7*). Using the three states of water as an example, ice (a solid) exerts pressure only in a downward direction. The same is true for all solid materials. As a liquid, water exerts pressure against all sides of the container in contact with it. As a gas (water vapor), it exerts pressure on all the surfaces of the container because it completely fills the container. In refrigeration work, the term fluid is generally

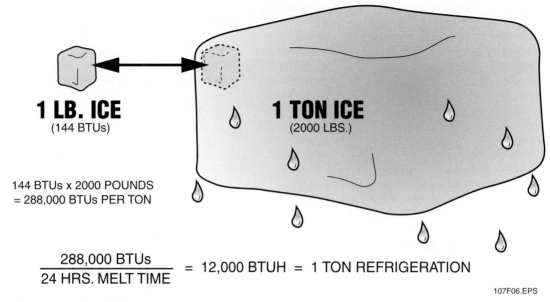

1 LB. ICE
(144 BTUs)

1 TON ICE
(2000 LBS.)

144 BTUs x 2000 POUNDS
= 288,000 BTUs PER TON

$$\frac{288,000 \text{ BTUs}}{24 \text{ HRS. MELT TIME}} = 12,000 \text{ BTUH} = 1 \text{ TON REFRIGERATION}$$

107F06.EPS

Figure 6 ◆ One ton of refrigeration.

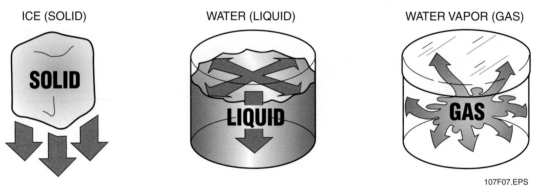

ICE (SOLID)

WATER (LIQUID)

WATER VAPOR (GAS)

SOLID

LIQUID

GAS

107F07.EPS

Figure 7 ◆ The directions of pressure.

used when describing pressure. Fluid means the liquid or gaseous state of a material such as a refrigerant. Gases tend to exert pressure equally in all directions like the water vapor given in the example.

2.3.1 Atmospheric Pressure

The Earth is surrounded by a blanket of air called the atmosphere. Air is matter consisting of oxygen, nitrogen, and water vapor. It has weight, and exerts a force called **atmospheric pressure** on all things on the Earth's surface. Atmospheric pressure can be measured with a barometer. For this reason, it is often referred to as barometric pressure.

Figure 8 shows a simple mercury tube barometer. The top of the barometer tube is sealed, while the open end at the bottom rests in a container of mercury. Air pressure pushing down on

the mercury in the container causes the column of mercury in the tube to rise. The extent of the rise is determined by the amount of pressure applied to the mercury.

Visualize a column of air with a cross-sectional area of one square inch and extending from the Earth's surface at sea level to the limits of the atmosphere. Also, assume that the temperature at sea level is 70°F. If this column of air is applied to the mercury tube barometer, the height of the mercury in the tube will be slightly less than 30" (29.92"). Its weight will be 14.7 pounds. This means that every square inch of any surface at sea level has 14.7 pounds of air pressure pushing down on it. These values of 29.92 inches of mercury and 14.7 pounds per square inch at sea level at 70°F are standards that are used frequently in refrigeration work. A pressure scale, called the **absolute pressure** scale, is based on the barometer measurements just described. On this scale,

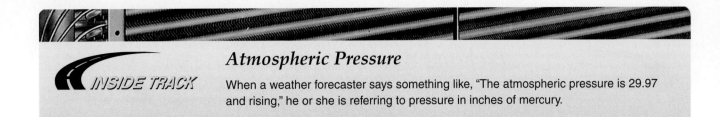

INSIDE TRACK

Atmospheric Pressure

When a weather forecaster says something like, "The atmospheric pressure is 29.97 and rising," he or she is referring to pressure in inches of mercury.

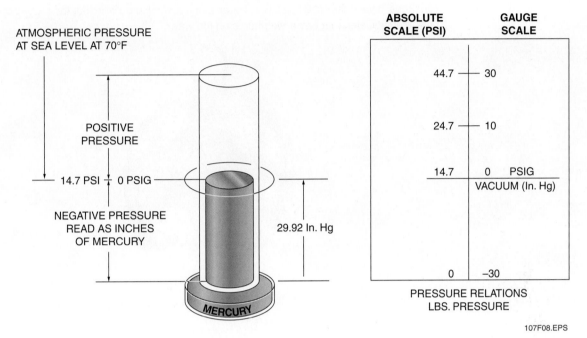

Figure 8 ◆ Absolute and gauge pressure scale comparison.

pressures are expressed as pounds per square inch (psi) or pounds per square inch absolute (psia), starting from zero, which represents a complete absence of pressure.

2.3.2 Gauge Pressure

Another scale, called **gauge pressure**, is normally used for refrigeration work. Gauge pressure scales use atmospheric pressure as their zero starting point. Positive gauge pressures, those above zero (14.7 psi), are expressed in pounds per square inch gauge or psig. Negative pressures, those below 0 psig, are expressed in inches of mercury vacuum or in Hg vac. Gauge pressures can easily be converted to absolute pressures by adding 14.7 to the gauge-pressure value. For example, a gauge pressure of 10 psig equals an absolute pressure of 24.7 (10 + 14.7). A comparison of the gauge and absolute pressure scales is shown in *Figure 8*.

2.3.3 Pressure/Temperature Relationships

Pressure and temperature have a special relationship. Two things are important to remember. First,

the temperature at which a liquid or gas changes state depends on the pressure. Second, the boiling temperature of a liquid will drop as the pressure on it decreases. It will rise as the pressure increases. Using our water example, water boils at 212°F at sea level (14.7 psia). With a lower atmospheric pressure of about 11.6 psia that exists at 5,000' above sea level, the same water would boil at the lower temperature of about 203°F. At the higher pressure of about 29.7 psia (15 psig + 14.7 atmospheric pressure), such as can be reached in a pressure cooker, the water boils at the higher temperature of about 250°F. *Figure 9* shows the temperature/pressure relationship of water.

If the pressure on a liquid can be lowered enough, its boiling point need not be a high temperature. This relationship is basic to your work in refrigeration. Each refrigerant used with cooling equipment has its own temperature/pressure relationship. Like that of water, the boiling temperature of refrigerants rises as the pressure is increased, and drops as the pressure is decreased.

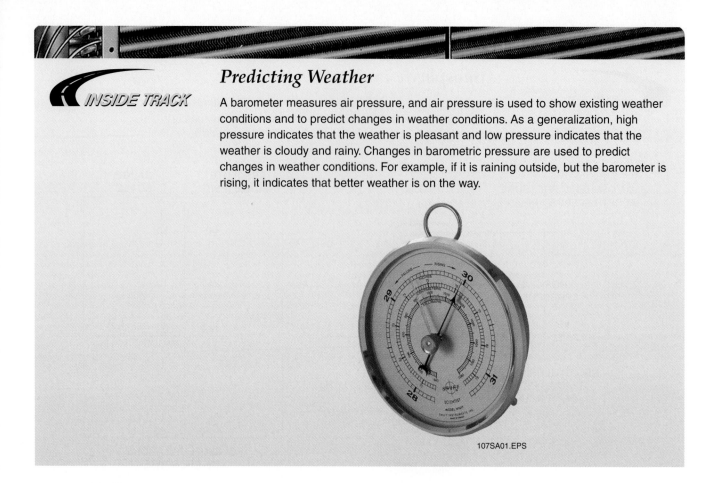

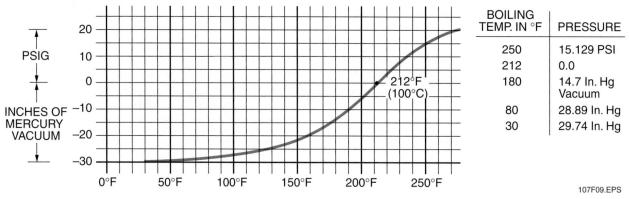

BOILING TEMP. IN °F	PRESSURE
250	15.129 PSI
212	0.0
180	14.7 In. Hg Vacuum
80	28.89 In. Hg
30	29.74 In. Hg

107F09.EPS

Figure 9 ◆ Temperature/pressure relationship of water.

2.3.4 Movement of Fluids

Differences in pressure cause the flow of fluids. This flow is always from a higher pressure to a lower pressure. Just as heat moves from a higher temperature to a lower temperature, so too does a liquid or gas move from a higher pressure to a lower pressure. The result is that liquids and gases can be made to move by adjusting or changing the pressure around them. In refrigeration, a compressor is used to create the pressure differential that causes refrigerant to flow in a system.

2.4.0 Instruments Used to Measure Temperature and Pressure

A variety of instruments are used to make temperature and pressure measurements.

2.4.1 Thermometer

The dial and electronic thermometers (*Figure 10*) are the two types of thermometers most commonly used for measuring temperatures in HVAC equipment. Many other special-purpose thermometers are also available. Thermometers come

INSIDE TRACK

Infrared Thermometers

Infrared thermometers allow the user to measure temperature without touching the thermometer to the measurement point. They work by detecting infrared energy emitted by the device at which they are pointing. Laser sighting devices allow the technician to aim the thermometer like a pistol at the surface to be measured. The temperature is displayed instantly. This type of thermometer is useful for taking temperature readings in hard-to-reach areas. However, it is not effective on reflective material such as sheet metal ductwork.

107SA02.EPS

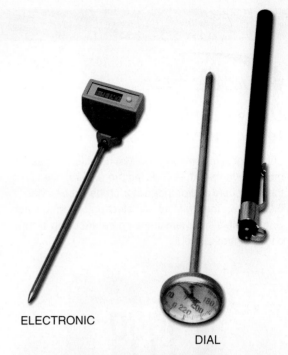

ELECTRONIC

DIAL

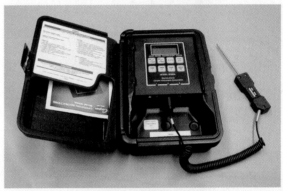

DIGITAL TEMPERATURE PROBE

107F10.EPS

Figure 10 ◆ Thermometers.

in a variety of temperature ranges based on their intended use.

Dial thermometers are available in various forms. They are popular because they are rugged, small, and inexpensive. They come in many stem lengths and dial sizes. Because pocket-style dial thermometers have smaller calibrated scales, it is sometimes difficult to get accurate readings. When accuracy is needed, electronic thermometers should be used.

Electronic thermometers display temperature on either an analog meter or a digital liquid crystal display (LCD) readout. They generally use either a **thermocouple** or **thermistor** type of temperature probe, or both, to sense the heat and generate the temperature reading. Thermocouple

probes use a sensing device (thermocouple) made of two dissimilar wires welded together at one end called a junction. When the junction is heated, it generates a low-level DC voltage that produces a temperature reading on the electronic thermometer indicator. Thermocouple probes tend to be rugged and inexpensive. Thermistor probes use a semiconductor electronic element (thermistor) in which resistance changes with a change in temperature.

Often, several different probes are used with the same electronic thermometer to allow temperature measurement over a wide range. Many electronic thermometers have two or more probes, so that measurements can be made at several locations within the equipment at the same time. Most thermometers of this type can calculate and display the difference in temperature between the locations being measured.

Using a Multimeter to Measure Heat

Many digital multimeters (DMMs) can also be used to measure temperature. This feature requires the use of thermocouple and/or thermistor probe accessories that convert the DMM into an electronic thermometer. Some DMMs can measure temperature using a non-contact infrared probe.

107SA03.EPS

Electronic thermometers are precise measuring instruments. Be sure to read and follow the manufacturer's instructions for operating electronic thermometers. Also, be sure to follow the manufacturer's instructions for calibration of the instrument.

2.4.2 Gauge Manifold Set

The gauge manifold set is the most frequently used item of HVAC service equipment. It is used to install and check refrigerant charges and to measure low-side and high-side pressures in an operating system in order to evaluate system performance. The gauge manifold set is regularly used to route and control the flow of refrigerant, refrigerant oil, or other acceptable gases to and from the system in support of other servicing tasks such as refrigerant charging.

Figure 11 shows a standard two-valve gauge manifold set. It consists of two pressure gauges mounted on a manifold assembly. A compound gauge is mounted on the left side of the manifold and a high-pressure gauge is mounted on the right. A compound gauge allows measurement of pressures both above and below atmospheric pressure. Most compound gauges used with the gauge manifold set can measure pressures above atmospheric pressure in the range from 0 to 150 psig. Below atmospheric pressure they can measure from 0 to 30 in Hg vac. Zero, which represents atmospheric pressure, is the starting point for both scales. The compound gauge is used to measure system low-side (suction) pressures, including any vacuum that exists in a system. The high-pressure gauge on the right side of the manifold is used to measure system high-side (discharge) pressures. High-pressure gauges can usually measure system pressures in the range from 0 to 500 psig. However, a high-pressure refrigerant such as R-410A requires a 0–800 psi high-pressure gauge and a –500 psi low-pressure gauge. The hoses must be rated for 800 psi.

As shown in *Figure 11*, a standard two-valve gauge manifold set has two hand valves and three

107F11.EPS

Figure 11 ◆ Gauge manifold set.

hose ports. The hand valves are adjusted to monitor system pressures on the compound and high-pressure gauges and/or route the flow of refrigerant to and from the system during servicing activities. The gauge manifold set hose ports are connected to the system being serviced and/or other service instruments through a set of environmentally safe, high-vacuum, high-pressure service hoses. These hoses must be equipped with fast, self-sealing fittings that immediately trap refrigerant in the hoses when disconnected. Use of these fittings helps to meet the clean air non-venting regulations and also greatly reduces the amount of air that can enter and contaminate the hose after it has been disconnected.

Most gauge manifold sets and service hoses are color-coded. The low-pressure compound gauge, hand valve, and low-pressure hose port are blue. A blue service hose is used to connect the mani-

fold low-pressure hose port to the equipment suction service valve. Red is the color used to mark the high-pressure gauge, hand valve, and hose port. A red service hose is used to connect the high-pressure hose port to the equipment discharge service valve or liquid line valve. The center hose port is the utility port. This port is normally connected through a yellow service hose to other service instruments or devices. When not in use, the utility port should be capped.

Gauge manifold sets are also available with four hand valves and related hose ports. Use of this type of manifold can reduce service time by eliminating the need to switch a single utility hose between different service devices. Four-valve manifolds and related service hose sets are color-coded as follows: blue (low pressure), red (high pressure), yellow (charging), and black (vacuum). Digital electronic gauge manifold sets that display system pressures on liquid crystal display (LCD) readouts are also available.

NOTE

The gauge manifold set is a precise measuring instrument. Its accuracy is critical for correct servicing. The technician must ensure that the gauge manifold set is always handled with care both during use and in transport. The calibration of the gauge manifold set should be checked regularly. If necessary, it should be calibrated according to the manufacturer's instructions.

2.4.3 Manometer

Another instrument that is used to measure pressure is the manometer. Both electronic and non-electronic manometers are in common use. Manometers can be used to measure pressures within a refrigeration system, but are seldom used for this purpose. Manometers are most often used to measure gas pressure in heating systems and velocity or static air pressure in air distribution systems. You will learn more about manometers and their uses later in your training.

3.0.0 ◆ MECHANICAL REFRIGERATION SYSTEM

This section covers the basic components and operation of the mechanical refrigeration system.

INSIDE TRACK

Digital Gauge Manifold

Like electrical test instruments, digital readout gauge manifolds have become popular. With a digital readout, there is no need to interpret the readings as there is with analog gauges.

LOW-SIDE **100.0**psi   HIGH-SIDE **180**psi

POWER/CAL   MODE/UNITS/SUPERHEAT   MAX/MIN/AVG/RESET   REFRIGERANT

ROBINAIR Model 41800

LOW   HIGH   TEMP

107SA04.EPS

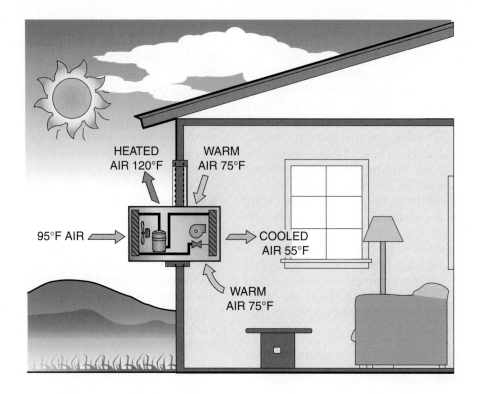

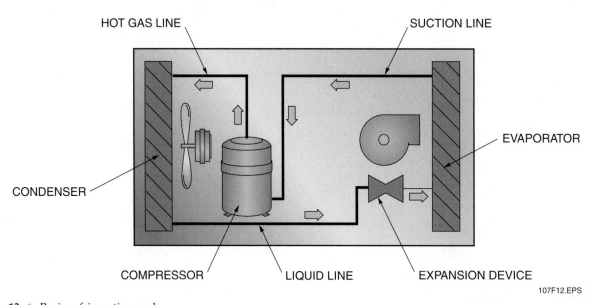

Figure 12 ◆ Basic refrigeration cycle.

3.1.0 System Components

There are many types of systems used to provide cooling for personal comfort, food preservation, and industrial processes. Each of these systems uses a mechanical refrigeration system. *Figure 12* shows a basic system and the following components:

- *Evaporator* – A heat exchanger that transfers the heat from the area or item being cooled to the refrigerant.
- *Compressor* – Creates the pressure differences in the system needed to make the refrigerant flow and the refrigeration cycle work.
- *Condenser* – A heat exchanger that transfers the heat absorbed by the refrigerant to the outdoor air or another substance.
- *Metering device* – Provides a pressure drop that lowers the boiling point of the refrigerant just before it enters the evaporator. This is also known as the expansion device.

Also shown in *Figure 12* is the piping, called lines, used to connect the basic components in order to provide the path for refrigerant flow. Together, the components and lines form a closed refrigerant system. The types of lines are:

- *Suction line* – The tubing that carries refrigerant gas from the evaporator to the compressor.
- *Hot gas line* (also called the discharge line) – The tubing that carries hot refrigerant gas from the compressor to the condenser.
- *Liquid line* – The tubing that carries the liquid refrigerant formed in the condenser to the metering device.

The arrows in *Figure 12* show the direction of flow through the system. The purpose of the refrigerant is to move heat. It is the medium by which heat can be moved into or out of a space or substance. A refrigerant is any liquid or gas that picks up heat by evaporating at a low temperature and pressure, and gives up heat by condensing at a higher temperature and pressure. Refrigerants often boil at extremely low temperatures. For example, R-410A boils at –60°F.

Remember that the operation of a mechanical refrigeration system is the same for all systems. Only the type of refrigerant used, the size and style of the components, and the installed locations of the components and the lines will change from system to system. Other devices, called accessories, may be used in some systems to gain the desired cooling effect and to perform special functions. The events that take place within the system happen again and again in the same order. This repeating series of events is called the refrigeration cycle.

3.2.0 Refrigeration Cycle

In order to effectively service HVAC systems, it is essential that you understand the basic refrigeration cycle.

3.2.1 Basic Operation

The refrigeration cycle is based on two principles:

- As liquid changes to a gas or vapor, it is capable of absorbing large quantities of heat.
- The boiling point of a liquid can be changed by altering the pressure exerted on the liquid.

As shown in *Figure 12*, the refrigerant flows through the system in the direction indicated by the arrows. We will begin with the evaporator. It receives low-temperature, low-pressure liquid refrigerant from the metering device. The evaporator is a series of tubing coils that expose the cool liquid refrigerant to the warmer air passing over the coils. Heat from the warm air is transferred through the tubing to the cooler refrigerant. This causes the refrigerant to boil and vaporize. It is important to realize that even though it has just boiled, it is still not considered "hot" because refrigerants boil at such low temperatures. So, it is a low-temperature, low-pressure refrigerant vapor that travels through the suction line to the compressor.

The compressor receives the low-temperature, low-pressure vapor and compresses it. It then becomes a high-temperature, high-pressure vapor. This travels to the condenser via the hot gas line.

The condenser is a series of tubing coils through which the refrigerant flows. As cooler air moves across the tubing, the hot refrigerant vapor gives up heat. As it continues to give up heat to the outside air, it cools to the condensing point, where it begins to change from a vapor into a liquid. As more cooling takes place, all of the refrigerant becomes liquid. Further cooling of the saturated liquid is known as subcooling. This high-temperature, high-pressure liquid travels through the liquid line to the input of the metering device.

The metering device regulates the flow of refrigerant to the evaporator. It also decreases its pressure and temperature. By the use of a built-in restriction, such as a tiny hole or orifice, it converts the high-temperature, high-pressure refrigerant from the condenser into the low-temperature, low-pressure refrigerant needed to absorb heat in the evaporator.

The evaporator is generally designed so that the refrigerant is completely vaporized before it reaches the end of the evaporator. Thus the refrigerant absorbs more heat from the warm air flowing over the evaporator coils. Because the refrigerant is totally vaporized by this time, the additional heat it absorbs causes an increase in sensible heat. This additional heat is known as superheat.

3.2.2 Refrigeration Cycle in a Typical Air Conditioning System

Figure 13 shows a typical air conditioning system. The components are divided into two sections based on pressure. The high-pressure side includes all the components in which the pressure of the refrigerant is at or above the condensing pressure. This is often referred to as the head pressure, discharge pressure, or high-side pressure. The low-pressure side includes all the components in which the pressure of the refrigerant is at or below the evaporating pressure. This is often called the suction pressure or low-side pressure. The dividing line between the sections cuts through the compressor and the metering device.

We will now discuss the refrigeration cycle in more detail. We will describe a typical air conditioner that uses HCFC-22 (R-22) as the refrigerant. R-22 boils at 40°F when under a pressure of 68.5 psig. This example will demonstrate the concepts and temperature/pressure relationships you have learned so far. For our example, assume an air temperature of 75°F for the room being cooled and an outdoor air temperature of 95°F. These values will vary due to equipment and load conditions. The numbers below correspond to the numbers shown in *Figure 13*. Follow along on the figure as this system is described.

1. A mixture (75 percent liquid, 25 percent vapor) of R-22 is supplied from the metering device to the evaporator. This mixture is at a pressure of 68.5 psig, which corresponds to the 40°F boiling point of R-22 refrigerant. (See the chart shown in *Figure 13*.) The 40°F boiling point is used here because it is typical of the temperatures normally used for evaporators in air conditioning systems.

2. Because the refrigerant flowing through the evaporator is cooler (40°F) than the warmer inside room air (75°F), it absorbs heat, causing the liquid refrigerant to boil and turn into a vapor. After traveling about 90 percent of the way through the evaporator tubing, all the refrigerant has boiled into a vapor known as the saturated vapor.

3. During the remaining 10 percent of travel through the evaporator, the saturated vapor continues to absorb heat from the warmer air, thus raising its temperature to 50°F. In other words, the saturated vapor is superheated 10°F (50°F − 40°F). This superheated vapor flows through the suction line and is drawn into the low-pressure side of the compressor. The cooled inside room air is recirculated by the evaporator fan back into the room at a temperature of about 55°F.

4. The superheated vapor applied at the suction input of the compressor typically picks up an additional 3 to 5 degrees of superheat because the vapor in the suction line absorbs more heat from the warmer surrounding air as it travels from the evaporator to the compressor.

5. After compression, the highly superheated gas from the compressor flows through the hot gas line to the condenser. This hot gas may be close to 200°F at 296.8 psig. Since the saturated temperature corresponding to 296.8 psig is 130°F (see the temperature/pressure chart), the hot gas line has gained about 70°F (200°F − 130°F) of discharge superheat. This superheat must be removed before the vapor can be condensed into a liquid. The 200°F refrigerant in the hot gas line easily gives up some of its superheat to the surrounding 95°F air. The hot gas line is normally not insulated and the tubing is a good conductor of heat.

6. Because the refrigerant in the condenser is still hotter than the warmer outside air passing over the condenser, it easily gives up the remaining superheat. This drops its temperature to 130°F. As heat continues to be transferred from the vapor to the cooler outside air, the vapor begins to cool and condenses into a liquid. After the refrigerant has traveled about three-quarters of the way through the condenser, all of the refrigerant has condensed into a liquid. The 130°F condensing temperature is set by the condenser design. A standard condenser is designed to have a condensing temperature about 35°F higher than the surrounding air. In this case, 95°F outside air is used to absorb the heat, so 95°F + 35°F = 130°F condensing temperature.

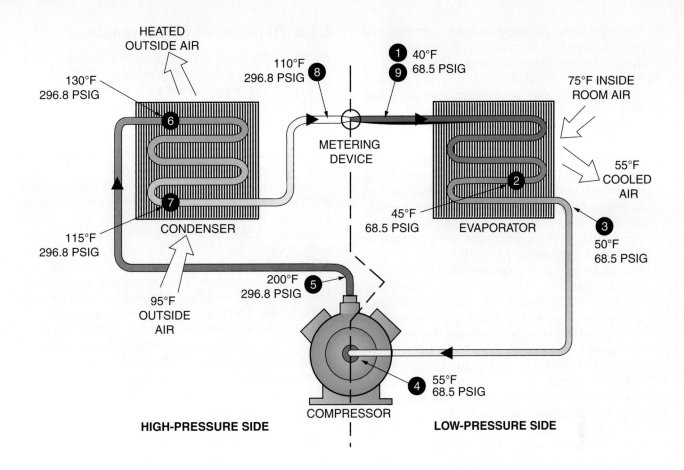

Temperature °F	Refrigerant/Pressure (PSIG) R-22
27	51.2
28	52.4
29	53.6
30	54.9
31	56.2
32	57.5
33	58.8
34	60.1
35	61.5
36	62.8
37	64.2
38	65.6
39	67.1
40	68.5
41	70.0
42	71.5
43	73.0
44	74.5
45	76.0
46	77.6
47	79.2
48	80.8
49	82.4

Temperature °F	Refrigerant/Pressure (PSIG) R-22
50	84.0
55	92.6
60	101.6
65	111.2
70	121.4
75	132.2
80	143.6
85	155.7
90	168.4
95	181.8
100	195.9
105	210.8
110	226.4
115	242.7
120	259.9
125	277.9
130	296.8
135	316.6
140	337.3
145	358.9
150	381.8
155	405.1

TEMPERATURE/PRESSURE CHART

107F13.EPS

Figure 13 ◆ Typical air conditioning cycle for HCFC-22 (R-22) refrigerant.

7. During the remaining travel through the condenser, the liquid refrigerant continues to drop in temperature (subcool). This lowers its temperature about 15°F to 115°F. In other words, the liquid refrigerant is subcooled 15°F (130°F − 115°F).

8. Subcooled liquid refrigerant from the condenser travels through the liquid line to the metering device. The liquid line is usually not insulated and may be long. Thus, the 115°F liquid refrigerant may be further subcooled as it gives up more heat to the cooler outside air. This drop could increase the subcooling by another 5°F, lowering the temperature of the liquid refrigerant to 110°F.

9. The metering device controls the flow of liquid refrigerant to the evaporator. Subcooled liquid from the condenser enters at the high temperature of 110°F and high pressure of 296.8 psig. It leaves the metering device at the low temperature of 40°F and low pressure of 68.5 psig, thereby lowering the boiling point of the liquid refrigerant supplied to the evaporator. In the metering device, the subcooled liquid refrigerant at 296.8 psig is passed through a small opening or orifice. This changes the pressure of the liquid refrigerant from 296.8 psig to 68.5 psig, causing some of it to "flash" into a vapor. This flash gas cools the remaining liquid to produce a mixture of about 75 percent liquid and 25 percent vapor. The pressure of this mixture is 68.5 psig, which corresponds to the 40°F boiling point needed for correct evaporator operation. This low-temperature, low-pressure mixture from the metering device then travels to the evaporator.

10. The refrigerant has now completed its cycle and is ready to start over again. It should be pointed out here that the above discussion is theoretical and does not take into account pressure drops across components and within piping that normally occur in an actual system.

4.0.0 ◆ REFRIGERANTS

Refrigerants are used in cooling systems to move heat into or out of a space or substance. This section briefly describes refrigerants, with the focus on their impact on the environment. You will learn more about their characteristics and specific uses later in your training. This section limits the discussion to ammonia and fluorocarbon refrigerants, since for all practical purposes they are the only ones in common use today.

4.1.0 Refrigerant Trade Names

Traditionally, each refrigerant had a trade name or an "R" (refrigerant) name. Names like R-11, R-123, R-500, etc., were assigned by the American Society of Heating, Refrigerating, and Air-Conditioning Engineers (ASHRAE). These names were substituted for the true chemical names. For example, R-22 describes the refrigerant with the chemical name of hydrochlorofluoromethane.

As a result of the *Clean Air Act* and the concerns about refrigerants and their effects on our environment, the way refrigerants are named has changed. ASHRAE has substituted acronyms such as CFCs, HCFCs, and HFCs for the "R" in the refrigerant name. These acronyms describe the way the refrigerants are chemically structured. Their meanings will be explained in the following paragraphs. The number previously assigned to a refrigerant by ASHRAE has been retained. Both old and new names are currently being used in the trade. Some examples of changed names for commonly used refrigerants are shown in *Table 2*.

4.2.0 Ammonia

Ammonia (R-717) has excellent heat transfer qualities and is used mainly in ice plants, ice skating rinks, and large food processing plants. Though not classified as poisonous, ammonia has a harsh effect on the respiratory system. Only very small quantities can be safely inhaled. Exposure for 5 minutes to 50 parts per million (ppm) is the maximum exposure allowed by the Occupational Safety and Health Administration (OSHA). Ammonia is hazardous to life at 5,000 ppm and is flammable at 150,000 to 270,000 ppm. Ammonia has an odor that can be smelled at 3 to 5 ppm. This odor gets very irritating at 15 ppm. Anyone working on an ammonia system must be specifically trained for that purpose. As far as our environment is concerned, ammonia is considered safe.

Table 2	Examples of Old and New Refrigerant Names	
Old Name	**New Name**
R-22	HCFC-22
R-123	HCFC-123
R-134a	HFC-134a
R-500	CFC-500
R-502	CFC-502

107T02.EPS

Refrigerant Certification

The U.S. Environmental Protection Agency (EPA) requires that all persons who install, service, repair, or dispose of equipment containing a refrigerant possess a certification card. The certification card is obtained by passing a test for one or more categories of work as identified by the EPA. The categories are:

- *Type I* – Small appliances containing less than 5 pounds of refrigerant, such as refrigerators and small air conditioners.
- *Type II* – Appliances that use high-pressure refrigerants such as R-22, R-500, and R-502.
- *Type III* – Appliances such as centrifugal chillers that use low-pressure refrigerants.
- *Type IV* – This is a universal certification for any of the above categories.

Refrigerant Compound Numbering

Each refrigerant has a specific chemical composition, which can be shown graphically. For example, the composition of the HCFC refrigerant R-22 appears like this:

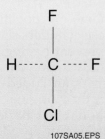

107SA05.EPS

The name of the refrigerant is derived from its components. You can see from this graphic that R-22 is an HCFC. The numbers—22 in this case—relate to the number of atoms of each component in a molecule of the refrigerant. Refrigerant blends have a much different numbering system.

4.3.0 Fluorocarbon Refrigerants

Man-made (synthetic) refrigerants in popular use today are all fluorocarbons or a mixture of fluorocarbon refrigerants. Included in this group are refrigerants such as CFC-11 (R-11), CFC-12 (R-12), HCFC-22 (R-22), HCFC-123 (R-123), and HFC-134a (R-134a). These refrigerants all stem from one of two base molecules, methane and ethane.

Methane and ethane are called **hydrocarbons** because they are organic compounds that contain only hydrogen and carbon atoms. When most or all of the hydrogen atoms in the methane or ethane molecule are replaced with elements such as chlorine, fluorine, and/or bromine, the changed molecule is called a **halocarbon**, short for halogenated hydrocarbon. Chlorine, fluorine, and bromine are chemically related elements called **halogens**. To halogenate means to cause some other element to combine with a halogen. When all the hydrogen atoms in a hydrocarbon molecule are replaced with chlorine or fluorine, the molecule is said to be fully halogenated. Halocarbons in which at least one or more of the hydrogen atoms have been replaced with fluorine are called fluorocarbons. Fluorocarbon refrigerants fall into three groups, CFCs, HCFCs, and HFCs, based on their chemical structure.

Refrigerant and the Law

It is unlawful to knowingly release CFCs, as well as other types of fluorocarbon refrigerants, into the atmosphere. If caught doing so, you can be subject to a stiff fine and possibly a prison term. Because of the damaging effects of these refrigerants, a new class of environmentally safe refrigerant blends, known as green refrigerants, has emerged.

There is evidence that the ozone layer surrounding the Earth is being destroyed by various chemicals, most notably chlorine. The ozone layer filters out harmful radiation from the sun that would otherwise reach the Earth's surface and damage life. The chlorine in CFC and HCFC refrigerants is now known to contribute to this damage. Since the passage of the Clean Air Act in 1990, these refrigerants have come under increasing government regulation and control. The U.S. Environmental Protection Agency (EPA) is responsible for making and enforcing laws pertaining to the use of these refrigerants.

It is important to note that CFCs have the greatest impact on the ozone layer because CFCs contain the greatest volume of chlorine. As a result, the production of new CFCs has been banned. HCFC refrigerants such as R-22 are also being phased out. Although they do contain chlorine, they are considered to be ozone-safe. However, there is evidence that they contribute to the problem of global warming when released to the atmosphere.

4.4.0 Refrigerant Containers

Refrigerants come in disposable, returnable, or refillable recoverable metal containers, which vary in shape and size (see *Figure 14*). Low-pressure refrigerants such as CFC-11, CFC-113, and HCFC-123 come in standard steel drums or cylinders. They have boiling points close to, or slightly above, ambient (room) temperature. The pressure they exert on the container is much less than that of medium and high-pressure refrigerants, such as CFC-12, HCFC-22, HFC-134a, CFC-500, and CFC-502. These refrigerants are liquefied compressed gases. If improperly handled, the pressurized containers that contain them can burst or leak, causing damage, injury, or even death.

4.4.1 Disposable Cylinders

Disposable cylinders are not manufactured for repeated use. These cylinders should be stored in dry locations to prevent rusting. They should be transported carefully to prevent abrasion of their painted surfaces. As an added protection, they should be kept in their original cartons. Disposable cylinders should not be left around with quantities of refrigerant in them. Over time, rough handling or excessive heat could cause them to explode, especially if weakened by rust or corrosion.

WARNING!
Disposable cylinders must never be refilled. Not only is it dangerous, it is also against the law. Violators can be fined up to $25,000 and can face up to five years in jail.

When empty, disposable cylinders are recycled as scrap metal, be sure that the cylinder pressure is zero pounds, then render the cylinder useless by puncturing the rupture disk or breaking off the shutoff valve.

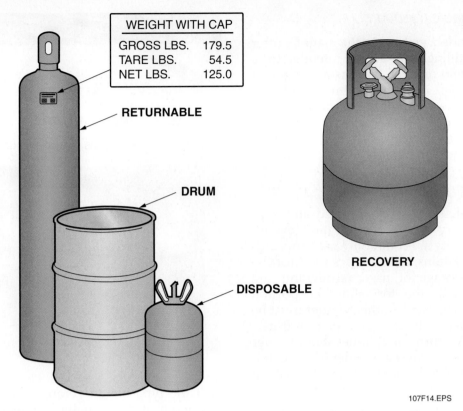

WEIGHT WITH CAP	
GROSS LBS.	179.5
TARE LBS.	54.5
NET LBS.	125.0

RETURNABLE

DRUM

RECOVERY

DISPOSABLE

107F14.EPS

Figure 14 ◆ Refrigerant containers.

Returnable Cylinder Labeling

INSIDE TRACK

Returnable cylinders and tanks are normally stamped with various weight values. These values include the following:

- *Tare weight* – The empty weight of the vessel.
- *Gross weight* – The combined weight of the vessel (tare weight) plus the weight of the refrigerant when the vessel is full. Be aware that the term full actually means a vessel that is filled to 80 percent of capacity. The remaining 20 percent of the volume must be available for expansion.
- *Net weight* – The weight of the contents in the vessel. For example, when ordering 50 pounds of refrigerant from a supplier, we are actually talking about the net weight. Manufacturers design vessels so that when the full net weight is reached, 20 percent of the volume remains for expansion.

Also stamped on the shoulder or collar of returnable cylinders is the date when the cylinder was tested. Returnable and reusable cylinders must be retested every five years.

5-15

4.4.2 Returnable Cylinders

Returnable cylinders go back to the manufacturer for reuse and refilling. They are not intended to be refilled in the field or to be used as a refrigerant recovery tank. These containers are not filled with more than 80 percent liquid. Excess liquid causes hydrostatic pressure that can result in an explosion. This pressure increases rapidly with even very small changes in temperature.

4.4.3 Recovery Cylinders

Recovery cylinders are typically used and supplied with recovery/recycle units. Cylinders with a 50-pound capacity are commonly used. According to EPA regulations, all cylinders used for the recovery and storage of used refrigerants are painted gray with the top shoulder portion painted yellow. The label on the cylinder must be marked to properly identify the type of refrigerant it contains. A returnable cylinder should never be substituted for a recovery cylinder. These cylinders must never be filled to more than 80 percent of capacity.

NOTE

The U.S. Department of Transportation (D.O.T.) governs the construction and labeling of cylinders used for refrigerant storage and recovery. Recovery cylinders are colored gray with a yellow top. However, cylinders used to recover R-410A, even though they use that color scheme, must be specifically manufactured and labeled for use with R-410A because of its higher pressures.

4.5.0 Identifying Refrigerants

Refrigerant containers are color-coded and marked with labels to identify the type of refrigerant they contain. These labels also include important health information about the contents of the container. *Table 3* lists the color codes for some common refrigerant containers.

Table 3	Color Codes Used for Some Common Refrigerant Containers		
CFC-11	Orange	CFC-12	White
CFC-13	Dark purple	CFC-500	Yellow
CFC-502	Purple	HCFC-22	Green
HCFC-123	Gray	HCFC-124	Dark green
HFC-410A	Pink	HFC-134A	Light blue
Refrigerant recovery cylinders—Gray with yellow top			

107T03.EPS

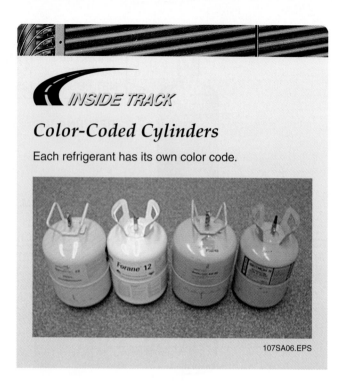

Color-Coded Cylinders

Each refrigerant has its own color code.

107SA06.EPS

If the type of refrigerant used in a system is unknown, it can usually be identified by using one of the following methods:

- Check the manufacturer's service literature for the equipment.
- Check the nameplate on the equipment.
- Check the data marked on the thermostatic expansion valve (metering device).

4.6.0 Refrigerant Safety Precautions

Butyl-lined gloves and safety glasses must be worn to avoid getting refrigerant on your skin or in your eyes. When accidentally released to the atmosphere, refrigerant can cause frostbite or burn the skin.

Refrigerants can cause suffocation if the amount and time of exposure is great enough. Always maintain ample ventilation. Refrigerant vapor is invisible, has little or no odor, and is heavier than air. Be especially careful of low places where it might accumulate.

Equipment rooms or other areas with large machines holding large amounts of refrigerant must have alarm systems that detect low oxygen levels and sound an alarm. When refrigerant is exposed to an open flame, a toxic gas is formed. A self-contained respirator must be available outside equipment rooms or other areas containing large equipment. Use the respirator if you must enter a contaminated area. Some equipment rooms have a mechanical ventilation system that can be used to clear contaminated air from the room.

Refrigerant Precautions

In addition to wearing protective clothing and equipment, follow these rules when handling and using refrigerants:

- Always double-check to be sure you are using the proper refrigerant. The containers are color-coded and labeled to identify their contents. Container labels also include product, safety, and warning information.
- Refer to technical bulletins and material safety data sheets available from the manufacturers for information important to your health. They describe the flammability, toxicity, reactance, and health problems that could be caused by a particular refrigerant if spilled or incorrectly used.
- Do not drop, dent, or abuse refrigerant containers. Do not tamper with safety devices.
- Always use a proper valve wrench to open and close the valve.
- Replace the valve cap and hood cap to protect the cylinder valve when not in use or empty.
- Secure containers in place to prevent them from becoming damaged when moved (especially in a van or truck). Strap or chain containers in an upright position.
- Do not store containers where the temperature can cause the pressure to exceed the cylinder relief valve settings.

5.0.0 ◆ COMPRESSORS

The compressor is the keystone of the refrigeration system. It creates the pressure difference that causes refrigerant to flow around the system. In the process, it takes refrigerant vapor at a low temperature and pressure and converts it to a higher temperature and pressure.

Compressors are usually driven by an electric motor. Very large compressors can be driven by internal combustion engines or steam turbines. Compressors are divided into three groups based on the way they are joined to their motors or engines (*Figure 15*).

- *Open-drive compressor* – The compressor is separate from its motor. One end (the shaft) extends outside the case. A mechanical seal is used with the rotating shaft to prevent leakage of the refrigerant. The compressor motor drives the compressor using a belt (belt drive) or flexible coupling (direct drive). Belt-driven arrangements allow the motor to run at one speed while the compressor can run at another. The proper combination of pulleys (also called drives) produces the desired speed of the compressor. Most direct-drive systems use an electric motor to drive the compressor. This means that the compressor also runs at the speed of the drive motor.

COMPRESSORS TAKE REFRIGERANT VAPOR AT A LOW TEMPERATURE AND PRESSURE AND RAISE IT TO A HIGHER TEMPERATURE AND PRESSURE.

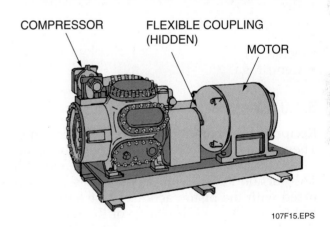

107F15.EPS

Figure 15 ◆ Open-drive compressor.

- *Hermetic (welded hermetic) compressor* – The compressor and motor have a common drive shaft. They are sealed in a welded steel enclosure or shell. Hermetic compressors (*Figure 16*) are more compact, less noisy, and require less maintenance than open-type compressors because they have no belts or couplings to break or wear out. Because they are sealed, the entire unit must be replaced when they fail.

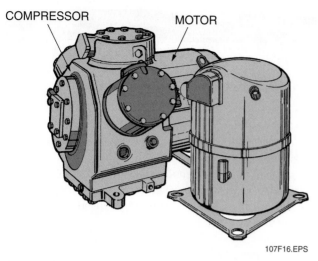

Figure 16 ◆ Semi-hermetic and hermetic compressors.

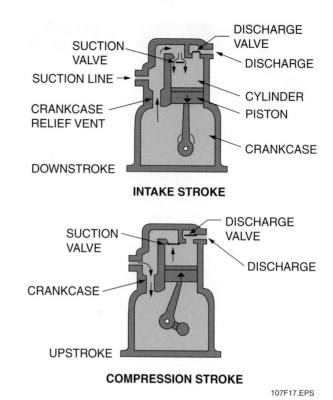

Figure 17 ◆ Reciprocating compressor.

- *Semi-hermetic (serviceable hermetic) compressor –* Similar to the hermetic compressor, the compressor and motor share the same housing and a common drive shaft. When they fail, access to the compressor or motor for repair is possible by removing the heads and/or the bottom and end.

Five types of compressors are commonly used in mechanical refrigeration systems:

- Reciprocating
- Rotary
- Scroll
- Screw
- Centrifugal

5.1.0 Reciprocating Compressors

Reciprocating compressors are very common. They use one or more pistons moving back and forth within a cylinder or cylinders (*Figure 17*). The suction and discharge valves are synchronized with the piston action. These valves control the intake and discharge of the refrigerant. Reciprocating compressors are typically used in refrigerators, air conditioners, and commercial processing equipment. Welded hermetic compressors below 10 tons are most popular, but the use of compressors in the 10 to 20 ton range is increasing. *Figure 18* shows a cutaway view of a hermetic reciprocating compressor. Serviceable semi-hermetic compressors are used in commercial air conditioning and heat pumps above 10 tons. Open reciprocating compressors are used mostly for refrigeration work and on industrial and large commercial air conditioning and heat pumps anywhere in the 5 to 150 ton range.

Figure 18 ◆ Cutaway view of a hermetic reciprocating compressor.

5.2.0 Rotary Compressors

Rotary compressors are usually welded hermetic compressors. They are frequently used on appliances, room air conditioners, and central air conditioning below five tons. There are two types of rotary compressors: stationary vane and rotary vane.

In the stationary vane compressor (*Figure 19*), a shaft with an attached off-center (eccentric) rotor rotates or rolls around the cylinder. A stationary vane mounted in the compressor housing slides in and out and follows the rotating motion of the rotor as it moves within the cylinder. This vane also separates the suction and discharge sides of the cylinder. As the shaft turns, the rotor rolls around the cylinder, drawing suction gas into the intake opening. At the same time, the gas is compressed against the cylinder wall on the discharge or compression side. A valve at the discharge keeps the compressed gas from leaking back into the cylinder and into the suction side during the off cycle. This process continues as long as the compressor is running.

Rotary vane compressors have a rotor centered on the drive shaft. However, the drive shaft is positioned off-center in the cylinder. Mounted on the rotor are two or more vanes that slide in and out to follow the shape of the cylinder. As the rotor turns, these vanes trap low-pressure suction gas and compress it against the cylinder wall, then force it out the discharge opening. The vanes also keep the compressed gas from mixing with the incoming low-pressure gas.

5.3.0 Scroll Compressors

Scroll compressors usually are welded hermetic compressors. Of all the compressor types, the scroll compressors have the fewest working parts. They operate efficiently even in applications that have large changes in refrigerant pressures, such as commercial refrigeration and heat pumps.

No suction or discharge valves are used in a scroll compressor (*Figure 20*). However, a valve is used on the discharge side to prevent reverse rotation. The scroll compressor achieves compression

INSIDE TRACK

Scroll Compressor Operating Noise

Scroll compressors produce unusual sounds when they start up and shut down. If you have never heard these sounds, you might think the compressor is defective. When you have the opportunity during your training, listen closely to these sounds so you can recognize them in the field. Some manufacturers of compressors and air conditioning equipment have training programs available on video or CD-ROM to help you identify these sounds.

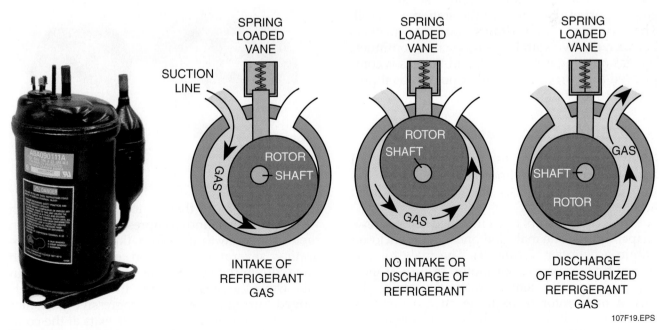

Figure 19 ◆ Stationary vane rotary compressor.

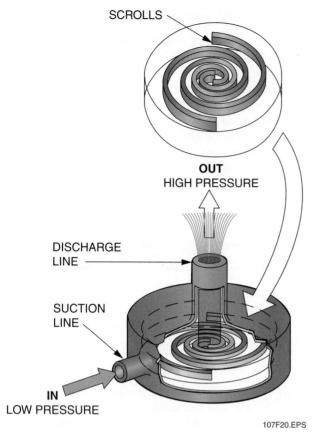

Figure 20 ◆ Scroll compressor operation.

by the use of two spiral-shaped parts called scrolls. One is fixed; the other is driven and moves in an orbiting action inside the fixed one. There is contact between the two. Refrigerant gas enters the suction port at the outer edge of the scroll, and after compression is squeezed out a separate discharge port at the center of the stationary scroll. The orbiting action draws gas into pockets between the two spirals. As this action continues, the gas opening is sealed off, and the gas is compressed and forced into smaller pockets as it progresses toward the center.

5.4.0 Screw Compressors

Screw compressors are used in large commercial and industrial applications requiring capacities from 20 to 750 tons. They are made in both open and hermetic styles.

Screw compressors use a matched set of screw-shaped rotors, one male and one female, enclosed within a cylinder (*Figure 21*). The male rotor is driven by the compressor motor. In turn, it indirectly drives the female rotor. Normally the driven male rotor turns faster than the female rotor because it has fewer lobes than the female rotor. Typically, the male has four lobes and the

female has six. As these rotors turn, they mesh with each other and compress the gas between them. The screw threads form the boundaries separating several compression chambers, which move down the compressor at the same time. In this way, the gas entering the compressor is moved through a series of progressively smaller compression stages until the gas exits at the compressor discharge in its fully compressed state.

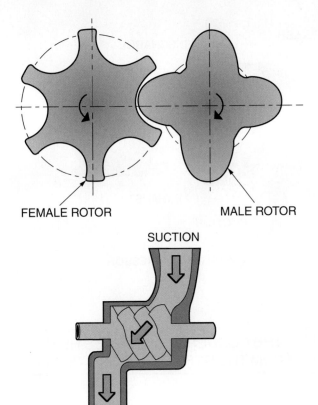

FEMALE ROTOR   MALE ROTOR

SUCTION

DISCHARGE

(SIDE VIEW)

107F21.EPS

Figure 21 ◆ Screw compressor.

5.5.0 Centrifugal Compressors

Centrifugal compressors are made in open and hermetic designs. They are typically used in commercial and industrial refrigeration and air conditioning systems with capacities larger than 100 tons. Standard models range up to 10,000 tons of capacity, with custom models exceeding 20,000 tons.

Centrifugal compressors use a high-speed impeller with many blades that rotate in a spiral-shaped housing (*Figure 22*). The impeller is driven at high speeds (typically 10,000 rpm) inside the compressor housing. Refrigerant vapor is fed into the housing at the center of the impeller. The impeller throws this incoming vapor in a circular path outward from between the blades and into the compressor housing. This action, called centrifugal force, creates pressure on the high-velocity gas and forces it out the discharge port. Often, several impellers are put in series to create a greater pressure difference and to pump a sufficient volume of vapor. A compressor that uses one impeller is called a single stage, one that uses two impellers is called a double stage, and so on. When more than one stage is used, the discharge from the first stage is fed into the inlet of the next stage.

Screw Compressor Rotors

This figure shows what the rotors in a screw compressor look like.

INSIDE TRACK

107SA09.EPS

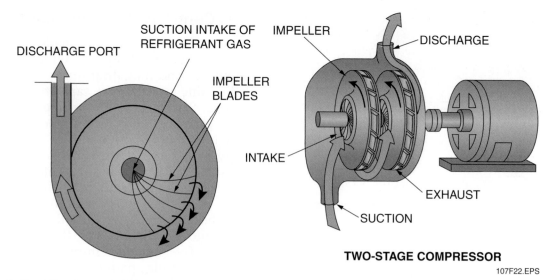

DISCHARGE PORT

SUCTION INTAKE OF REFRIGERANT GAS

IMPELLER

DISCHARGE

IMPELLER BLADES

INTAKE

EXHAUST

SUCTION

TWO-STAGE COMPRESSOR

107F22.EPS

Figure 22 ◆ Centrifugal compressor.

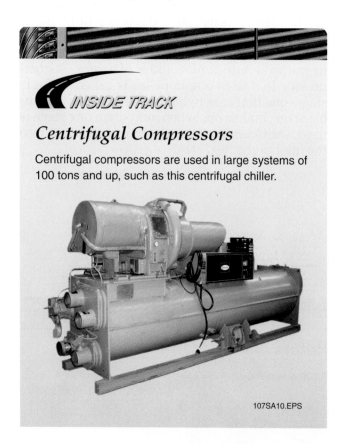
6.0.0 ◆ CONDENSERS

Condensers (*Figure 23*) are used for removing heat from the refrigeration system. They take in high-pressure, high-temperature refrigerant gas from the compressor and change it into a high-temperature, high-pressure liquid. They do this by transferring the heat from the refrigerant to the air, to water, or both. As the refrigerant flow progresses through the condenser, it first rejects the superheat and then fully condenses into a

THE CONDENSER REMOVES HEAT FROM THE REFRIGERATION SYSTEM.

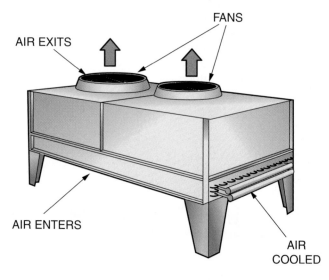

FANS

AIR EXITS

AIR ENTERS

AIR COOLED

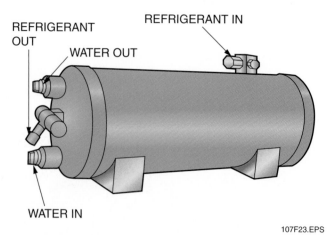

REFRIGERANT OUT

REFRIGERANT IN

WATER OUT

WATER IN

107F23.EPS

Figure 23 ◆ Condensers.

subcooled, high-temperature, high-pressure liquid. For the condenser to operate properly, the condensing medium of air or water must always be at a lower temperature than the refrigerant it is condensing. Condensers are classified according to the medium used to carry the heat away from the refrigerant vapor. The following sections describe these types of condensers:

- Air-cooled
- Water-cooled
- Evaporative

6.1.0 Air-Cooled Condensers

Air-cooled condensers (*Figure 24*) reject the heat absorbed by the system directly to the outdoor air. At normal design (peak load) conditions, the refrigerant flowing through the condenser is about 25°F to 35°F warmer than the outside air to which it is rejected. This means that a saturation temperature of 120°F to 130°F is typical in the condenser when the outdoor air is 95°F. Because the medium is outdoor air, this temperature tends to be greater than is needed for condensers used in water-cooled systems.

Propeller (axial) fans are used with most air-cooled condensing units to increase the amount of air being circulated across the condenser. This increases its capacity to reject heat. Because air-cooled condensers require the circulation of air over their surfaces, their location and the temperature of the surrounding air are very important to proper operation. The higher the temperature of the condensing air, the more work the system compressor must do to raise the refrigerant temperature to a level that allows efficient heat transfer. This causes the compressor to use more power. Air-cooled condensers are typically used in residential air conditioning up to about 5 tons and in commercial air conditioning up to about 50 tons. Air-cooled condensers are generally of two types: fin- and tube-condensers and plate condensers (*Figure 24*).

6.1.1 Fin-and-Tube Condensers

In the fin-and-tube condenser, the refrigerant vapor passes through the rows of tubing. The tubing is encased in metal ribs or fins. These condensers provide increased exposure to the surrounding air. Cooler air passing over the fins and tubing absorbs heat from the refrigerant.

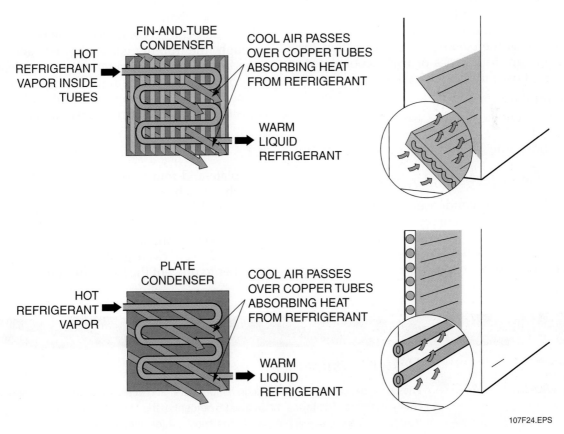

107F24.EPS

Figure 24 ◆ Air-cooled condensers.

6.1.2 Plate Condensers

The plate condenser operates the same as the fin-and-tube condenser. Basically, it is formed by two sheets of metal that have been pressed or stamped into the correct shape and then welded together. The plates provide a larger surface area for the transfer of heat to the surrounding air.

6.2.0 Water-Cooled Condensers

Water-cooled condensers are more complicated, more expensive, and require more maintenance than air-cooled condensers. However, they are more efficient and operate at much lower condensing temperatures (about 15°F lower) than air-cooled condensers. This allows the system compressor to run at lower head pressures, requiring the use of less power. Depending on the type, the velocity of the water flowing in a water-cooled condenser should be between three and ten feet per second. If it flows too fast, the tubing may become pitted. If it flows too slowly, scaling will occur. In areas where water is plentiful, the water that flows through the condenser may be used once and then drained into a waste system. Most often, the water portion of a water-cooled condenser is connected via piping to a cooling tower, which can be located on the roof of the building. In the cooling tower, the heat absorbed by the water in the condenser is rejected from the system into the atmosphere by evaporation. The cooled water is then returned to the system for reuse. There are four types of water-cooled condensers, as shown in *Figure 25*:

- Tube-in-tube
- Shell-and-tube
- Shell-and-coil
- Plate-and-frame

6.2.1 Tube-in-Tube Condensers

In the tube-in-tube condenser, the water flows through one or more curved tubes. The high-temperature, high-pressure refrigerant supplied from the compressor flows in the opposite direction. In the condenser, the hot refrigerant gives up heat to the cooler water and condenses into a sub-cooled liquid.

6.2.2 Shell-and-Tube Condensers

Shell-and-tube condensers may be either horizontally or vertically mounted. Vertical condensers contain straight, vertical tubes encased in a metal shell. High-temperature, high-pressure refrigerant supplied from the compressor enters the top of the condenser metal shell containing the tubes. The water also enters at the top and travels down through the vertical tubes where it absorbs heat from the surrounding refrigerant, then leaves at the bottom. As the refrigerant gas condenses on the cooler water tubes, liquid refrigerant falls to the bottom of the condenser metal shell where it collects and leaves the condenser at the output. The horizontal condenser allows water flowing in the tubing to make several passes before leaving the condenser. This and the use of fins make for greater cooling efficiency, since the refrigerant is exposed to more than one column of moving water.

6.2.3 Shell-and-Coil Condensers

Construction of the shell-and-coil condenser is similar to that of the shell-and-tube condenser, except that the tubing is wound around inside the shell, rather than being a straight length. As with other water-cooled condensers, the water flowing through the tubing cools the refrigerant surrounding it. As the refrigerant gas condenses on the cooler water tubes, liquid refrigerant falls to the bottom of the condenser metal shell, where it collects and leaves the condenser at the output.

6.2.4 Plate-and-Frame Condensers

The plate-and-frame condenser, also known as a plate heat exchanger, consists of a series of metal plates held in place by a metal frame. Within the heat exchanger, the cooling medium flows through one channel, while the warm fluid flows through another. The plates are gasketed, welded, or brazed together to prevent the fluids from mixing.

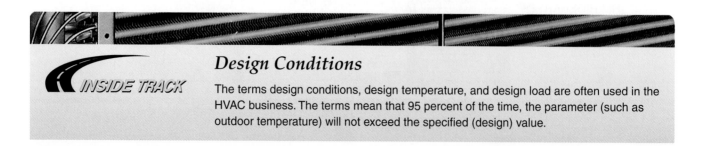

INSIDE TRACK

Design Conditions

The terms design conditions, design temperature, and design load are often used in the HVAC business. The terms mean that 95 percent of the time, the parameter (such as outdoor temperature) will not exceed the specified (design) value.

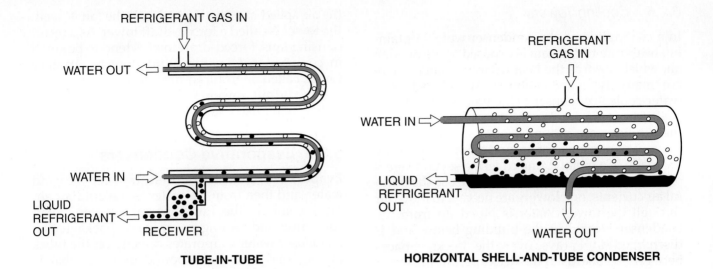

TUBE-IN-TUBE

HORIZONTAL SHELL-AND-TUBE CONDENSER

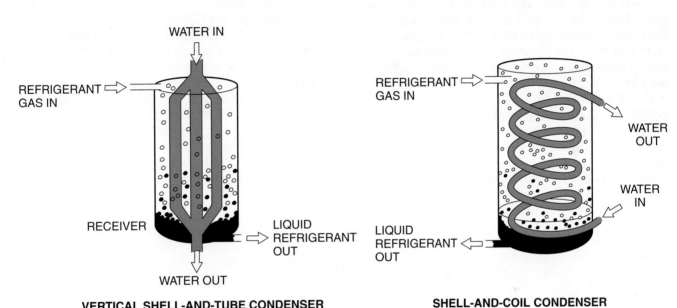

VERTICAL SHELL-AND-TUBE CONDENSER

SHELL-AND-COIL CONDENSER

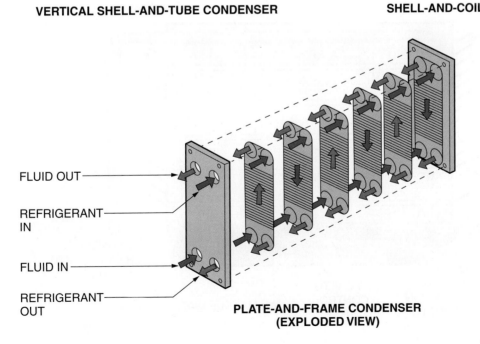

**PLATE-AND-FRAME CONDENSER
(EXPLODED VIEW)**

107F25.EPS

Figure 25 ◆ Water-cooled condensers.

6.2.5 Cooling Towers

In a cooling tower, the condenser water containing heat from the system is exposed to the outside air, which absorbs the heat from the water. There are many types of cooling towers used with water-cooled condensers. One type is the natural-draft tower (*Figure 26*). Built to system capacity, these towers are mounted outdoors, usually on the roof to make use of natural air currents. They are made of a metal frame covering several layers or tiers of wooden decks. Since they use the natural air currents, no blowers are needed to move air through the tower. Water is piped up from the condenser located in the building below and is discharged in sprays over the decks. Spaces between the boards in the decks permit the water to drip or run from deck to deck, while being spread out and exposed to air breezes that enter the tower from the open sides. The cooled water is collected in a catch basin at the bottom of the tower where it is pumped back to the condenser for reuse.

Because cooling towers work partly on evaporation, any water lost due to evaporation must be replaced in order to maintain the system. This is done using a float that senses the water level in the catch basin and adds water as needed. Mechanical fans may be used to push and increase the air speed in a cooling tower. If the fan is used, the tower is called a forced-draft tower. As a result of using fans, forced-draft towers tend to be small in comparison to natural-draft towers. Another tower, called an induced-draft tower, is similar to a forced-draft tower, but the fans pull the air rather than push it across the wet deck surfaces.

6.3.0 Evaporative Condensers

Evaporative condensers first transfer heat to water, and then from the water to the outdoor air. They combine the functions of a water-cooled condenser and cooling tower in one package. The condenser water evaporates directly off the tubes of the condenser. Each pound of water that is evaporated removes about 1,000 Btus from the refrigerant flowing through the tubes.

Air enters the bottom of the unit (*Figure 27*) and flows by convection upward over the condensing coil filled with refrigerant. At the same time, water is sprayed over the coil. Both the air and the water absorb heat from the refrigerant in the coil. Water eliminators, located above the water spray, remove water from the rising air. The air is then moved out of the top of the unit using one or more fans. Cooled by both air and water, the refrigerant in the coil condenses into a subcooled liquid at the output of the coil.

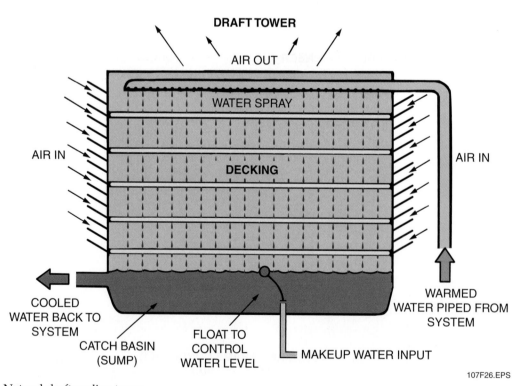

107F26.EPS

Figure 26 ◆ Natural-draft cooling tower.

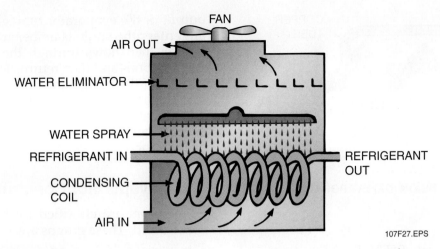

FAN

AIR OUT

WATER ELIMINATOR →

WATER SPRAY →

REFRIGERANT IN →

CONDENSING COIL →

AIR IN →

REFRIGERANT OUT

107F27.EPS

Figure 27 ◆ Evaporative condenser.

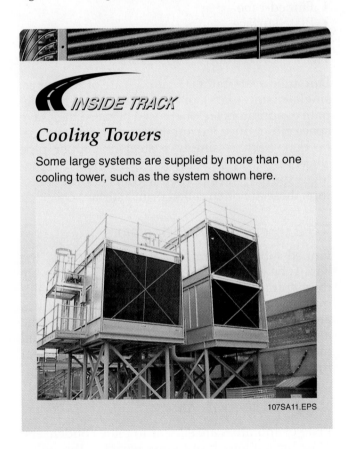

INSIDE TRACK

Cooling Towers

Some large systems are supplied by more than one cooling tower, such as the system shown here.

107SA11.EPS

7.0.0 ◆ EVAPORATORS

Evaporators (*Figures 28* and *29*) are used to extract heat from the conditioned space. They take in low-temperature, low-pressure liquid refrigerant from the metering device and change it into a low-temperature, low-pressure gas. This is done by transferring heat from either air or water to the refrigerant. As the refrigerant flow progresses through the evaporator, the heat in the warmer medium (air or water) causes it to boil (evaporate) and change into a vapor. When this occurs, the refrigerant absorbs heat from the medium being

THE EVAPORATOR ABSORBS HEAT IN THE REFRIGERATION SYSTEM.

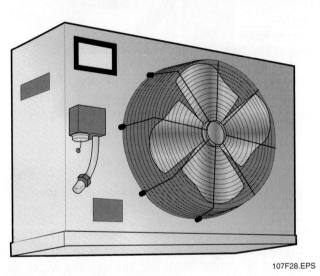

107F28.EPS

Figure 28 ◆ Forced-draft evaporator.

cooled. The amount of heat absorbed depends on how much heat is lost by the medium. The heat gain must equal the heat loss. For example, if the air passing over an evaporator gives up 800 Btus of heat, then the refrigerant in the evaporator must gain 800 Btus. Evaporators are of two types: the direct-expansion type and the flooded type (*Figure 29*).

7.1.0 Direct Expansion (DX) Evaporators

Direct expansion (DX) evaporators are the most widely used. DX evaporators have one continuous tube, or coil, through which the liquid refrigerant flows. The refrigerant, with a small amount

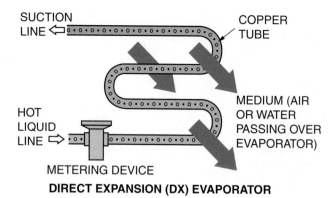

DIRECT EXPANSION (DX) EVAPORATOR

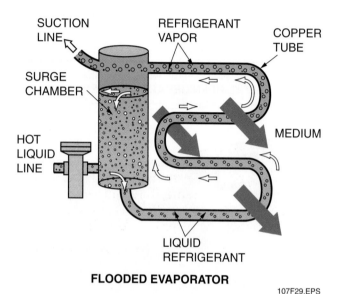

FLOODED EVAPORATOR

107F29.EPS

Figure 29 ◆ Direct expansion and flooded evaporators.

of gas mixed in, enters the evaporator and is gradually warmed by the medium until it boils and becomes a vapor near the outlet. The flow of refrigerant into the DX coil is controlled by the metering device at its input. It supplies just the right amount of refrigerant to the evaporator so that it is all transformed into vapor by the time it reaches the evaporator output (actually, the metering devices used with most DX evaporators are designed to produce between 10°F and 20°F of superheat of the refrigerant vapor at the evaporator output). Movement of the medium (air or water) over the evaporator can be by natural draft, or it can be enhanced using fans or pumps (forced draft).

7.2.0 Flooded Evaporators

In a flooded evaporator, refrigerant can be circulated through the evaporator more than once (*Figure 29*). A special receptacle known as a surge chamber is connected between the input and output of the evaporator tubing. Liquid refrigerant enters the surge chamber from the metering device, then flows through the evaporator coil, where it boils and then returns to the surge chamber. All of the refrigerant vapor exits through the suction line for input to the compressor. Any refrigerant not changed into a vapor collects in the surge chamber for recycling back through the evaporator.

7.3.0 Evaporator Construction

Evaporators are classified by the way they are constructed. These groups are:

- Bare-tube
- Finned-tube
- Plate-surface

7.3.1 Bare-Tube Evaporators

Bare-tube evaporators (*Figure 30*) are of two types, single-circuit (path) or multiple-path. Multiple-path evaporators are often used because they save space and reduce the number of metering devices needed. Each is simply a steel or copper pipe shaped in a way that best matches the job. The piping is the only surface used to transfer heat; therefore, these are often called prime-surface coils.

7.3.2 Finned-Tube Evaporators

Finned-tube evaporators are a variation of the bare-tube evaporator. Attached to the tubing are thin spiral-wound or rectangular fins of aluminum or copper, like those used with the fin-and-tube condenser. These fins increase the amount of surface area exposed to the heated medium. This in turn increases the amount of heat that can be transferred to the refrigerant.

7.3.3 Plate Evaporators

Plate evaporators are similar to plate condensers. They have a length of tubing weaving through a metal plate. The plate provides greater surface area for heat transfer. This type of evaporator is typically used as a shelf in older upright freezers.

7.3.4 Chilled Water System Evaporators

In many large refrigeration and air conditioning installations, cooling coils are installed at some distance apart in the building or complex. Because of the expense and the problems involved with long runs of refrigerant piping, cooling in these remote areas is done with chilled water rather than refrigerant as the medium. This chilled water

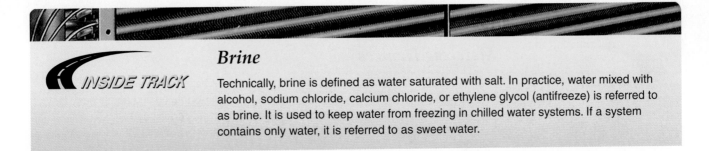

Brine

Technically, brine is defined as water saturated with salt. In practice, water mixed with alcohol, sodium chloride, calcium chloride, or ethylene glycol (antifreeze) is referred to as brine. It is used to keep water from freezing in chilled water systems. If a system contains only water, it is referred to as sweet water.

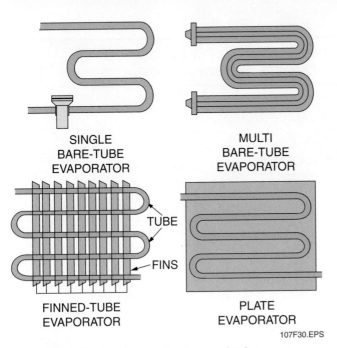

SINGLE BARE-TUBE EVAPORATOR

MULTI BARE-TUBE EVAPORATOR

TUBE

FINS

FINNED-TUBE EVAPORATOR

PLATE EVAPORATOR

107F30.EPS

Figure 30 ◆ Evaporator construction methods.

is called a secondary refrigerant. When temperatures are below freezing, brine is used in place of water. A refrigeration system, called the primary system, is used to cool the secondary water or brine refrigerant. The primary system generally uses shell-and-tube or shell-and-coil evaporators called chillers to absorb heat from the water or brine. Like other evaporators, chillers can be of the direct expansion or flooded type. Their construction is similar to that used for shell-and-tube or shell-and-coil condensers.

In the direct-expansion chiller, the colder liquid refrigerant runs through the tubing while the warmer secondary refrigerant (water or brine) circulates within the shell around the outside of the tubes. As a result, heat contained in the water or brine is transferred to the primary refrigerant flowing through the tubes. The refrigerant leaves the evaporator as a gas.

In the flooded-type chiller, the water or brine is circulated through the tubing, which is located on the bottom of the evaporator shell. This tubing is submerged below the liquid refrigerant level

within the shell, which is controlled by a float valve. The evaporator shell acts as a surge chamber. The top portion is left vacant so the refrigerant vapor can be properly separated from the liquid refrigerant for output to the suction line as it evaporates.

8.0.0 ◆ METERING (EXPANSION) DEVICES

The metering device is located between the condenser outlet and the evaporator inlet. High-pressure, high-temperature liquid refrigerant from the condenser enters the metering device. It leaves as a low-pressure, low-temperature mixture of liquid and vapor. Regardless of type, the metering device performs two functions:

- It allows the liquid refrigerant to flow into the evaporator at a rate that matches the rate at which the evaporator boils liquid refrigerant into a vapor.
- It provides a pressure drop that lowers the boiling point of the refrigerant.

There are many types of metering devices. They can be divided into two categories, fixed and adjustable (*Figure 31*). Fixed metering devices are used mainly in domestic refrigerators and freezers and residential air conditioning units. Adjustable metering devices are used most often on systems with variable load requirements. This section briefly describes the main types of fixed and adjustable metering devices. You will study these devices in greater detail later in your training.

8.1.0 Fixed Metering Devices

Fixed metering devices have a fixed restriction or fixed opening (orifice) size. The capillary tube (*Figure 31*) is the simplest metering device. It is a fixed length, small-diameter copper tube, usually with an inside diameter of ⅟₁₆" to ⅛". Because of its high resistance to refrigerant flow, it restricts the flow of liquid refrigerant from the condenser to the evaporator. The greater its length or the

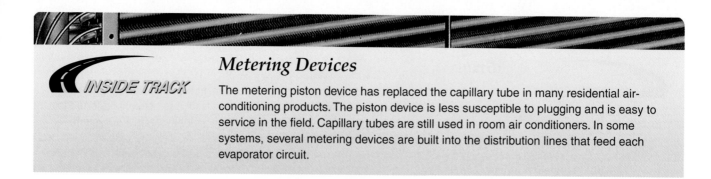

Metering Devices

The metering piston device has replaced the capillary tube in many residential air-conditioning products. The piston device is less susceptible to plugging and is easy to service in the field. Capillary tubes are still used in room air conditioners. In some systems, several metering devices are built into the distribution lines that feed each evaporator circuit.

> THE METERING DEVICE CONTROLS THE AMOUNT OF REFRIGERANT SUPPLIED TO THE EVAPORATOR AND PROVIDES A PRESSURE DROP THAT LOWERS THE REFRIGERANT BOILING POINT.

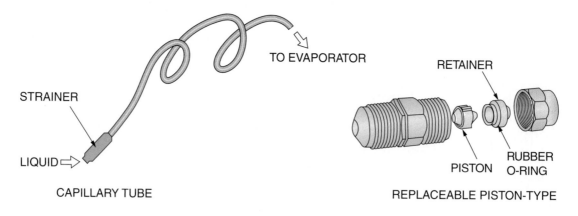

FIXED METERING DEVICES

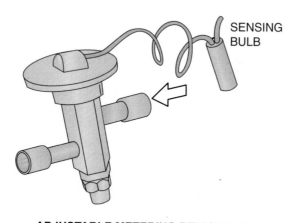

ADJUSTABLE METERING DEVICE (TXV)

107F31.EPS

Figure 31 ◆ Typical metering devices.

smaller its diameter, the greater the pressure drop. Capillary tubes are often coiled to conserve space and protect them from damage.

Another type of fixed metering device similar to the capillary tube is the fixed orifice device. This device is a compact and rugged assembly that is installed at the evaporator inlet. It contains a piston. Pistons are made with different-sized orifices to match the capacities of different equipment. The smaller the orifice, the greater the pressure drop. Often, the piston is installed in this type of metering device at the time of system installation in order to match the metering device to the condensing unit. The amount of refrigerant flow through the fixed metering device is controlled by the size of the orifice opening, the length of tubing, and the condenser pressure. As the outdoor temperature increases, the pressure also increases, driving more refrigerant through the fixed metering device.

8.2.0 Adjustable Metering Devices

Adjustable metering devices differ in their mechanisms and how they are controlled. They all work to regulate refrigerant flow so that the evaporator capacity matches the cooling load. There are six types of adjustable metering devices in common use:

- Hand-operated expansion valve
- Low-side float valve
- High-side float valve
- Automatic expansion valve
- Thermostatic expansion valve
- Electric and electronic expansion valves

Each of these devices is covered in detail in a later module.

8.2.1 Thermostatic Expansion Valve

The thermostatic expansion valve or TXV controls the amount of refrigerant flowing through the evaporator by sensing the level of superheat in the suction line at the evaporator output. It is designed to maintain a constant superheat. Like the automatic expansion valve, a TXV has a diaphragm with a valve and seat below it. The difference between the two is that the adjustment spring is below the diaphragm rather than above it. The pressure exerted on the underside of the diaphragm is a combination of evaporator and adjustment spring pressures. The spring is adjusted to exert a pressure that represents the desired level of evaporator superheat. Pressure on top of the diaphragm is applied via tubing from a remote sensing bulb used to monitor the superheat at the evaporator outlet. This bulb contains a refrigerant charge, or other suitable chemical charge, as the sensing fluid. If the superheat increases or decreases as a result of changing load, a corresponding pressure change is transmitted

---

⟪ INSIDE TRACK

TXVs

This photo shows a thermostatic expansion valve installed in a system. Note that the sensing bulb is wrapped in insulation to make sure that it senses only the temperature of the refrigerant leaving the evaporator, and is not influenced by the external temperature. This TXV is externally equalized, as indicated by the connection from the bottom of the TXV to the suction line. This design, which is common in larger systems, allows the superheat to be accurately maintained regardless of the pressure drop through the evaporator.

Improper installation of a TXV can prevent the system from operating correctly. The sensing bulb must be securely fastened to a clean, straight section of the suction line close to the evaporator outlet. In addition, the bulb must be thoroughly insulated with waterproof insulation to prevent it from being influenced by the surrounding air. Because the bulb is sensing the temperature of the refrigerant vapor, it should be installed on the top of the suction line to avoid the possibility of it reacting to any liquid refrigerant that might enter the line.

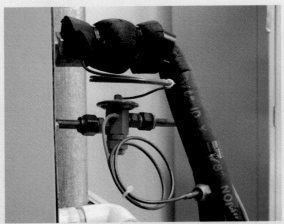

107SA12.EPS

TXVs and Troubleshooting

It can be very tempting to tweak the adjustment screw on a TXV when you have a hard-to-find problem. Don't do it before you check the superheat. If the superheat is within range (usually between 5° and 15°), the problem is not likely to be in the TXV, and you should look elsewhere. You may do more harm than good by trying to adjust the metering device.

---

from the bulb to the valve. This pressure change counteracts the combined evaporator and adjustment spring pressures to increase or decrease the valve opening, allowing more or less refrigerant flow through the valve.

9.0.0 ◆ OTHER COMPONENTS

Additional components can be added to the basic refrigeration system in order to improve safety, endurance, efficiency, or servicing (*Figure 32*). Some of these components are factory-installed, while others may be installed in the field. This section briefly describes the most commonly used components. They include the following:

- Filter-drier
- Sight glass/moisture liquid indicator
- Suction line accumulator
- Crankcase heater
- Oil separator
- Heat exchangers
- Receiver
- Service valves
- Compressor muffler

9.1.0 Filter-Drier

The filter-drier or strainer-drier combines the functions of a refrigerant filter and a refrigerant drier in one device (*Figure 33*). The filter protects metering devices and the compressor from foreign matter such as dirt, scale, or rust. The drier removes moisture from the system and traps it where it can do no harm. Filter-driers are normally installed in the liquid line ahead of the metering device. Filter-driers are replaced periodically during system maintenance or immediately after a system repair, such as a compressor burnout.

9.2.0 Sight Glass and Moisture Liquid Indicator

The sight glass is like a window that allows the technician to view the condition of the system refrigerant. It is typically used when checking the refrigerant charge. One common location for a sight glass is at the condenser outlet to view the condition and flow of refrigerant leaving the condenser. Another location is near the inlet of the metering

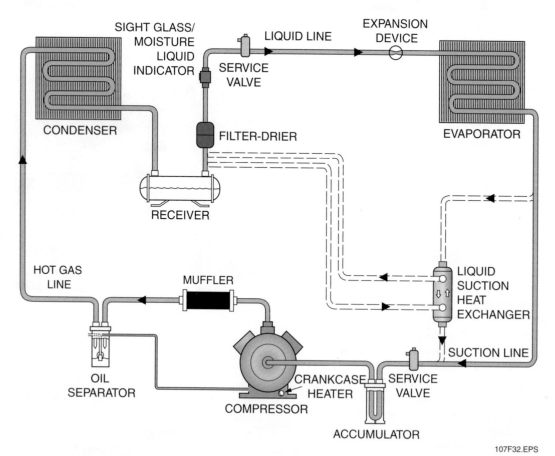

Figure 32 ◆ Other refrigeration system components.

Filter-driers are most often installed in the liquid line. They are usually installed/replaced whenever the system is opened to the air, because air contains moisture and other contaminants that can damage the compressor.

If a severe compressor burnout has occurred, a special liquid line filter-drier designed for one-time use is installed, then replaced with a clean one after the refrigerant has circulated for a while. A suction line filter-drier is also installed in such cases.

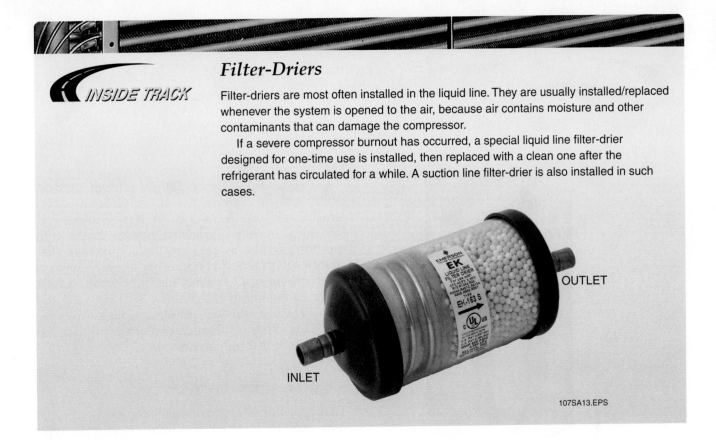

107SA13.EPS

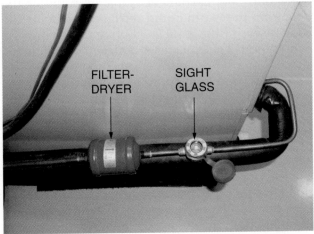

107F33.EPS

Figure 33 ◆ Filter-drier and moisture-indicating sight glass.

device, where the refrigerant condition can be viewed as the liquid refrigerant arrives at its destination. A moisture liquid indicator (*Figure 33*) is a sight glass with a small moisture-indicating device installed in it. This moisture indicator is exposed to the refrigerant and changes color depending on the amount of moisture in the refrigerant. When the moisture is within the limits set by the manufacturer, the indicator is one color. If too much moisture is present, the device will change color.

9.3.0 Suction Line Accumulator

The suction line accumulator (*Figure 34*) is a trap designed to prevent liquid **floodback** or slugs of liquid refrigerant from entering the compressor cylinders. The compressor cannot compress liquid. If liquid refrigerant is allowed to enter the compressor, noisy operation, high power consumption, and compressor damage may result. Accumulators are installed in the suction line as near the compressor suction inlet as possible. At this location, any liquid refrigerant or oil will be trapped temporarily in the accumulator. Some accumulators have heaters that help to vaporize refrigerant liquid. Refrigerant vapor is drawn from the top of the accumulator to be returned to the compressor. A small orifice at the bottom of the internal U-tube allows tiny amounts of both liquid refrigerants and oil to return to the compressor. In larger systems, trapped oil is piped back to the compressor.

9.4.0 Crankcase Heater

Crankcase heaters are installed on compressors to prevent liquid refrigerant from migrating to the compressor crankcase and causing damage. These heaters work by evaporating refrigerant from the oil. They are usually fastened to the bottom of the

Suan Line Accumulator

INSIDE TRACK

Suction line accumulators are widely used in residential heat pumps. During low-temperature operation, the accumulator may become covered with frost. This is normal operation under these circumstances.

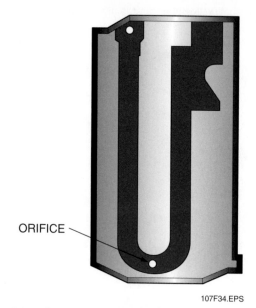

ORIFICE

107F34.EPS

Figure 34 ◆ Suction line accumulator.

crankcase or inserted directly into the compressor crankcase (immersion type). Wrap-around or bellyband heaters that encircle the outside shell of welded hermetic compressors are also used.

9.5.0 Oil Separator

Oil is used in a refrigeration system for four purposes:

• It lubricates the compressor.
• It helps seal the system.
• It dampens compressor noise.
• It acts as a coolant for the compressor and compressor motor.

Because there is oil in the compressor, it mixes with the refrigerant and travels with it to other areas of the system. Oil separators minimize the amount of oil that circulates through the system. Oil coats the inside of every component through which it passes. It reduces the heat-transfer ability and efficiency of the evaporator and condenser. Another reason for the oil separator is to slow down the accumulation of oil in places from which oil return is difficult.

Oil separators (*Figure 32*) are seldom used in residential or commercial air conditioning systems. Their use is mainly in refrigeration and industrial systems. Typically, they are installed in the hot gas line as close to the compressor discharge as practical. Separators usually have a reservoir (sump) to collect the trapped oil. A float valve in the sump maintains a seal between the high-pressure and low-pressure sides of the system. This valve automatically returns the oil to the compressor through an orifice.

9.6.0 Heat Exchangers

Two types of heat exchangers can be used with refrigeration systems: the liquid-to-suction type and the refrigerant water pre-heater. The liquid-to-suction heat exchanger transfers some of the heat from the warm liquid refrigerant leaving the condenser to the cool suction gas leaving the evaporator. This increases efficiency and helps subcool the liquid refrigerant. In some applications, it is used to evaporate the small amount of liquid refrigerant expected to return from the evaporator in the suction line to the compressor. Operation of the heat exchanger is similar to that of a water-cooled condenser. The liquid refrigerant leaving the condenser, and the cool suction gas leaving the evaporator, flow in opposite directions through the heat exchanger. The amount of heat that can be exchanged between the gas and liquid is determined by the temperature difference between the two, the amount of surface area, and how much time there is for heat exchange to occur.

The refrigerant water pre-heater is used to pre-heat the water supplied at the input of a hot water heater. In this heat exchanger, heat is transferred from the compressor hot gas line to the water. This reduces energy consumption whenever heat needs to be rejected from the system and helps to de-superheat the discharge gas leaving the compressor. Instead of rejecting heat outdoors, the heat is transferred to the hot-water system. Shell-and-coil and tube-in-tube type heat exchangers are typically used as water pre-heaters.

9.7.0 Receiver

The receiver is a tank or container used to store liquid refrigerant in the system (*Figure 32*). This storage is needed in some systems to accommodate changes during operation, to freely drain the condenser of refrigerant, and to provide a place to store the system charge during system service procedures or prolonged shutdowns. The receiver is installed in the liquid line between the condenser and the metering device. Receivers are used to store the excess refrigerant created by varying cooling loads in many systems that use self-adjusting metering devices. Note that some residential and commercial air conditioning equipment store this excess in the condenser. In these cases, the condenser is large enough to hold the excess while delivering less than peak capacity and subcooling.

9.8.0 Service Valves

Service valves are access ports that installers and service technicians can use to measure system pressures and perform servicing procedures such as charging, evacuation, and dehydration. There are several types of service valves. The most basic type of service valve, known as a piercing valve or line tap, can be temporarily installed on tubing to test pressures. The piercing valve is clamped to the tubing and a needle-like point pierces the tubing as the valve is tightened.

Most systems have factory-installed service valves like the ones shown in *Figure 35*. Note that there are two service valves, one for the system high side and the other for the low side. In this type of valve, a wrench is used to turn the valve stem and change the position of the valve. The valve has three positions. With the stem turned all the way in, or front-seated, the service port is closed off from the system. This position is used when connecting pressure gauges or other service equipment to the port. With the stem fully backed out, or back-seated, the refrigerant can readily flow to the service port. This position is used when charging, evacuating, or recovering refrigerant. When the stem is slightly open, or cracked, the service port is open to the system. This position is used when making pressure readings.

Schrader valves are similar to the valves on automobile tires, but they are not identical or interchangeable. The Schrader valve opens when the stem is depressed. Manifold gauge hoses are equipped with fittings designed to depress the stem as the fitting is tightened. When the Schrader valve is used for charging, evacuation, recovery,

etc., the core is usually removed using a special tool designed for this purpose (*Figure 36*) to decrease the pressure drop. This tool can also be used to replace a defective core.

9.9.0 Compressor Muffler

Mufflers are used most often in systems with open or semi-hermetic reciprocating compressors. Reciprocating compressors generate sound that can be transmitted along the piping. A muffler installed in the discharge line, as near the compressor as practical, is used to remove or dampen these pulsations. The muffler lowers the system noise and prevents possible damage from vibration.

107F35.EPS

Figure 35 ◆ Service valves.

107F36.EPS

Figure 36 ◆ Schrader valve core removal tool.

10.0.0 ◆ CONTROLS

Controls are the devices used to start, stop, regulate, and/or protect the components of the mechanical refrigeration system. They can be divided into two groups: primary and secondary. Primary controls start or stop the refrigeration cycle either directly or indirectly by sensing temperature, humidity, or pressure, or by measuring time. Secondary controls regulate and protect the cycle and its components. You have already been introduced to many controls in the HVAC module *Basic Electricity*. This section will introduce you to other types of control devices. You will study these devices again in greater detail later in your training.

10.1.0 Primary Controls

Primary controls start or stop the refrigeration cycle either directly or indirectly as a result of sensing temperature, humidity, or pressure, or by measuring time. Primary controls include the following:

- Thermostat
- Pressurestat
- Humidistat
- Time clock

10.1.1 Thermostats

All thermostats sense and respond to the temperature in a conditioned space. They switch the system on or off at a preset temperature called the setpoint by opening a set of contacts in the system control circuit. This may be done in several different ways. Some are activated by the warping of a bimetal strip (*Figure 37*). A bimetal device operates on the principle that different metals expand or contract at different rates when heated or cooled. A familiar form of the bimetal thermostat is the mercury bulb room thermostat. Its contacts are enclosed in an airtight glass bulb containing a small amount of mercury. Expansion and contraction of the bimetal element tilts the bulb in different directions. In one position, the mercury rolls to the low end of the bulb and closes or makes the electrical circuit; in the other, it rolls to the high end of the bulb and opens or breaks the electrical circuit. If the thermostat makes on a temperature rise and breaks on a temperature drop, it is a cooling thermostat. If it makes on a temperature drop and breaks on a temperature rise, it is a heating thermostat.

Another type of thermostat is activated by pressure applied from a bellows attached to a chemical-filled sensing bulb with a capillary tube

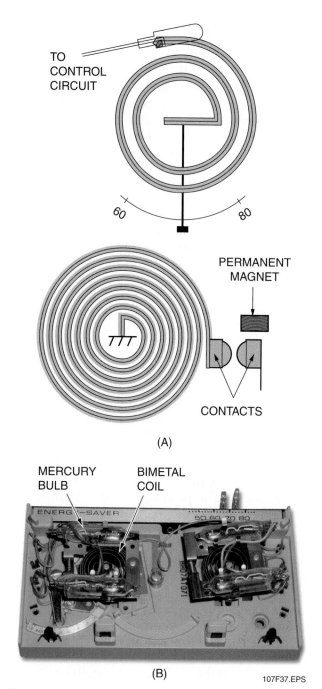

Figure 37 ◆ Bimetal strip used as a thermostat.

(*Figure 38*). When filled with a refrigerant, the bulb pressure increases or decreases as the temperature rises or drops. An increase in bulb pressure causes the bellows to expand, mechanically closing or opening the electrical contacts, depending on whether it is a heating or cooling thermostat. Cooling thermostats close on a pressure rise, while heating thermostats open on a pressure rise. The action is the opposite for a pressure drop. This type of thermostat is sometimes called a remote bulb thermostat because the bulb can be located at a different location from the rest

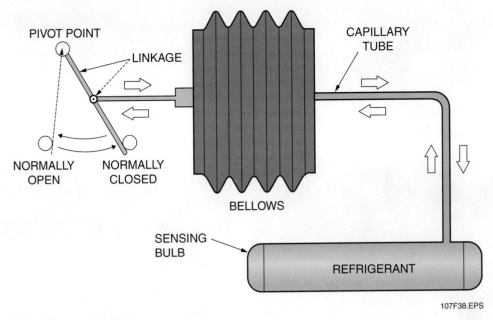

Figure 38 ◆ Remote bulb thermostat.

of the thermostat. Another pressure-actuated type of thermostat is a diaphragm thermostat. In this type of thermostat, the bulb pressure moves a diaphragm rather than a bellows to open and close the electrical contacts.

Electronic thermostats (*Figure 39*) use electronic components to sense temperature changes and perform switching functions. These thermostats generally use either a thermocouple or thermistor to sense the temperature. A thermocouple sensor is made of two dissimilar wires welded together at one end called a junction. When the junction is heated, it generates a low-level DC voltage that is applied to the switch circuits in the thermostat. A thermistor sensor is a semiconductor device in which the resistance changes with a change in temperature. As the temperature being measured varies, the thermistor resistance varies, causing a change in the current applied to the thermostat switch circuits. When the setpoint is reached, the switches open or close to control the related components. Electronic thermostats are more accurate and reliable than other types. Many contain microprocessor chips that allow them to be programmed for automatic startup, shutdown, or setpoint changes.

10.1.2 Pressurestat

Pressurestats (*Figure 40*) control systems using variations in pressure. They work on both the high-pressure and low-pressure sides of the compressor. Pressurestats allow operation of the compressor when low-side pressure has reached preset conditions. They also act as a cutout to

Figure 39 ◆ Programmable electronic thermostat.

open the circuit if the high-side pressure exceeds a safe level. Two versions of pressurestats are the bellows type and the Bourdon type. The bellows type is directly connected to the refrigeration system through a capillary tube. As the pressure within the system changes, the pressure in the bellows also changes. The bellows expand with a pressure increase and contract with a pressure decrease. The movement in and out of the bellows makes or breaks an electrical circuit as contacts mechanically linked to the bellows open or close.

The Bourdon pressurestat uses a thin C-shaped tube that is closed at one end and connected through a capillary tube to the refrigeration system. A mercury bulb switch like that used in a room thermostat is linked to the Bourdon tube

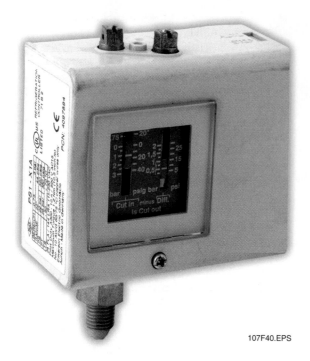

Figure 40 ◆ Pressurestat.

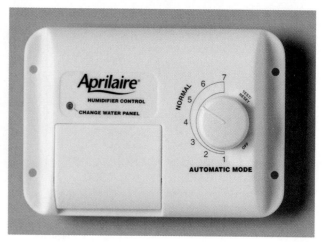

107F41.EPS

Figure 41 ◆ Humidistat.

through a spring-loaded lever. As the pressure inside the Bourdon tube increases, the tube straightens and, by the action of the spring-loaded lever, opens or closes the contacts in the mercury switch. If a pressure rise opens the switch contacts and a drop closes the contacts, the device is a high-pressure switch. If a pressure rise closes the contacts and a drop opens the contacts, the device is a low-pressure switch.

10.1.3 Humidistat

Humidistats (*Figure 41*) sense moisture or humidity. Electromechanical humidistats use materials such as human hair or nylon that expand when they absorb moisture from the air (hygroscopic materials). As the humidity increases, the hygroscopic material becomes moist and expands. This movement, applied through a spring-loaded lever, opens or closes the humidistat electrical contacts. A reduction in humidity causes the hygroscopic material to dry and contract. If a rise in humidity closes the switch contacts and a drop opens the contacts, the device being controlled turns on with a rise in humidity and cycles off with a drop in humidity. If a rise in humidity opens the switch contacts and a drop closes the contacts, the device being controlled turns on with a humidity drop and cycles off with a humidity rise.

Electronic humidistats use electronic sensing elements like lithium salts with hygroscopic properties. Another method uses carbon particles embedded in a hygroscopic material. These sensing elements work like a thermistor. Changes in humidity affect the resistance of the material and thereby change the current in an electronic circuit.

10.1.4 Time Clock

Time clocks are often used to start and stop a refrigeration system or selected components within a system. Time clocks typically are built into thermostats. When a time clock is used, the system thermostat, pressurestat, or humidistat acts as the primary controls during normal operation. However, when a timed event occurs, such as startup or shutdown, the time clock circuit overrides the other controls. Timers are used to activate defrost circuits in heat pumps and low-temperature refrigeration systems.

10.2.0 Secondary Controls

Secondary controls regulate and/or protect the refrigeration system while it is in operation. They are often referred to as operating controls because they keep the cycle adjusted and running properly. Operating controls include the following:

- Metering devices
- Relays, contactors, and starters
- Condenser water valve
- Refrigerant solenoid valve
- Evaporator pressure regulators
- Check valve
- Timed devices

Controls that protect the system and its components from damage are called safety controls. Safety controls include the following:

- Electrical overloads
- Current/temperature devices
- Thermostats
- Pressurestats
- Fusible plug
- Rupture disc
- Oil safety switches
- Flow switches

You have already been introduced to the electrical controls listed above in *Basic Electricity*. Most of the other ones listed have been covered previously in this module. The following paragraphs introduce you to those devices not discussed previously.

10.2.1 Condenser Water Valve

The condenser water valve is used in systems with water-cooled condensers. It regulates the head pressure at a constant level or at a preset minimum level. It is usually a self-contained, pressure-actuated bellows valve.

10.2.2 Evaporator Pressure Regulator

The evaporator pressure regulator (EPR) maintains a constant pressure and, therefore, a constant saturation temperature in the evaporator. It does this regardless of changes in pressure elsewhere in the system. It is typically a self-contained, pressure-actuated bellows valve. Its operating pressure is adjusted using a spring-tension screw.

10.2.3 Check Valve

Check valves are used to ensure single-direction flow. They prevent reverse flow. They use either a movable flapper or a ball that moves away from its seat to allow flow in the desired direction and seals on the seat to stop the flow in the wrong direction. They are also used in heat pumps and water circuits.

10.2.4 Pressure Relief Devices

Fusible plugs and rupture discs protect the refrigeration system or specific components from damage caused by an over-pressure condition. A fusible plug has a soft metal core with a low melting point. In the case of extreme pressure or fire, the high temperature melts the core. This allows the gas to escape before hazardous pressures build up. The rupture disc is a thin graphite disc that ruptures at pressures above the maximum desired pressure.

10.2.5 Oil Safety Switches

An oil safety switch is a pressure-actuated, electrical safety control used to protect the compressor against damage caused by a loss of oil pressure.

10.2.6 Flow Switches

Flow switches are used to shut down the system when either air or water flow to the evaporator or condenser is inadequate. They also prevent the system from starting with inadequate flow.

11.0.0 ◆ PIPING

Most piping used in refrigeration systems is ACR copper tubing. Aluminum, steel, stainless steel, and plastic tubing may also be used for certain applications. The piping layout is usually made by the system designer. The HVAC technician is interested in piping mainly from a servicing viewpoint.

There are four requirements for a good piping layout:

- It provides refrigerant paths.
- It avoids excessive pressure drops.
- It returns oil to the compressor crankcase.
- It protects compressors.

The purpose of the piping system is to provide a path for the flow of refrigerant from one component to another. Refrigerant flow must be accomplished without excessive pressure drops, such as those caused by friction, long risers, restrictions, and other piping conditions.

Some oil circulates in all refrigeration systems. The piping layout, therefore, must return the oil to the compressor crankcase. A **slug** of liquid refrigerant or large amounts of oil entering the compressor in the suction line can seriously damage the compressor. Good piping practices minimize the potential for damage.

In *Figure 42*, the major pipelines of the refrigeration cycle are identified. The suction line carries cold low-pressure gas from the evaporator to the compressor. The hot gas line carries hot high-pressure gas from the compressor to the condenser. Where separate receivers are used, a condensate line is installed to drain refrigerant from the condenser to the receiver. The liquid line carries the liquid refrigerant from the receiver to the metering device.

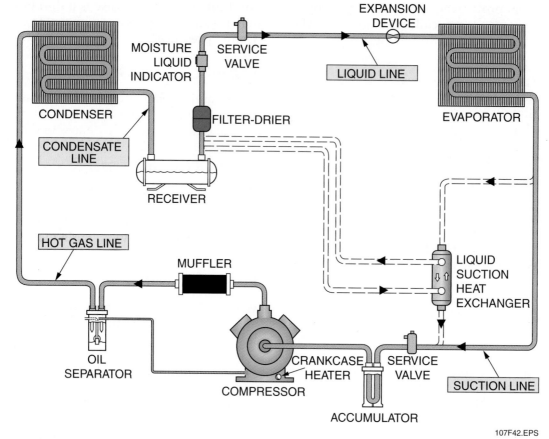

Figure 42 ◆ Refrigeration system major pipelines.

11.1.0 Basic Principles

Certain basic principles apply to all refrigerant lines: Keep them simple and pitch horizontal lines in the direction of flow. This helps to maintain the flow of oil in the right direction, prevent oil traps, and avoid backward flow during shutdown.

Unnecessary oil pockets (*Figure 43*) should be avoided. Poor layout or complicated routing can result in pockets in which oil can collect. Also, allow room for expansion and vibration. All lines expand and contract with changes in temperature.

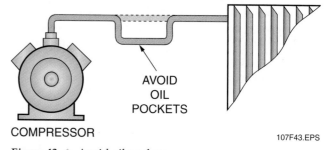

Figure 43 ◆ Avoid oil pockets.

11.2.0 Suction Line

The suction line is the first line to be considered. It is the most critical from a layout standpoint. The pressure drop at full load must be within practical limits and oil return must be maintained under minimum load conditions. Its design must prevent liquid refrigerant from draining to the compressor during shutdown. It must also prevent oil and refrigerant from returning to the compressor in slugs during operation.

In suction risers (*Figure 44*), oil is carried upward by the refrigerant gas. A minimum gas velocity must be maintained to keep the oil moving. The

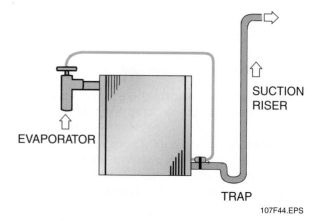

Figure 44 ◆ Suction riser.

trap at the bottom of the riser promotes free drainage of liquid refrigerant away from the TXV bulb, so that the bulb senses suction gas superheat instead of evaporating liquid temperature. In addition, the trap creates turbulence in the line as the vapor turns the corners. This helps to mix, or entrain, the accumulated oil, and assists in lifting it to the top of the riser.

When the compressor is at a higher level than the evaporator, it is more difficult to keep oil entrained with the vapor refrigerant leaving the evaporator, because gravity will affect the flow of oil. In such cases, the size of the suction line may be reduced to compensate.

Where system capacity is varied through compressor capacity control or some other arrangement, a short riser will usually be sized smaller than the remainder of the suction line to ensure oil return up the riser (*Figure 45*). Where system capacity is variable over a wide range, it may not be possible to find a pipe size for a single suction riser that will ensure oil return and still have a reasonable pressure during maximum conditions. In that case, a double suction riser (*Figure 46*) should

be used. The design of the double suction riser is such that under low demand, refrigerant flows up only one riser. Oil held in the trap blocks the other riser. At full load, when the compressor is operating on all cylinders, refrigerant and oil will travel up both risers.

When the compressor is on the same level or below the evaporator (*Figure 47*), a rise to at least the top of the evaporator must be placed in the suction line. This is to prevent liquid draining from the evaporator into the compressor during shutdown. The suction line loop can be left out and the piping simplified, if the system is on pump-down control.

Pump-down control is accomplished by placing a solenoid valve in the liquid line, and ahead of the metering device. A simple suction line, draining by gravity directly to the compressor and without traps, is allowed in this situation.

It is important to prevent liquid refrigerant from draining from the evaporator to the compressor during shutdown, but it is equally important to avoid unnecessary traps in the suction line near the compressor. Such traps collect oil, which can be

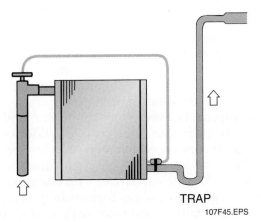

107F45.EPS

Figure 45 ◆ Reduced riser.

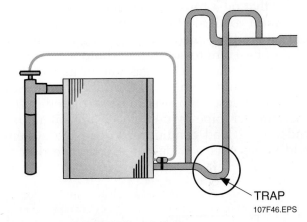

107F46.EPS

Figure 46 ◆ Double riser.

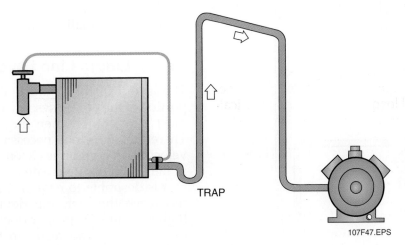

107F47.EPS

Figure 47 ◆ Inverted loop.

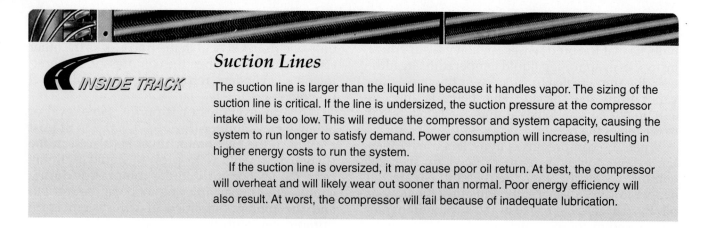

Suction Lines

The suction line is larger than the liquid line because it handles vapor. The sizing of the suction line is critical. If the line is undersized, the suction pressure at the compressor intake will be too low. This will reduce the compressor and system capacity, causing the system to run longer to satisfy demand. Power consumption will increase, resulting in higher energy costs to run the system.

If the suction line is oversized, it may cause poor oil return. At best, the compressor will overheat and will likely wear out sooner than normal. Poor energy efficiency will also result. At worst, the compressor will fail because of inadequate lubrication.

carried to the compressor in the form of slugs during startup, thereby causing serious damage. The part of the suction line near the compressor should be free-draining into the compressor.

11.3.0 Hot Gas Line

Considerations for the hot gas lines are similar to those for suction lines. They must also ensure that the pressure drop at full load is maintained within limits; that oil return or circulation is kept under minimum load conditions; and that refrigerant and oil are prevented from draining back to the head of the compressor during shutdown.

Figure 48 shows a hot gas line in its simplest form. It should pitch down from the compressor to the condenser. If a riser must be placed in the line, smaller pipe is used to ensure oil return under light load conditions. Where the system load varies over a wide range, a double riser is required.

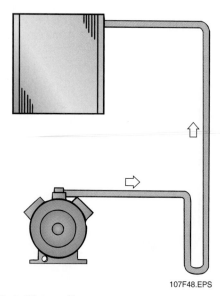

107F48.EPS

Figure 48 ◆ Hot gas line.

Since the hot gas line connects to the head of the compressor, provisions must be made to prevent oil or condensed refrigerant from flowing back through the line and into the compressor during shutdown. A loop to the floor between the compressor and the riser will normally provide an adequate reservoir to trap and hold the oil or condensed refrigerant.

If the compressor is located where its temperature can be lower than the temperature at the condenser or receiver, a check valve should be used. The preferred location for a muffler is in the downflow side of the hot gas loop, as close to the compressor as possible. If the muffler is placed in a horizontal section of the hot gas line, position it vertically so that the outlet connection comes off at the bottom to avoid trapping oil.

The piping between the condenser and the receiver (condensate line) must provide for the drainage of condensed refrigerant to the receiver and the venting of the gas generated in the receiver back to the condenser. This line should be large enough to allow gas formed in the receiver to flow back to the condenser without restricting the drainage of liquid refrigerant from the condenser. When the horizontal distance between the condenser and the receiver is more than six feet, a separate gas equalizer line is required.

11.4.0 Liquid Line Layout

The layout of the liquid line is the least critical portion of the piping system. Oil mixes freely with most refrigerants when they are in liquid form. Therefore, it is not necessary to provide high velocities in liquid lines to ensure oil return. Traps in the liquid do not create oil return problems.

It is desirable to have a slightly subcooled liquid reach the metering device at a sufficiently high pressure for proper operation. An excessive pressure drop can result in the loss of capacity at the metering device; a pressure drop without

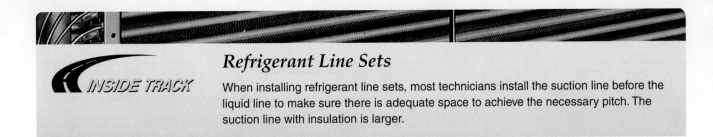

Refrigerant Line Sets

INSIDE TRACK

When installing refrigerant line sets, most technicians install the suction line before the liquid line to make sure there is adequate space to achieve the necessary pitch. The suction line with insulation is larger.

sufficient subcooling will cause some of the refrigerant to flash back into the vapor state. The pressure drop in liquid lines can be due to pipe friction, vertical rise, and accessories. Liquid pipe sizes are normally chosen for pressure drops; vertical rise in the liquid line is normally dictated by the job conditions. The total pressure drop through accessories should not exceed 4 psi.

11.5.0 Insulation

Liquid lines are not normally insulated, except where the surrounding temperature is higher than the liquid refrigerant. Hot gas lines are generally above the surrounding temperature and need only be insulated for personnel protection. Suction lines should be insulated to prevent condensation. Some heat absorption is desirable to evaporate any slop-over, but excessive heat gain by the suction gas must be avoided. Suction line insulation must be covered with a vapor barrier and weatherproofed when outdoors.

The three major considerations in refrigerant piping layout are compressor protection, oil return, and pressure drop. *Figure 49* illustrates the piping layouts and components by which these objectives are achieved.

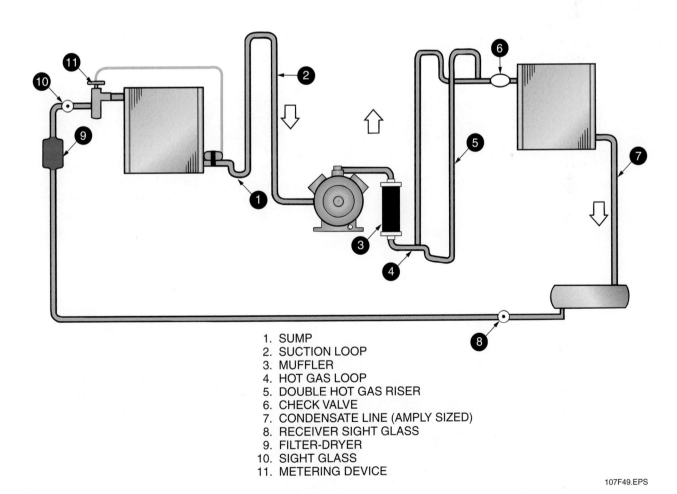

1. SUMP
2. SUCTION LOOP
3. MUFFLER
4. HOT GAS LOOP
5. DOUBLE HOT GAS RISER
6. CHECK VALVE
7. CONDENSATE LINE (AMPLY SIZED)
8. RECEIVER SIGHT GLASS
9. FILTER-DRYER
10. SIGHT GLASS
11. METERING DEVICE

107F49.EPS

Figure 49 ◆ Piping layout.

Summary

Air conditioning makes it possible to change the condition of the air in an enclosed area. It is a process that heats, cools, cleans, and circulates air, and controls its moisture content. The cooling portion of air conditioning depends on refrigeration. Refrigeration is also used for the preservation of food and in many industrial processes. The refrigeration cycle is based on the following concepts:

- Heat always flows from a warmer substance or location to a cooler substance or location.
- Heat must be added to or removed from a substance before a change in state can occur.
- The flow of a gas or liquid is always from a higher pressure area to a lower pressure area.
- The temperature at which a liquid or gas changes state is dependent on pressure.
- Cold is merely the absence of heat.
- Heat is always present in some degree.

Notes

1. The Fahrenheit scale is based on boiling water having a sea level temperature of _____ .
   a. 100°
   b. 180°
   c. 212°
   d. 459°

2. Five pounds of water heated to raise the temperature 2° requires _____ .
   a. 5 Btus
   b. 10 Btus
   c. 25 Btus
   d. 100 Btus

3. Sensible heat is heat that _____ .
   a. produces a change in state without a change in temperature
   b. changes a liquid to a vapor
   c. can be sensed by a thermometer
   d. changes a vapor to a liquid

4. Superheat is added _____ .
   a. after all of a solid changes to a liquid
   b. in changing vapor to liquid
   c. after all liquid has been changed to vapor
   d. in changing liquid to a vapor

5. The transfer of heat from one object to another by direct contact is called _____ .
   a. radiation
   b. change of state
   c. convection
   d. conduction

6. A ton of refrigeration is equal to _____ .
   a. 2,880,000 Btus per day
   b. 144 Btus per hour
   c. 12,000 Btus per hour
   d. 180 Btus per hour

7. Zero gauge pressure corresponds on the absolute scale to _____ .
   a. 10.7 psi
   b. 14.7 psi
   c. 27.4 psi
   d. 44.7 psi

8. The boiling temperature of a liquid will be lower as _____ .
   a. the pressure is increased
   b. the pressure is decreased
   c. it changes state

9. A gas or liquid always flows _____ .
   a. from a lower pressure to a higher pressure
   b. from a higher temperature to a lower temperature
   c. from a higher pressure to a lower pressure
   d. in a straight line

10. In a refrigeration system, the high-pressure, high-temperature vapor is converted into a high-pressure, high-temperature liquid by the _____ .
    a. compressor
    b. evaporator
    c. expansion (metering) device
    d. condenser

11. In a refrigeration system, the low-pressure, low-temperature liquid is converted into a low-pressure, low-temperature vapor by the _____ .
    a. compressor
    b. evaporator
    c. expansion (metering) device
    d. condenser

12. In a refrigeration system, the low-pressure, low-temperature vapor is converted into a high-pressure, high-temperature vapor by the _____ .
    a. compressor
    b. evaporator
    c. expansion (metering) device
    d. condenser

13. The low side of the refrigeration system includes the _____ .
    a. suction side (input) of the compressor
    b. muffler
    c. discharge side of the compressor
    d. condenser

14. Head pressure refers to the pressure in the _____ .
    a. low side of the system
    b. evaporator
    c. expansion (metering) device
    d. high side of the system

15. If a gas is superheated 15° to a temperature of 87°F, we can determine that its _____ .
    a. boiling point is 72°F
    b. freezing point is 32°F
    c. saturation point is 32°F
    d. saturation point is 102°F

16. The difference between a halocarbon and a fluorocarbon refrigerant is that _____ .
    a. there is no difference
    b. the fluorocarbon has fluorine in it while the halocarbon still has carbon
    c. a halocarbon never has chlorine in it while the fluorocarbon does
    d. the fluorocarbon always has fluorine in it but not all halocarbons do

17. The fluorocarbon refrigerants considered least harmful to the environment are the _____ .
    a. CFHs
    b. CFCs
    c. HCFCs
    d. HFCs

18. What is the color code for refrigerant recovery cylinders?
    a. Orange
    b. Gray
    c. Gray with a yellow top
    d. Yellow with a gray top

19. A compressor with a piston that travels back and forth in a cylinder is a _____ compressor.
    a. reciprocating
    b. rotary
    c. centrifugal
    d. scroll

20. The main purpose of a condenser is to _____ .
    a. store liquid refrigerant
    b. remove heat from the refrigerant
    c. remove water from the refrigerant
    d. add heat to the refrigerant

21. The main purpose of an evaporator is to _____ .
    a. store liquid refrigerant
    b. remove heat from the refrigerant
    c. remove water from the refrigerant
    d. add heat to the refrigerant

22. A TXV regulates the flow of refrigerant to the evaporator in order to maintain a constant _____ .
    a. subcooling
    b. superheat
    c. discharge pressure
    d. airflow

23. Components (accessories) most often found in the refrigeration system liquid line are _____ .
    a. receiver, sight glass/moisture indicator, filter-drier
    b. liquid-suction heat exchanger, suction line accumulator
    c. muffler, oil separator
    d. water pre-heater

24. A secondary control is used to _____ .
    a. regulate the cycle
    b. protect the cycle
    c. regulate and/or protect the cycle
    d. control conditions within the cycle

25. Horizontal piping runs used in refrigeration systems should _____ .
    a. pitch toward the compressor
    b. pitch away from the compressor
    c. pitch in the direction of flow
    d. be level

Trade Terms Quiz

1. A pressure measurement that represents pressure measured from zero pressure is known as _____.

2. Compounds containing carbon and hydrogen atoms are known as _____.

3. The movement of heat in the form of invisible rays is _____.

4. The total heat content of a refrigerant is called its _____.

5. If an object has less heat energy than another object, it is said to be _____.

6. The physical process in which heat is moved from one material to another by direct contact is _____.

7. If heat moves freely in a material, it is known as a(n) _____.

8. The heat gained in changing to or from a liquid to a gas is called the _____.

9. When a large amount of liquid refrigerant enters the compressor cylinder, it is called a(n) _____.

10. The transfer of heat from a space where it is not wanted to a space where it is not objectionable is known as _____.

11. The term used to define the pressure at sea level is _____.

12. A form of energy that raises the temperature of a substance is _____.

13. Chlorine, fluorine, and bromine are chemically related elements called _____.

14. The amount of heat needed to raise the temperature of one pound of water 1°F is called a(n) _____.

15. The transfer of heat by flow of a liquid or gas caused by a temperature difference is called _____.

16. Pressure that is measured in reference to atmospheric pressure is known as _____.

17. A halocarbon in which at least one of the hydrogen atoms has been replaced with fluorine is a(n) _____.

18. A hydrocarbon in which most of the atoms have been replaced with fluorine, chlorine, bromine, astatine, or iodine is called a(n) _____.

19. Heat that can be measured with a thermometer is called _____.

20. An electrical device made of two different metals that generates a current in response to a heat difference is called a(n) _____.

21. A device that changes resistance in response to a temperature change is called a(n) _____.

22. The amount of heat energy in a substance is expressed as _____.

23. The amount of heat, expressed as Btu/lb/°F, needed to raise the temperature of one pound of a substance 1°F is called a(n) _____.

24. The heat given up or removed from a gas in changing to a liquid is called _____.

25. A liquid or gas that picks up heat by evaporating at a low temperature and pressure and gives up heat by condensing at a higher temperature and pressure is a(n) _____.

26. A substance that is not a good conductor of heat is referred to as a(n) _____.

27. 12,000 Btuh is equal to a(n) _____.

28. The heat energy absorbed or rejected without changing temperature is known as _____.

29. The sensible heat added to a substance after it has fully vaporized is called _____.

30. The term that represents force per unit of area is _____.

31. The cooling of a liquid below its condensing temperature is called _____.

32. The return of liquid refrigerant to the compressor is called _____.

33. The heat gained or lost in changing to or from a solid is called _____.

Trade Terms

Absolute pressure	Fluorocarbons	Latent heat of condensation	Sensible heat
Atmospheric pressure	Gauge pressure	Latent heat of fusion	Slug
British thermal unit (Btu)	Halocarbons	Latent heat of vaporization	Specific heat
Cold	Halogens		Subcooling
Conduction	Heat	Pressure	Superheat
Conductor	Heat content	Radiation	Thermistor
Convection	Hydrocarbons	Refrigerant	Thermocouple
Enthalpy	Insulators	Refrigeration	Ton of refrigeration
Floodback	Latent heat		Total heat

Cedric Brown

Commercial Service Technician
Total Comfort Service Center, Inc.
Columbia, South Carolina

How did you choose a career in HVAC?
I chose the HVAC field in the 11th grade and attended the Heyward Career Center HVAC-R program. I was hired as an intern at Total Comfort during my senior year.

What kinds of work have you done in your career?
I've been a service apprentice to senior technicians and a preventive maintenance technician. Now I'm a commercial service technician.

What do you like about your present job?
It's never the same day to day. There's always different work to do. There is freedom to learn how to do the job better. I've also become a spokesman for Total Comfort at public events, such as school board meetings, teacher and guidance counselor advisory meetings, and career days at middle schools and high schools.

What factors have contributed most to your success?
I owe a lot to my parents. They helped me become goal oriented and I continue to have a strong desire to continue to grow my skills in this field. Right now, I'm attending the University of South Carolina night program for a mechanical engineering degree while working days for Total Comfort.

What types of training have you been through?
I studied HVAC-R for two years in high school at the Heyward Career Center. Then I obtained an associate's degree in HVAC from Midlands Technical College. Total Comfort has provided me with four years of on-the-job technical training.

What advice would you give to those who are new to the HVAC field?
Be ready to learn. Set high goals for yourself.

Trade Terms
Introduced in This Module

Absolute pressure: Positive pressure measurements that start at zero (no pressure at all). Also gauge pressure plus the pressure of the atmosphere (14.7 psi at sea level at 70°F).

Atmospheric pressure: The pressure exerted on all things on the earth's surface as a result of the weight of the atmosphere. It is 14.7 psi at sea level at 70°F.

British thermal unit (Btu): The amount of heat needed to raise the temperature of one pound of water one degree Fahrenheit.

Cold: A relative term for temperature. Cold means having less heat energy than another object against which it is being compared.

Conduction: A means of heat transfer in which heat is moved from one material to another by means of direct contact.

Conductor: A material in which the transfer of heat by conduction occurs easily.

Convection: The transfer of heat by the flow of liquid or gas caused by a temperature differential.

Enthalpy: The total heat content (sensible and latent) of a refrigerant or other substance.

Floodback: Refrigerant returning to the compressor in the liquid state.

Fluorocarbons: Halocarbons in which at least one or more of the hydrogen atoms has been replaced with fluorine.

Gauge pressure: The pressure measured on a gauge, expressed as pounds per square inch gauge (psig) or inches of mercury vacuum (in Hg vac.). Also pressure measurements that are made in comparison to atmospheric pressure.

Halocarbons: Hydrocarbons, like methane and ethane, that have most or all of their hydrogen atoms replaced with the elements fluorine, chlorine, bromine, astatine, or iodine.

Halogens: Substances containing chlorine, fluorine, bromine, astatine, or iodine.

Heat: A form of energy. It causes molecules to be in motion and raises the temperature of a substance. Other forms of energy like electricity, light, and magnetism deteriorate into heat.

Heat content: The amount of heat energy contained in a substance. Measured in Btus.

Hydrocarbons: Compounds containing only hydrogen and carbon atoms in various combinations.

Insulators: Materials that resist heat transfer by conduction.

Latent heat: The heat energy absorbed or rejected when a substance is changing state (solid to liquid, liquid to gas, or vice versa) but maintaining its measured temperature.

Latent heat of condensation: The heat given up or removed from a gas in changing back to a liquid state (steam to water).

Latent heat of fusion: The heat gained or lost in changing to or from a solid (ice to water or water to ice).

Latent heat of vaporization: The heat gained in changing from a liquid to a gas (water to steam).

Pressure: Force per unit of area.

Radiation: The movement of heat in the form of invisible rays or waves, similar to light.

Refrigerant: A liquid or gas that picks up heat by evaporating at a low temperature and pressure, and gives up heat by condensing at a higher temperature and pressure.

Refrigeration: The transfer of heat from a space or object where it is not wanted to a space or object where it is not objectionable.

Sensible heat: Heat that can be measured by a thermometer or sensed by touch. The energy of molecular motion.

Slug: A large amount of liquid refrigerant and/or oil entering a compressor cylinder.

Specific heat: The amount of heat required to raise the temperature of one pound of a substance one degree Fahrenheit. Expressed as Btu/lb/°F.

Subcooling: Cooling a liquid below its condensing temperature.

Superheat: The measurable heat added to the vapor or gas produced after a liquid has reached its boiling point and completely changed into a vapor.

Thermistor: A semiconductor device that changes resistance with a change in temperature.

Thermocouple: A device made of two different metals that generates electricity when there is a difference in temperature from one end to the other.

Ton of refrigeration: Large unit for measuring the rate of heat transfer. One ton is defined as 12,000 Btus per hour or 12,000 Btuh.

Total heat: Sensible heat plus latent heat.

This module is intended to present thorough resources for task training. The following reference works are suggested for further study. These are optional materials for continued education rather than for task training.

Air Conditioning Systems, Principles, Equipment, and Service, Latest Edition. Upper Saddle River, NJ: Prentice Hall.

Basic Refrigeration (Slides and Student Handbook), Latest Edition. York, PA: York International Corporation, Publications Distribution Center.

General Training Air Conditioning (Fundamentals)— GTAC-I, Latest Edition. Syracuse, NY: Carrier Corporation, Literature Services.

NCCER makes every effort to keep these textbooks up-to-date and free of technical errors. We appreciate your help in this process. If you have an idea for improving this textbook, or if you find an error, a typographical mistake, or an inaccuracy in NCCER's Contren® textbooks, please write us, using this form or a photocopy. Be sure to include the exact module number, page number, a detailed description, and the correction, if applicable. Your input will be brought to the attention of the Technical Review Committee. Thank you for your assistance.

Instructors – If you found that additional materials were necessary in order to teach this module effectively, please let us know so that we may include them in the Equipment/Materials list in the Annotated Instructor's Guide.

Write: Product Development and Revision
National Center for Construction Education and Research
3600 NW 43rd St., Bldg. G, Gainesville, FL 32606

Fax: 352-334-0932

E-mail: curriculum@nccer.org

Craft

Module Name

Copyright Date

Module Number

Page Number(s)

Description

(Optional) Correction

(Optional) Your Name and Address

Introduction to Heating
03108-07

03108-07
Introduction to Heating

Topics to be presented in this module include:

Overview

Just about every home in the United States has some type of heating system. Gas-fired forced-air furnaces are the most common, but oil-fired furnaces and electric heat are also used. Installing and servicing furnaces is a big responsibility. Because flame and combustible fuels are involved, there is a potential for fire or explosion. In addition, furnaces that develop internal leaks can kill building occupants. On the other hand, there are tens of millions of furnaces being safely used today. If furnaces are properly installed according to the manufacturer's instructions, and periodically inspected and serviced by qualified technicians, they will operate satisfactorily for many years.

Objectives

When you have completed this module, you will be able to do the following:

1. Explain the three methods by which heat is transferred and give an example of each.
2. Describe how combustion occurs and identify the byproducts of combustion.
3. Identify the various types of fuels used in heating.
4. Identify the major components and accessories of an induced draft and condensing gas furnace and explain the function of each component.
5. State the factors that must be considered when installing a furnace.
6. Identify the major components of a gas furnace and describe how each works.
7. With supervision, use a manometer to measure and adjust manifold pressure on a gas furnace.
8. Identify the major components of an oil furnace and describe how each works.
9. Describe how an electric furnace works.
10. With supervision, perform basic furnace preventive maintenance procedures such as cleaning and filter replacement.

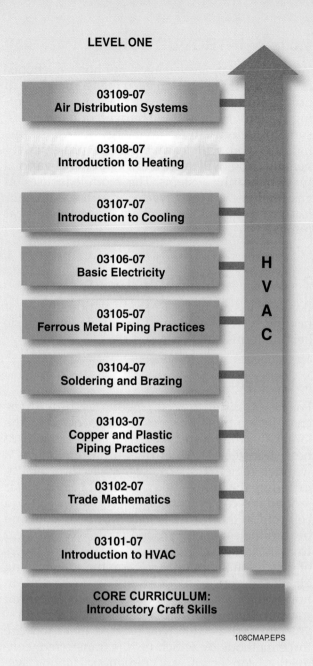

03109-07
Air Distribution Systems

03108-07
Introduction to Heating

03107-07
Introduction to Cooling

03106-07
Basic Electricity

03105-07
Ferrous Metal Piping Practices

03104-07
Soldering and Brazing

03103-07
Copper and Plastic Piping Practices

03102-07
Trade Mathematics

03101-07
Introduction to HVAC

CORE CURRICULUM:
Introductory Craft Skills

H V A C

108CMAP.EPS

Trade Terms

Annual Fuel Utilization Efficiency (AFUE)
Atomize
Combustion
Condensing furnace
Electrode
Flame rectification
Heat exchanger
Hot surface igniter
Induced-draft furnace
Infiltration
Manometer

Natural-draft furnace
Oil burner
Orifice
Piezoelectric
Primary air
Redundant gas valve
Relative humidity
Safety pilot
Secondary air
Standing pilot
Spud
Thermocouple

Prerequisites

Before you begin this module, it is recommended that you successfully complete *Core Curriculum*; and *HVAC Level One*, Modules 03101-07 through 03107-07.

This course map shows all of the modules in *HVAC Level One*. The suggested training order begins at the bottom and proceeds up. Skill levels increase as you advance on the course map. The local Training Program Sponsor may adjust the training order.

Required Trainee Materials

1. Paper and pencil
2. Appropriate personal protective equipment

1.0.0 ◆ INTRODUCTION

We rely on heating systems in our homes, schools, and places of business to keep the temperature in our comfort range. There are many types of heating systems. Most of them burn oil or natural gas; some use electricity as an energy source. In this module, you will learn the basic principles of heating. You will also learn about the various types of heating systems. This module focuses on gas-fired and oil-fired warm air furnaces. In later lessons, you will learn about boilers and heat pumps.

2.0.0 ◆ HEATING FUNDAMENTALS

Before you can begin to study various types of heating systems, you must first understand basic heating terms and processes.

2.1.0 Heat Transfer

For heat transfer to occur, there must be a difference in temperature between two objects. The larger the difference, the greater the amount of heat transfer. Heat always transfers from a warmer region to a cooler region. Any object, including the human body, gives off heat if the air around it is cooler than the object. The reason you feel cold in the winter is that your body is losing heat to the cooler air around it. To keep the heat from escaping, you have two choices: wear layers of insulation (clothing), or warm the surrounding air so that your body stops giving away heat. In hot weather, your body feels hot because it cannot transfer as much heat to the surrounding air.

In the last module, you learned that there are three methods by which heat is transferred: conduction, convection, and radiation. Heating systems use one or more of these methods to warm the air in the conditioned space (see *Figure 1*).

2.1.1 Conduction

Conduction is the flow of heat from one part of a material to another part or to a substance in direct contact with it. The rate at which a material transmits heat is known as its conductivity. The amount of heat transmitted by conduction between two surfaces is determined by the surface area, thickness, and conductivity of the materials, and the temperature difference between the two. An example of conduction is the transfer of heat from the gas burners to the **heat exchangers** in a furnace (*Figure 1A*).

2.1.2 Convection

Convection is air motion due to the warmer portions rising and the denser, cooler portions falling. For convection to occur, there must be a difference in temperature between the source of heat and the surrounding air. The greater the difference in temperature, the greater the movement of air by convection. The greater the movement of air, the greater the transfer of heat.

Hot water heating elements, such as the one shown in *Figure 1B*, rely on convection. The room air at the baseboard heater is heated, causing it to rise. As it rises, it gives up its heat to the cooler room air and becomes colder. The cold air then falls to the floor and the process repeats.

2.1.3 Radiation

Radiation is the transfer of heat through space by wave motion. Heat passes from one object to another without warming the space between them. The amount of heat transferred by radiation depends upon the area of the radiating body, the temperature difference between the two objects, and the distance between the heat source and the object being heated.

The heat from the sun or a fireplace is radiated heat (*Figure 1C*). Another example of radiated heat is a plug-in electric space heater. The heat from radiation heat sources tends to be intense near the source and to heat only solid objects that are directly in its path. When you are sitting at a campfire, for example, the front of your body can be hot while your back is cold.

One of the most common and effective applications of radiant heat is the gas-fired radiant tube

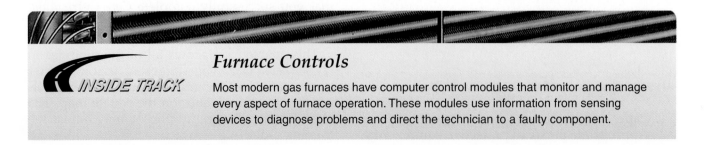

Furnace Controls

INSIDE TRACK

Most modern gas furnaces have computer control modules that monitor and manage every aspect of furnace operation. These modules use information from sensing devices to diagnose problems and direct the technician to a faulty component.

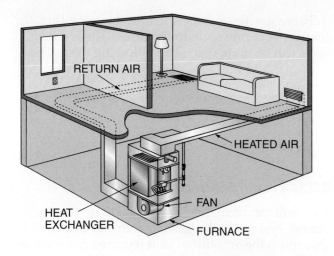

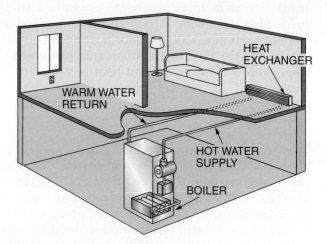

(A) CONDUCTION

(B) CONVECTION

(C) RADIATION

108F01.EPS

Figure 1 ◆ Heat transfer methods.

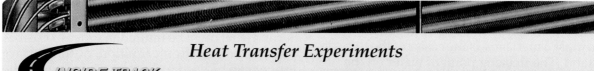

Heat Transfer Experiments

INSIDE TRACK

Here are some simple experiments you can do to test heat transfer theories:

Conduction: Take a piece of copper pipe and attach a thermometer to one end. When the temperature reading has stabilized, apply heat to the other end of the pipe and watch the temperature rise.

Convection: Obtain a light plastic bag. While holding the bag upside down, fill it with hot air from a hair dryer. Let the bag go and watch it rise. As the air cools, the bag will begin to descend.

Radiation: Obtain two thermometers. Place one in the shade and one in direct sunlight. Note how much higher the temperature reading is on the thermometer in direct sunlight.

heating system in which heat is obtained from heated liquid flowing through tubes embedded in floors or ceilings. Radiant floor heating is gaining popularity in both residential and commercial applications. In-floor radiant heat provides both radiant and convective heating.

2.1.4 Humidity

Heat transfer is affected by the moisture content of the air, which is known as humidity. You will most often hear the term **relative humidity**. This term describes the capacity of air to hold water. A relative humidity of 50 percent means that the air contains half the moisture it is capable of holding. The body loses heat more rapidly when the humidity is low. Therefore, homes in many climates are equipped with humidifiers that add moisture to the air during winter months when the air is dry. Humidifiers are frequently added to forced-air furnaces.

2.2.0 Temperature

Temperature is defined as the degree of hotness or coldness measured on a numerical scale. It can be accurately measured using a thermometer. There are two thermometer scales in frequent use. One type is the inch-pound system known as the Fahrenheit scale; the other is the metric system called the Celsius (centigrade) scale. It is important to understand both of these scales, because some temperature information is given in Fahrenheit degrees and some in Celsius (for example, 32°F or 0°C).

NOTE

The Celsius scale is required on many government projects.

The Celsius thermometer, though not generally familiar to most Americans, is the easier of the two to understand. The freezing point of water on the Celsius scale is 0° and the boiling point of water is 100°. The freezing point of water on the Fahrenheit scale is 32° and the boiling point of water is 212°. To reduce confusion about which scale is involved, a capital C is placed after a Celsius degree indication and a capital F is placed after the number of Fahrenheit degrees.

There are times when it may be necessary to change from one scale to the other. These conversions can be done using the following formulas:

$$\text{Degrees Celsius} = \tfrac{5}{9} \times (°F - 32°)$$
$$\text{Degrees Fahrenheit} = (\tfrac{9}{5} \times °C) + 32°$$

For example:

$$32°F = 0°C\ [\tfrac{5}{9} \times (32 - 32) = \tfrac{5}{9} \times 0 = 0]$$

2.3.0 Heat Measurement

The unit of measurement of heat in the inch-pound system is the British thermal unit (Btu). One Btu is the amount of heat required to raise the temperature of one pound of water 1°F.

It is also important to be aware of the rate of heat change, that is, how fast or how slowly a piece of heating equipment produces heat. The term for this is Btus per hour (Btuh). For example, the capacity of a furnace may be stated as 110,000 Btuh.

Furnaces are selected based on their Btu rating. Actually, there are two ratings. The input rating is the actual amount of heat in Btus produced by the furnace, while the output rating is the number of usable Btus. The difference is a function of the efficiency rating of the furnace. For example, if a furnace with an input rating of 150,000 has an efficiency rating of 90 percent, its output rating will be 135,000 Btus, or 90 percent of 150,000.

2.4.0 Combustion

Combustion is the burning of fuel to create heat. During combustion, oxygen is combined with a fuel to release the stored energy in the form of heat. There are three conditions necessary for combustion to take place.

* First, there must be fuel. The fuel can be gas, such as natural gas; liquid, such as fuel oil; or solid, such as coal. Two elements that all fuels have in common are carbon and hydrogen.
* Second, fuel must be ignited in order to burn. A pilot burner, electronic ignition, or **hot surface igniter** can be used to ignite a gas burner, while an electric spark is used to ignite fuel oil.
* Third, oxygen must be present for the burning of fuel to take place.

Combustion is a chemical reaction between fuel, heat, and oxygen; therefore, all three components must be present in order for combustion to

---

occur. Furthermore, the fuel must be in a gaseous state to burn. Natural gas, because it is a gas, will burn in its natural state as long as heat and oxygen are present. Fuel oil is **atomized** (converted to a fine spray) before it is burned.

2.4.1 Complete Combustion

Complete combustion takes place when carbon combines with oxygen to form carbon dioxide (CO_2). Carbon dioxide is nontoxic and can be exhausted to the atmosphere. Hydrogen combines with oxygen to form water vapor (H_2O), which is also harmless and can be exhausted to the atmosphere.

2.4.2 Incomplete Combustion

Incomplete combustion results from a lack of oxygen and causes undesirable products to form. These undesirable products are carbon monoxide, pure carbon (soot), and aldehyde. Both carbon monoxide and aldehyde are toxic. Soot causes coating of the heating surfaces of the furnace and reduces heat transfer.

Fuel-burning devices must be adjusted so that complete combustion of the fuel always takes place. Sufficient air must be provided for proper combustion to take place and to eliminate the hazards of incomplete combustion. Furnaces are normally adjusted to provide from 5 to 50 percent excess air in order to guard against the possibility of incomplete combustion.

2.4.3 Combustion Efficiency

When fuel is burned in a furnace, a certain amount of heat is lost in the hot gases, known as flue gases, that are vented through the chimney. This function is necessary for disposal of the products of combustion, but the loss must be kept down to allow the furnace to operate at its peak efficiency. Air entering the furnace at room temperature or lower is heated to flue gas temperatures that range from 100°F to 600°F, depending upon the design and adjustments of the furnace.

If the amount of heat lost is 20 percent, the furnace efficiency would be 80 percent. Charts are available for determining the efficiency of a furnace based on the temperature and the carbon dioxide content of the flue gases. Knowing the efficiency of the furnace makes it possible for a technician to calculate the furnace output.

The current industry standard for defining furnace efficiency is **Annual Fuel Utilization Efficiency (AFUE)**. AFUE takes into account operating efficiency as well as combustion efficiency. The *National Appliance Energy Conservation Act of 1987* requires that all gas furnaces built after 1991 have an AFUE of no less than 78 percent. Most existing

Carbon Monoxide and Flue Gases

Carbon monoxide is a deadly gas that results from incomplete combustion of the fuel supply to the furnace burners. Incomplete combustion can be caused by insufficient air to the burners or excessive cooling of the flame, which can result from the flame touching the heat exchanger (impingement).

Aldehydes will also be emitted by incomplete combustion. While carbon monoxide is odorless, aldehydes have a distinct sharp odor that will irritate the membranes of the nose and throat. If aldehydes are present, you can be pretty sure that carbon monoxide is also present. However, the absence of aldehydes does not mean that carbon monoxide is absent.

The products of combustion that are vented to the outdoors include carbon dioxide, nitrogen, and hydrogen. If the furnace heat exchanger or vent pipe is damaged or corroded, carbon dioxide and other flue gases may enter the building. While these gases are not poisonous themselves, they do replace air in the building. Eventually, occupants may suffocate from lack of air, or the reduced air supply will result in incomplete combustion, producing deadly carbon monoxide.

For all these reasons, proper venting of the furnace, annual inspection of the heat exchangers and flue, and proper adjustment of the flame are essential parts of the installation and service technician's job.

One way you can help customers is by encouraging the installation of carbon monoxide detectors.

High-Efficiency Furnaces

Condensing gas furnaces extract so much heat from the products of combustion that they can be vented with plastic (PVC) pipe. During furnace operation, the vent pipe will be cool to the touch, unlike standard-efficiency and mid-efficiency furnaces.

When installing a condensing gas furnace to replace a gas furnace that is common-vented with a gas water heater, the venting arrangement for the water heater must be reconsidered. For example, the existing chimney or vent will now only be venting the water heater. In all probability, this existing vent will now be oversized for the water heater, which may prevent the water heater from venting properly. This situation can be corrected by lining the chimney with a properly sized liner or by installing a new and smaller metal vent.

PVC FLUE VENT

108SA01.EPS

natural-draft furnaces that rely on convection for combustion air have an AFUE of less than 78 percent. With the addition of special features such as electronic ignition and vent dampers, it is possible to bring a natural-draft furnace up to an AFUE of 80 percent. Natural-draft furnaces have been replaced by induced-draft furnaces, which use a fan to draw in and exhaust combustion air. Induced-draft furnaces have an AFUE that ranges from 78 to 85 percent, depending on the kinds of efficiency options installed. Condensing gas furnaces have the highest efficiency—greater than 90 percent. These furnaces use a condensing coil as a secondary heat exchanger to extract the latent heat from the combustion by-products before they are vented outdoors.

2.4.4 Flames

The type of flame and the intensity with which it burns have a direct relationship to the efficiency of the heating unit. Pressure-type oil burners burn with a yellow flame, while gas burners burn with a blue flame that has a slight orange tip. The difference is mainly due to the manner in which air is mixed with the fuel.

A yellow flame in a gas burner denotes incomplete combustion. In this case, the burner should be checked. A blue flame is produced when approximately 50 percent of the air (primary air) is mixed with the gas before ignition. The balance of air, called secondary air, is supplied during combustion to the exterior of the flame. Improper gas flames are the result of inefficient or incomplete combustion. They can be caused by too much primary air, a lack of secondary air, or by contact between the flame and a cool surface.

2.5.0 Fuels

Fuels are available in three forms: gases, liquids, and solids. Gases include natural gas, manufactured gas, and liquified petroleum (LP). Liquids include fuel oils. There are six grades of fuel oil, with No. 2 being the most common. Solids include coal and wood. This module focuses on gas and oil, which are the types most likely to be encountered in the HVAC trade. *Table 1* shows the heating values of common fuels.

Table 1 Heating Values of Common Fuels

Fuel	Heat Released
Coal	
Bituminous	12,000 to 15,000 Btus/lb
Anthracite	13,000 to 14,000 Btus/lb
Oil*	
Grade 1	137,000 Btus/gallon
Grade 2	140,000 Btus/gallon
Grade 4**	141,000 Btus/gallon
Grade 5	148,000 Btus/gallon
Grade 6	152,000 Btus/gallon
Gas	
Natural***	900 to 1,200 Btus/cubic ft
Manufactured	500 to 600 Btus/cubic ft
Liquified Petroleum (LP)	1,500 to 3,200 Btus/cubic ft
Wood	6,200 Btus/lb (avg)

*Grades are determined by the American Society for Testing and
 Materials (ASTM).
**This grade is not commonly used.
***Check with local gas company for specific values.

2.5.1 Gaseous Fuels

There are three types of gaseous fuels: natural gas, manufactured gas, and liquified petroleum.

- *Natural gas* – Natural gas comes from the earth and usually accumulates in the upper part of oil wells. It is colorless and nearly odorless. An odorant, such as a sulfur compound, is added so that leaks can be detected. The content of this gas varies by locality and has a bearing on the Btu content, which varies from 900 to 1,200 Btus/cubic ft depending on the locality, but is usually in the range of 1,000 to 1,050 Btus/cubic ft. The chief component of natural gas is methane. It also includes other hydrocarbons.
- *Manufactured gas* – Manufactured gases are combustible gases that are normally produced from solid or liquid fuel and are used mostly in industrial processes. They are produced mainly from coal, oil, and other hydrocarbons, and are comparatively low in Btus per cubic foot (500 to 600). They are not considered to be economical space-heating fuels.
- *Liquified petroleum* – LP is a by-product of the oil refining process. It is stored in liquid form, but is vaporized when used. There are two types of LP gas, propane and butane. Propane is more useful as a space-heating fuel because it boils at −40°F and can be readily vaporized for heating in a northern climate. Butane vaporizes at about 32°F. Propane has a heating value of about 2,500 Btus per cubic foot, whereas butane has a heating value of approximately 3,200 Btus

per cubic foot. LP gas is usually propane with a small amount of butane added. When LP gas is used as a heating fuel, the equipment must be designed or modified for its use.

WARNING!

Propane and butane vapors are considered to be more dangerous than those of natural gas because they have a higher specific gravity. The vapor is heavier than air and tends to accumulate in low spots and near the floor, increasing the danger of an explosion at ignition.

LP gas is an alternative in areas where natural gas is not available. Manufacturers of gas appliances make LP conversion kits to adapt natural gas furnaces for LP. However, the use of LP gas is heavily regulated by state and local laws in some locales.

2.5.2 Fuel Oils

Fuel oils are rated according to their Btu per gallon content and the American Petroleum Institute (API) gravity. The API gravity is an index related to the heating values for standard grades of fuel oil. There are six common grades of oil: Nos. 1, 2, 4, 5 (light), 5 (heavy), and 6. The lighter-weight oils have a higher API gravity.

- *Grade 1* – A light-grade distillate for use in vaporizing-type oil burners.
- *Grade 2* – A heavier distillate for use in domestic pressure-type oil burners.
- *Grade 4* – A light residue or heavy distillate used for higher-pressure commercial oil burners.
- *Grade 5* (light) – A medium-weight, residual-type fuel used in commercial oil burners that are specifically designed for it.
- *Grade 5* (heavy) – A residual-type fuel for commercial oil burners. Usually requires preheating.
- *Grade 6* – A heavy residue used for commercial burners. Preheating is necessary in the tank to permit pumping; additional preheating is needed at the burner to convert the fuel into a fine spray or atomize it. Grade 6 is also known as Bunker C.

Grade 1 fuel oil (kerosene) has a slightly lower heat content than Grade 2 fuel oil. In northern climates, Grade 1 is used if the oil is to be stored in an outside tank. Grade 2 fuel oil has a tendency to thicken when exposed to cold, which makes it hard to pump from the storage tank to the burner. Fuel oil must be atomized in order to burn.

3.0.0 ◆ FORCED-AIR FURNACES

In a forced-air furnace, cooled return air from the space being heated is passed over a heat exchanger where heat is transferred to the air before the air is returned to the space through the supply ductwork. In most furnaces of this type, a fuel such as natural gas or heating oil is burned to create the heat that warms the heat exchangers.

In a fuel-burning furnace, the products of combustion are exhausted to the atmosphere through the flue passages and the vent system. In an electric furnace, air passes directly over the heated elements without the use of a heat exchanger, since no products of combustion are formed. A forced-air furnace uses a fan to move the air over the heat exchanger and to circulate the air through the distribution system (ductwork). If cooling is installed with a forced-air furnace, the cooling coil is normally placed in the airstream at the outlet of the furnace.

3.1.0 Types of Forced-Air Furnaces

There are four different designs of forced-air furnaces in common use. Each requires a different arrangement of the basic components.

* The upflow design (*Figure 2*) is used in the basement or in a first-floor equipment room. The blower is located below the heat exchanger. Air enters at the bottom or lower sides of the unit and exits through the top into the duct plenum.

INSIDE TRACK

Blower Speeds

It is common practice to use a forced-air furnace as the air handler for a split system cooling application. A cooling coil is added to the furnace as shown here. Ideally, the blower for a furnace or air handler that supplies cooling as well as heat will operate at different speeds for the two modes. A higher speed is used for cooling than for heating because larger quantities of air are required for cooling.

If the furnace is being adapted to provide cooling as well as heating, it may be necessary to replace the existing blower with a two-speed blower and a control circuit that automatically selects the fan speed based on the mode of operation. Modern furnaces are designed with that need in mind, so the fan installed at the factory is suitable for cooling as well as heating.

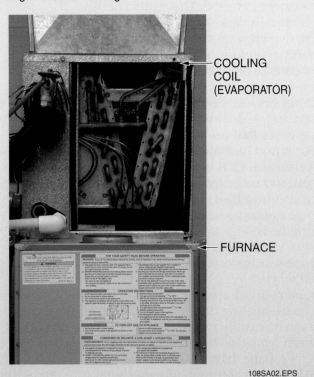

COOLING COIL (EVAPORATOR)

FURNACE

108SA02.EPS

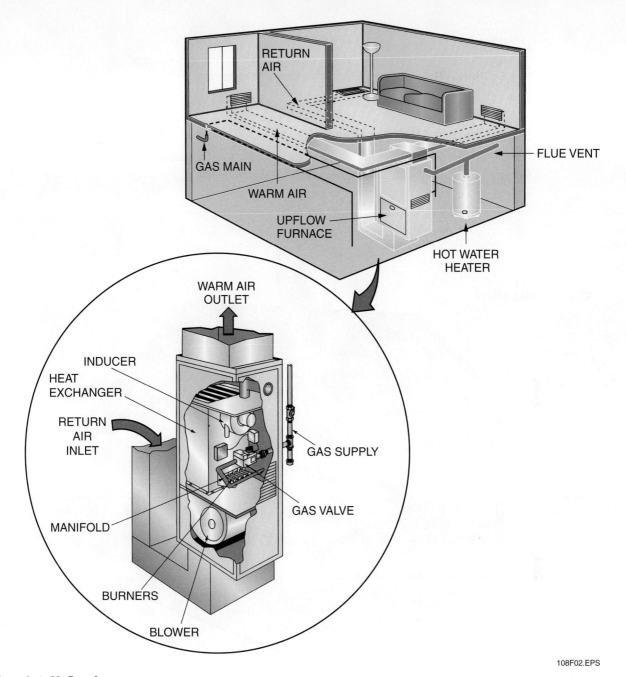

Figure 2 ◆ Upflow furnace.

108F02.EPS

- The horizontal furnace (*Figure 3*) is used in attics or crawlspaces where the height of the furnace must be kept at a minimum. Air enters at one end of the unit through the blower and filter compartment and is forced horizontally over the heat exchanger, exiting at the opposite end.

NOTE

When high-efficiency furnaces are installed in attics, it may be necessary to provide freeze protection.

- The low-boy furnace (*Figure 4*) occupies more floor space and is lower in height than the upflow design. It is best suited for basement installations with limited headroom. The blower is placed alongside the heat exchanger, the return air plenum is built above the blower compartment, and the warm air outlet or heat supply plenum is built above the heat exchanger.
- The downflow or counterflow furnace (*Figure 5*) is used in houses with an under-the-floor distribution system. The blower is located above the heat exchanger and the return air plenum is connected to the top of the blower area. The warm air supply plenum is connected to the bottom of the enclosure cabinet.

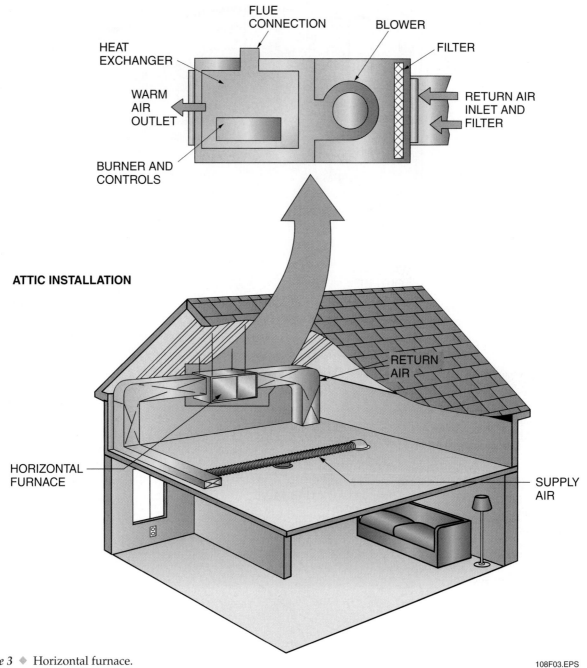

ATTIC INSTALLATION

FLUE CONNECTION

BLOWER

FILTER

HEAT EXCHANGER

WARM AIR OUTLET

RETURN AIR INLET AND FILTER

BURNER AND CONTROLS

RETURN AIR

HORIZONTAL FURNACE

SUPPLY AIR

Figure 3 ◆ Horizontal furnace.

108F03.EPS

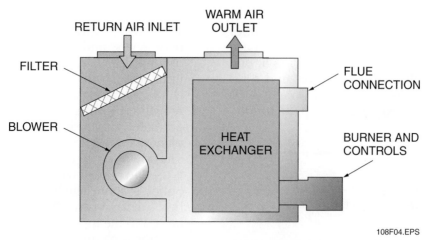

RETURN AIR INLET

WARM AIR OUTLET

FILTER

FLUE CONNECTION

BLOWER

HEAT EXCHANGER

BURNER AND CONTROLS

108F04.EPS

Figure 4 ◆ Low-boy furnace.

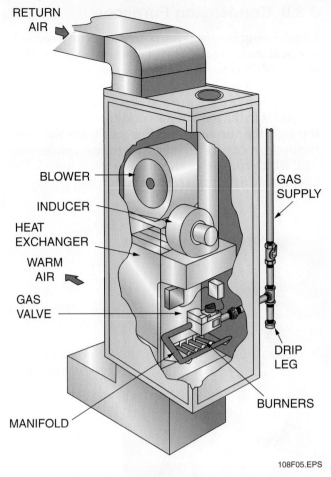

RETURN
AIR →

BLOWER

INDUCER

HEAT
EXCHANGER

WARM
AIR

GAS
VALVE

MANIFOLD

GAS
SUPPLY

DRIP
LEG

BURNERS

108F05.EPS

Figure 5 ◆ Downflow furnace.

Most furnace manufacturers offer a multi-poise furnace, which can be installed in some or all of the listed configurations. Check the manufacturer's instructions, because a slight modification is usually necessary in order to change from the design configuration. Failure to precisely follow the manufacturer's installation instructions can be hazardous because flue gas venting and condensate drainage can be affected by improper installation. Improper installation can lead to hazardous conditions for occupants or to furnace failure and expensive rework later.

3.2.0 Heat Exchangers

The heat exchanger is the part of the furnace where combustion takes place. It is usually made of a stamped or rolled aluminized steel product. There are several types of heat exchangers. The types shown in A, B, and C of *Figure 6* are commonly found in gas-fired furnaces. In these three types, the burners fire directly into the heat exchanger and the hot flue gases flow through the heat exchanger and out the other end into a collector box that directs the gases to the vent system.

The heat exchangers shown in D and E of *Figure 6* are typical of those found in oil-fired furnaces. The cylindrical heat exchanger (D) is a primary surface. The primary surface is in direct contact with the flame; the secondary surface (E) extracts

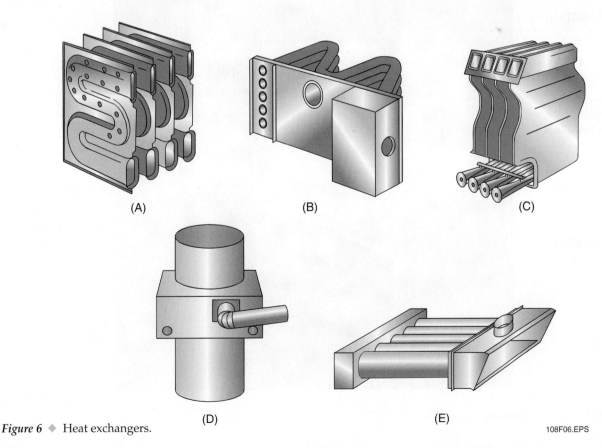

(A)

(B)

(C)

(D)

(E)

Figure 6 ◆ Heat exchangers.

108F06.EPS

heat from the hot flue gases. Oil does not burn as cleanly as gas. Therefore, the long, thin heat exchangers used for gas heating would get plugged up if used in oil-fired furnaces.

The number of heat exchanger sections depends on the furnace capacity. In a gas furnace, each section has its own burner. A low-capacity furnace might have two sections fed by two burners; a high-capacity furnace might have four or more sections. The heat exchanger sections terminate in a collector box that directs the flue gases into the flue pipe.

3.3.0 Condensing Furnaces

Condensing furnaces are equipped with a secondary heat exchanger made of stainless steel or plain steel with a corrosion-resistant coating that often looks like a refrigeration condensing coil. A condensing furnace is shown in *Figure 7*. The condensing coil is in the path of the conditioned air flowing through the furnace. The flue gas from the combustion process is piped through the condensing coil, which condenses the flue gas into a liquid. The latent heat removed by the condensing process is

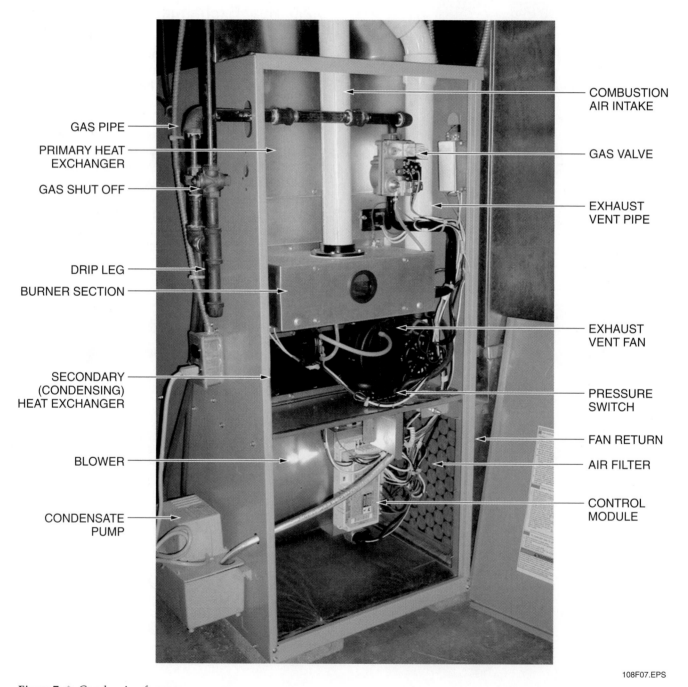

108F07.EPS

Figure 7 ◆ Condensing furnace.

transferred to the conditioned air. Condensing furnaces can reach efficiencies of 95 percent and more. Even though they may be more expensive initially, they save money over the long term.

3.4.0 Fans and Motors

The fan and fan motor are responsible for circulating the air through the heating system.

3.4.1 Induced-Draft Fan

As previously mentioned, the induced-draft fan is used on products in the mid- and high-efficiency classes. This fan draws combustion air into the

heat exchangers to support combustion. It also forces the combustion products out through the vent system. *Figure 8* shows the flow of combustion air and flue gases.

3.4.2 Blower

Much of the successful performance of a heating system depends upon the proper operation of the fan and fan motor, called the blower. The fan and motor circulate the air through the system, obtaining it from return air, and forcing it over the heat exchanger and through the supply distribution system to the space to be heated. Outside air can be introduced into the return air, but this is not

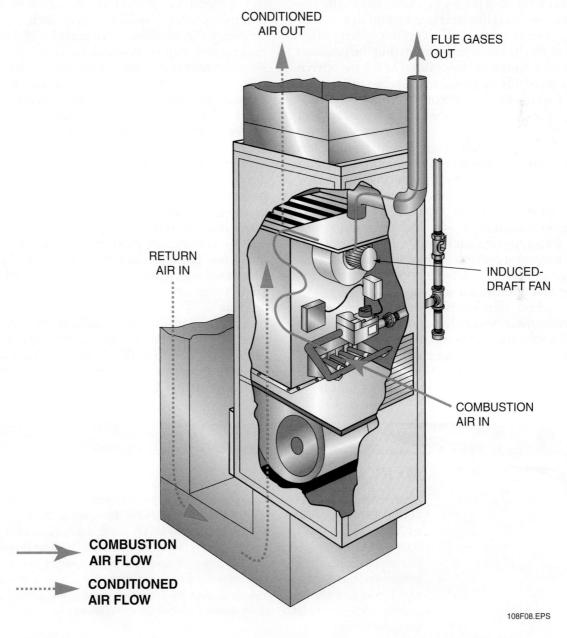

108F08.EPS

Figure 8 ◆ Furnace air flow.

always the case. Centrifugal fans with forward-curved blades are generally used. Air enters through both ends of the wheel and is compressed at the outlet by centrifugal force.

The volume of air is measured in cubic feet per minute (cfm).

Fans and motors must be matched to each other and to the system in order to deliver the required amount of air against the system resistance. It is usually necessary to make some adjustment of the air quantities at the time of installation because the resistance of the system cannot be accurately determined prior to installation.

The two types of fan motor arrangements in use are the belt drive and the direct drive. The belt-drive arrangement uses a fixed pulley on the fan shaft and a variable pulley on the motor shaft. The speed of the fan is in direct proportion to the ratio of the pulley diameters. The variable-pitch pulley diameter can be changed by adjusting the position of the outer flange of the pulley. On the belt-drive arrangement, the belt tension must be sufficient to prevent slippage. The motor is mounted on rubber isolators to prevent sound transmission and reduce vibration.

The direct-drive fan arrangement has the fan mounted on an extension of the motor shaft. Fan speeds are changed only by altering the motor speed. In most cases, the motor speed is altered by the use of extra windings on the motor. This is sometimes called a tap-wound or multi-speed motor. This type of motor has a series of electrical taps (connection points), each of which provides a different speed. Often, one speed is selected for heating and another for cooling.

Direct-drive blowers have replaced belt-drive blowers in most residential and many light commercial applications. Direct-drive blowers make it easier for the service technician to adjust the fan speed for optimum operation. Eliminating the drive belt also removes a likely point of failure.

Variable-speed, electronically commutated motors (ECM) are coming into more widespread use in furnaces, especially in higher efficiency, deluxe models. When combined with microprocessor-based room thermostats and furnace controls, they provide more precise control of air volume that translates into quieter and more efficient operation and enhanced indoor comfort.

3.5.0 Air Filters

Air filters remove dust, pollen, molds, and other particles from the air circulating through the building. They are most commonly located at the return air entrance to the furnace. Filters come in different efficiency levels and their capacity to capture particles is measured in microns. A micron is $\frac{1}{1000}$ of a millimeter. There are 25,400 in an inch. Standard-efficiency filters will remove particles of dust, dirt, pollen, and molds as small as 10 microns. That sounds pretty good, but it really means that 50 percent or more of the particles in the air are getting through the filter. The replaceable fiberglass filters you can buy at a hardware store are in this class. So are most of the permanent washable filters, which are made of metal or polyurethane foam.

High-efficiency filters will capture particles down to about half a micron. That covers mold, pollen, and much of the dust in the air. This class of filters includes self-magnetizing electrostatic filters and bag filters.

An electronic air cleaner is needed to capture really fine particles, including smoke and vapors. Electronic air cleaners can be obtained as stand-alone units, or may be added to a furnace as an accessory.

A dirty filter can block airflow, causing a furnace to lose efficiency. Filters need to be replaced or cleaned periodically, depending on the type of filter material. The furnace manufacturer's literature should specify the filter servicing

INSIDE TRACK

MERV Rating

When buying a furnace filter today, you will likely see a minimum efficiency rating value (MERV) on the filter label. MERV is part of *ASHRAE Standard 52.2* that addresses indoor air quality (IAQ). The higher the MERV rating, the more efficient the filter. However, don't assume that the highest MERV rating will give the best service. Higher ratings often mean the filter is more restrictive to airflow, allowing it to capture smaller particles. These more restrictive filters often cause the furnace fan motor to work harder as it attempts to draw air through the filter. Always follow the furnace manufacturer's MERV rating recommendation when selecting an air filter for a particular furnace.

frequency. Filters delivered with furnaces are not good-quality filters.

Generally, if the filter looks dirty or is damaged, it should be serviced. Turn off the furnace before removing the filter. The blower should not be allowed to operate with the filter removed.

Some thermostats are equipped with a check filter warning. In some cases, a sensor that detects a pressure drop across the filter activates the warning. In other cases, a timer records the elapsed running time and activates the warning when a specified time has elapsed.

Disposable filters must be replaced with filters of the same size and type. If the filter is too small, dirty air can bypass the filter. If a substitute filter has higher resistance than the original filter, it can cause the furnace to lose efficiency, and may even result in damage.

3.6.0 Automatic Vent Damper

The automatic vent damper is an energy-saving device. It is optional equipment on some heating units. One type of vent damper can be connected to the spark ignition control of the gas furnace (*Figure 9*). This type of system prevents the heated air from going up the flue when the furnace is not operating. It also provides immediate venting for products of combustion.

WARNING!
Automatic vent dampers must be installed only on heating units for which they have been designed. If installed on other furnaces, they would not only be inefficient, but could be dangerous. They are not used in induced-draft furnaces.

The vent damper installation package consists of a damper assembly, a damper operator, a flue adapter, and a damper cable. The damper can be installed in a vertical flue or in a flue that is not inclined more than 20 degrees from vertical. The damper operator opens and closes the vent damper upon demand of the space thermostat. When the thermostat calls for heat, the damper motor or operator is energized and the damper will open. The vent damper will remain open while the main burner is operating.

Another type of vent damper is the thermally actuated vent damper (*Figure 10*). One type of thermally actuated automatic vent damper uses a bimetal strip that transforms heat energy into mechanical energy. As the bimetal strip twists due to the heat from the furnace, the damper plate opens.

NOTE
Vent dampers enjoyed popularity as energy saving devices in the 1970s and 1980s. Advances in furnace technology and government-mandated minimum energy efficiency requirements for gas furnaces have made thermal vent dampers obsolete. However, you may still find them in older installations.

3.7.0 Humidifiers

Humidifiers are used to add moisture to indoor air. The amount of humidity inside a structure depends upon the outside temperatures, the building construction, and the relative humidity

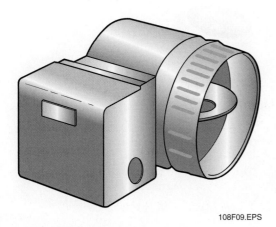
108F09.EPS

Figure 9 ◆ Automatic vent damper.

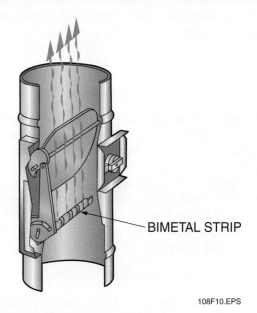
BIMETAL STRIP

108F10.EPS

Figure 10 ◆ Thermal vent damper.

that the interior of the house will withstand without a condensation problem. It is commonly held that a relative humidity of 30 to 50 percent is desirable. Too much humidity can cause condensation problems and too little humidity can cause static electricity problems and may cause furniture to crack or come apart at glued joints.

There are several types of humidifiers. In this discussion, we are mainly concerned with humidifiers that can be added to heating systems. Most humidifiers use some type of medium to pick up water from a reservoir and expose it to the airstream in the ductwork.

3.7.1 Plate-Type Humidifier

The plate-type evaporative humidifier has a series of porous plates mounted in a rack. The lower section of the plates extends down into water that is contained in a pan. A float device regulates the supply of water to maintain a constant level in the pan. The pan and plates are mounted in the warm air (supply) plenum.

3.7.2 Rotating-Drum Humidifier

The rotating-drum evaporative humidifier has a slowly revolving drum that is covered with a polyurethane pad. The drum is partially submerged in a supply of water, which is controlled by a float system. As the drum rotates, it absorbs water. The humidifier is mounted so that as the air passes over the wet surface, it picks up moisture that is carried through the supply duct throughout the structure.

3.7.3 Rotating-Disk Humidifier

The rotating-disk evaporative humidifier (*Figure 11*) is similar to the drum type in that the water-absorbing material revolves. It is normally mounted on the underside of the main warm air supply duct.

3.7.4 Fan-Powered Humidifier

The fan-powered evaporative humidifier (*Figure 12*) is mounted on the warm air plenum. Air is drawn in by the fan and forced over the wet core, and is then delivered back into the supply air plenum. The water flow over the core is controlled by a valve. A humidistat senses the humidity level and is used to turn the humidifier on and off. It controls both the fan and the water supply valves. The humidifier operates only when the furnace fan is running.

3.7.5 Bypass Humidifier

Bypass humidifiers (*Figure 13*) are usually installed on either plenum of any type of forced-air furnace.

They operate on the bypass principle. That is, air movement is accomplished by the static pressure differential between the supply and return plenums. Water is supplied to the water panel evaporator. Dry air that is forced through the wet panel picks up the available moisture and distributes the humidity through the conditioned space. Minerals and solid residues not trapped by the panel evaporator are flushed down the drain. For situations in which water hardness is not a problem, some bypass humidifiers incorporate a water-circulating system.

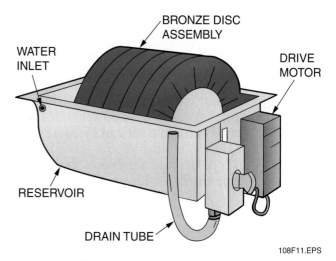

Figure 11 ◆ Rotating-disk humidifier.

Figure 12 ◆ Fan-powered humidifier.

Figure 13 ♦ Bypass humidifier.

3.7.6 Atomizing Humidifier

The atomizing humidifier consists of a metal enclosure containing a stainless steel water-atomizing nozzle. A finely dispersed water mist is produced that is capable of instant evaporation in the furnace ductwork. A minimum of 40 psi normal household water supply is required for efficient atomization. The humidifier operates only when the furnace fan is running.

3.7.7 Vaporizing Humidifier

The vaporizing humidifier uses an electrical heating element immersed in a water reservoir to evaporate moisture into the furnace supply air plenum. A timed flush cycle is included so that the accumulated solids from the water will not remain in the reservoir. A humidistat is connected into the system so that it not only starts the water heater, but also turns on the furnace fan if it is not already running.

3.7.8 Ultrasonic Humidifier

The ultrasonic humidifier contains a crystal known as a **piezoelectric** crystal. The crystal vibrates at a high frequency when an electric current is applied to it. Water dripping onto the vibrating crystal is atomized and injected into the airstream.

3.7.9 Steam Humidifier

In this type of humidifier, water is heated and converted into steam. Steam humidifiers (*Figure 14*) have been in use for many years. Due to their light weight, one worker can usually install them.

The unjacketed discharge manifold cools down completely when not in operation, but instant start-up is also considered to be one of the positive factors of steam humidifiers.

Problems can arise in the use of humidifiers when the mineral content (hardness) of the water is too great. When the water hardness is more than 10 grains of dissolved mineral particles per gallon, a water treatment unit should be added to the humidifying system.

3.8.0 Installation

Proper installation of a furnace is important for the safety of building occupants and efficient operation of the furnace. Consider the following factors during the installation of any furnace: location, safety controls, clearances, venting, and combustion air.

Applicable national, state, and local codes should be followed when installing a furnace. You will find, however, that manufacturers' instructions are often more stringent than the codes. This is done to make sure that codes are not inadvertently violated.

Proper handling of furnaces is especially important. They are made of sheet metal, which can be damaged if the furnace is dropped or mishandled. Also, some components, such as ceramic igniters, can be easily damaged.

3.8.1 Location

A furnace should be installed in a central location to reduce the length of duct runs. There should be plenty of space around the furnace to permit easy access for servicing and repair. Minimum clearances will be specified by the manufacturer.

Figure 14 ♦ Steam humidifier.

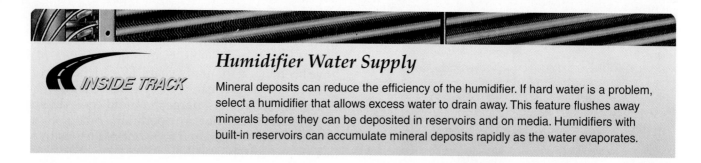

Humidifier Water Supply

INSIDE TRACK

Mineral deposits can reduce the efficiency of the humidifier. If hard water is a problem, select a humidifier that allows excess water to drain away. This feature flushes away minerals before they can be deposited in reservoirs and on media. Humidifiers with built-in reservoirs can accumulate mineral deposits rapidly as the water evaporates.

Furnaces should never be located near a source of aerosol sprays, bleaches, detergents, air fresheners, or cleaning solvents. Even small concentrations of such materials can corrode a furnace. In such environments, it is necessary to bring combustion air in from the outdoors. Manufacturers may refuse to honor warranties if a furnace is exposed to corrosive materials.

3.8.2 Safety Controls

Furnaces come equipped with the required safety controls. It is important to make sure that all controls are installed and working properly. Some of the furnace safety controls include the following:

- A flame rollout switch to shut off the furnace if the flame escapes the fire box
- A high limit switch to shut off the furnace if the flue is blocked or a fan stops working
- A control to shut off gas flow if ignition does not occur
- A control to shut off the furnace if fuel flow is interrupted

3.8.3 Clearances

Manufacturer recommendations or local codes specify the required distance between the furnace and combustible materials. Flammable materials must not come into contact with the heat exchangers, burners, or any other hot surfaces, such as the flue vent.

3.8.4 Furnace Venting

The flue gases leaving the furnace normally contain carbon dioxide and water vapor. Carbon monoxide, a poisonous gas, may be present if combustion is incomplete.

Carbon dioxide results if complete combustion occurs. It is not poisonous, but it displaces oxygen. A large concentration of carbon dioxide can cause the body to suffer from lack of oxygen. In extreme cases, it can cause asphyxiation. Because of the potential danger from products of combustion such as carbon monoxide, flue gases must be vented in accordance with local and national codes.

Standard-efficiency furnaces and mid-efficiency furnaces both produce hot flue gases. These gases must be vented through a special metal vent or a masonry chimney (*Figure 15*). It is often necessary

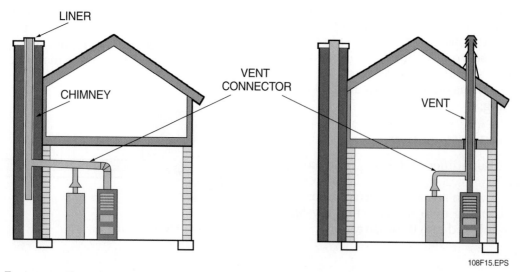

Figure 15 ◆ Furnace venting.

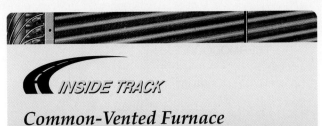

to use a double-wall metal vent for the vent connector. If a masonry chimney is used, it must be lined and have the correct dimensions. Even if the chimney has a tile liner, it is sometimes necessary to add a metal liner to meet code requirements. This is especially true if the chimney runs up the outside of the building, because a tile liner takes longer to warm up than a metal liner. The flue gases discharged by the furnace contain water vapor. When this water vapor enters a cold masonry chimney, it will condense and react with other combustion products to form compounds that will attack the mortar in the chimney, causing it to deteriorate. In addition, the condensate can return to the vent system, causing corrosion of the vent connector and possibly the heat exchangers.

When sizing and selecting a vent system, a lot depends on such factors as the distance from the furnace to the vent, the amount of heat in the flue gases, and whether the furnace is common-vented with another gas appliance such as a hot water heater.

The Gas Appliance Manufacturers Association (GAMA) provides tables and instructions that will help installers determine how to vent a particular furnace. Some manufacturers have simplified the tables and instructions for their product lines and provide the information in their installation instructions.

High-efficiency furnaces, such as condensing furnaces, can usually be direct-vented through an outside wall using PVC pipe.

3.8.5 *Combustion Air*

Many furnaces obtain the air for combustion from the air supply around the furnace. This approach is common with natural-draft and fan-assisted gas furnaces because the air used for combustion can usually be made up by **infiltration**. If the furnace is installed in a confined space such as an equipment closet, the room must be vented to allow for sufficient airflow.

High-efficiency furnaces need a lot of combustion air. For that reason, they must have combustion air piped in from the outdoors. The same is true for fan-assisted furnaces in tightly sealed buildings where there is not enough infiltration to provide makeup air. If these furnaces try to use indoor air for combustion, the pressure differential it creates can cause flue gases to be drawn into the building. It can also cause problems like difficulty in closing doors.

 WARNING!
Installation of combustion air piping requires special training in local codes. Improper installation can result in incomplete combustion, a dangerous condition.

4.0.0 ◆ GAS FURNACES

In a gas furnace, gas fuel is supplied at low pressure into a burner head, where it is mixed with the air required for combustion. The combustion forms hot gases that pass through the furnace heat exchanger or boiler into the vent pipe and chimney. The rate at which gas is supplied to the burner is controlled by a gas valve. The pressure is controlled by an automatic pressure regulator, which is usually built into the gas valve assembly.

The function of a gas burner is to produce a proper fire at the base of the heat exchanger. In order to do this, the equipment must control and regulate the flow of gas, ensure the proper mixture of gas with air, and ignite the gas under safe conditions. To accomplish these functions, the gas burner assembly (*Figure 16*) needs four major parts or sections: gas valve, ignition device, manifold and **orifice**, and burners.

4.1.0 Flame Ignition

Combustion begins when the gas is ignited. There are several ways that gas ignition can be accomplished. (See *Figure 17*.)

4.1.1 *Standing Pilot*

Older furnaces use **standing pilots**, which produce a small gas flame that remains on all the time. When gas begins to flow, it is immediately ignited by the pilot light. The **safety pilot** igniter

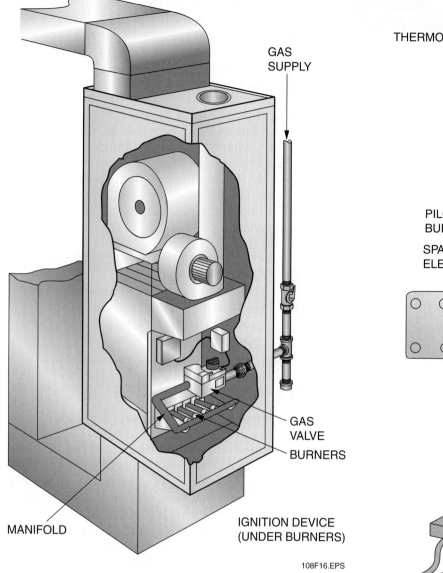

GAS
SUPPLY

GAS
VALVE

BURNERS

IGNITION DEVICE
(UNDER BURNERS)

MANIFOLD

108F16.EPS

Figure 16 ◆ Gas burner assembly.

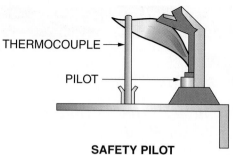

THERMOCOUPLE

PILOT

SAFETY PILOT

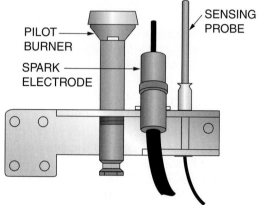

PILOT
BURNER

SPARK
ELECTRODE

SENSING
PROBE

ELECTRIC SPARK IGNITION

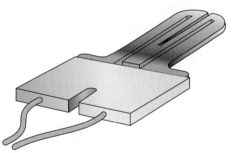

HOT SURFACE IGNITER

108F17.EPS

Figure 17 ◆ Gas furnace ignition devices.

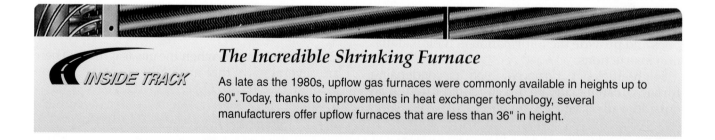

The Incredible Shrinking Furnace

INSIDE TRACK

As late as the 1980s, upflow gas furnaces were commonly available in heights up to 60". Today, thanks to improvements in heat exchanger technology, several manufacturers offer upflow furnaces that are less than 36" in height.

contains a **thermocouple** that converts the heat from the pilot flame into a small electrical current. The thermocouple is connected to the gas valve control circuit. If the pilot flame stops burning, the current from the thermocouple stops flowing and the gas valve is shut off.

Some safety pilots use a bimetal switching device instead of a thermocouple. It reacts to the heat from the pilot and keeps the switch closed as long as the pilot is lit.

4.1.2 Re-Ignition Pilot (Electric Spark Ignition)

The re-ignition pilot, or intermittent igniter, saves energy because, unlike the standing pilot, it does not require continuous gas flow to the pilot. When the thermostat calls for heat, a special transformer in the control circuit produces a high-voltage spark (10,000 volts or more) that ignites the pilot gas. These pilots often use a process known as **flame rectification** for pilot safety. This process is based on the fact that the pilot flame will produce a tiny current in the microampere range. As long as this current is sensed, it means that the pilot is on. If the current stops, the control circuit shuts off the gas supply to the furnace.

4.1.3 Direct Ignition

Many modern furnaces use a device known as a hot surface igniter (HSI). It is placed next to a burner in place of the pilot. When the thermostat calls for heat, a current flows through the igniter, causing it to become extremely hot. The heat ignites the gas. A flame sensor must be placed near the burners as a safety shutoff device. The flame sensor may be separate or, in some cases, integrated into the igniter. Hot surface igniters are made of ceramic and are therefore very fragile. They must be handled carefully.

CAUTION
Never touch the ceramic part of a hot surface igniter. The oils will cause damage to the HSI.

4.2.0 Gas Valve Assemblies

Older systems used a gas supply arrangement in which the gas valve and pressure regulator were separate components (*Figure 18*). On residential and small commercial systems manufactured

Combustion Air Sources

Induced-draft (fan-assisted) furnaces can draw combustion air from inside the building under certain circumstances. However, even if the available space and building construction meet the standards for using indoor air for combustion, there are other factors that must be considered. If corrosive chemicals such as bleach, solvents, construction adhesives, paint stripper, and other household products are used near the furnace, they can contaminate the combustion air, causing corrosion of the heat exchangers and creating a dangerous situation for building occupants. If the furnace cannot be sealed off from the contaminants, it will be necessary to duct combustion air from outdoors. Some furnace manufacturers will not warrant their heat exchangers if indoor air is used for combustion.

For the same reasons, care must be taken when locating the combustion air source for a condensing furnace. For example, it cannot be located near a swimming pool because there can be heavy concentrations of chlorine in the air around a pool.

A Rule of Thumb: A fan-assisted gas furnace installed in an "unconfined space," as defined by the *National Fuel Gas Code*, does not require that combustion air be ducted from the outdoors. The "1/20 rule" serves as a rule of thumb for making this determination. This rule is derived from the code requiring not less than 50 cubic feet of open space for every 1,000 Btuh of the combined input rating for all gas appliances (furnace, water heater, clothes dryer, etc.) in the room. This converts to 1 cubic foot for every 20 Btuh, or 1/20. The open space is considered to include adjoining rooms that cannot be closed off with doors. Be aware, however, that there is a trend in the industry to use outdoor air regardless of the space available. This trend stems from safety concerns, and is discussed later in the module.

Gas Furnace Installation

In years past, most natural-draft and fan-assisted mid-efficiency furnaces drew their combustion air from the space around the furnace. The air was supplied by infiltration through cracks in the building, as well as from the opening of doors and windows. In a modern building with tight construction, however, there may not be enough infiltration air to supply the combustion process. This will create the potential for flue gases to be drawn into the home. There are news articles every winter about individuals and entire families being killed by carbon-monoxide poisoning.

Although the *National Fuel Gas Code,* allows a fan-assisted furnace in an unconfined space to use indoor air for combustion, many people in the industry think it is best to err on the side of caution and always duct-in outside combustion air.

Even if a building has sufficient combustion air when the furnace is installed, circumstances can change. Homeowners often caulk around windows and doors and use other sealing methods that improve heating efficiency, but reduce the amount of infiltration air available for combustion. People can also add gas appliances after the furnace is installed, which creates additional demand for combustion air.

Local codes may impose special requirements for combustion air and venting beyond those in the National Fuel Gas Code or the manufacturer's instructions. Your company may have its own special requirements that are more demanding than the codes. In fact, this is quite common, because many companies take the position that the safety of current and future occupants overrides any other consideration.

When you are installing any gas appliance, thoroughly read the manufacturer's instructions, check local codes, and discuss it with your supervisor before proceeding.

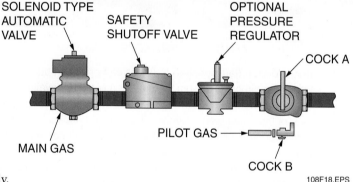

Figure 18 ◆ Older gas supply.   108F18.EPS

since the 1970s, however, these functions, along with pilot safety control (on pilot-type furnaces) have been combined into a single assembly. The modern gas valve is a **redundant gas valve**; that is, it contains two independent gas valves in series. If one fails, the other will shut off the gas flow when required.

The gas valve assembly shown in *Figure 19* is just one of the many types available. It has a gas outlet for the pilot, as well as a pilot control, which would not be found on the gas control for a direct-ignition furnace. Gas valves on some high-efficiency furnaces are designed to control two levels, or stages, of heat. These gas valves supply full or partial gas pressure, depending on the heat demand.

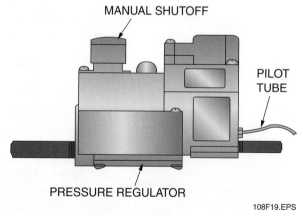

Figure 19 ◆ Redundant gas valve.

108F19.EPS

4.2.1 Automatic Gas Valve

The principal function of the automatic gas valve is to control the gas flow. Some of these valves control the gas flow directly; others regulate the pressure on a diaphragm, which in turn regulates the flow of gas. There are five principal types of automatic gas valves:

- The solenoid-operated valve (*Figure 20*) uses electromagnetic force to operate the valve plunger. When the thermostat calls for heat, the plunger is raised, opening the gas valve. These valves are often filled with oil to eliminate noise and lubricate the unit.
- The diaphragm-operated valve (*Figure 21*) uses gas pressure above and below the diaphragm to control the device. When the coil is energized, the gas supply to the upper section (above the diaphragm) is cut off and the pressure in the upper section is then bled off to the atmosphere. The pressure of the gas in the lower section then bends the diaphragm up, allowing gas to flow through the valve, as shown in *Figure 21*. When the coil de-energizes, the upper section is pressurized. With equal pressure above and below the diaphragm, the weight of the diaphragm forces it down to shut off the gas flow.
- Bimetal valve operators have a high-resistance wire wrapped around the diaphragm. When the thermostat calls for heat, current is supplied to the wire, causing it to heat and warp the diaphragm. The warping action opens the valve. The action of this valve is slow and the unit is sometimes referred to as a delayed-action valve.

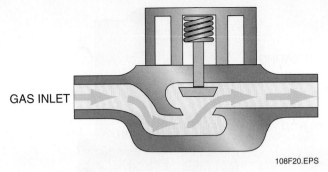

GAS INLET

108F20.EPS

Figure 20 ◆ Solenoid-operated gas valve.

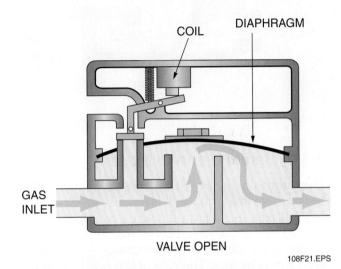

COIL   DIAPHRAGM

GAS INLET

VALVE OPEN

108F21.EPS

Figure 21 ◆ Diaphragm-operated gas valve.

- Bulb-type valve operators depend upon the expansion of a liquid-filled bulb to provide the operating force. The sensing bulb is attached to a bellows that expands and contracts, activating the gas flow valve.
- The heat motor valve (*Figure 22*) uses an electric heating coil to give off heat, which moves an expandable rod. The force necessary to operate the valve is provided by the movement of the expandable rod. When the rod is contracted and the valve is in the de-energized mode, the spring below the valve disc keeps the valve closed. The rod expands to open the valve when the heater produces enough heat.

In the energized mode, the expansion of the rod moves the valve stem downward, forcing the disc off the seat and allowing the gas to flow through the valve. Due to the time needed to heat the rod, a delay of about 20 seconds takes place. There is a delay of about 40 seconds when the valve is de-energized.

INSIDE TRACK

Improved HSI

The HSI with a ceramic element is very fragile. If it accidentally touches a hard surface while being removed or installed, it is likely to shatter. In 2005, manufacturers developed an HSI with a silicon-nitride element. This element is much more durable than the earlier types and is therefore less likely to shatter.

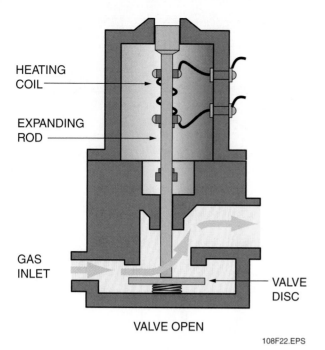

Figure 22 ◆ Heat motor valve.

HEATING
COIL

EXPANDING
ROD

GAS
INLET

VALVE
DISC

VALVE OPEN

108F22.EPS

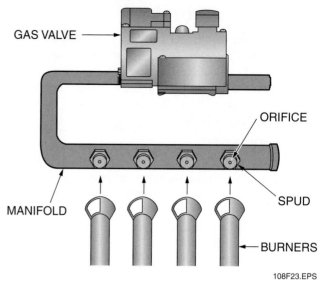

Figure 23 ◆ Gas manifold.

GAS VALVE

ORIFICE

MANIFOLD

SPUD

BURNERS

108F23.EPS

4.3.0 Manifold and Orifices

The manifold (*Figure 23*) delivers gas equally to all the burners in a furnace. It is usually made of ½" to 1" black iron pipe. A **spud** is screwed into the manifold. The orifice, which is the size of the opening in the spud, determines how much gas is delivered to the burner. The size of the orifice selected depends upon the type of fuel gas, the pressure in the manifold, and the gas input required for each burner. A number on the spud shows the orifice size. If the number is not visible, a numbered drill can be used to determine the size.

As discussed earlier, most gas burners require that some air be mixed with the gas before combustion; this air is called primary air. Primary air (*Figure 24*) amounts to approximately one-half of the total air required for proper combustion. Too much primary air causes the flame to lift off of the burner surface. Too little primary air causes a yellow flame.

Air supplied to the burner at the time of combustion is called secondary air. If too little secondary air is present, carbon monoxide forms. As discussed earlier, most gas burner units are designed to operate on approximately 50 percent excess secondary air supply. For correct flame generation, the proper ratio between primary and secondary air must be maintained.

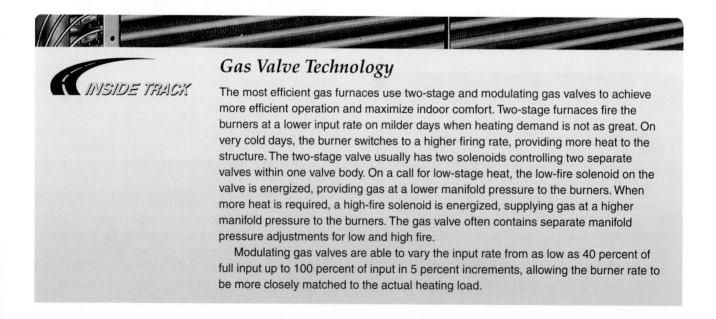

Gas Valve Technology

INSIDE TRACK

The most efficient gas furnaces use two-stage and modulating gas valves to achieve more efficient operation and maximize indoor comfort. Two-stage furnaces fire the burners at a lower input rate on milder days when heating demand is not as great. On very cold days, the burner switches to a higher firing rate, providing more heat to the structure. The two-stage valve usually has two solenoids controlling two separate valves within one valve body. On a call for low-stage heat, the low-fire solenoid on the valve is energized, providing gas at a lower manifold pressure to the burners. When more heat is required, a high-fire solenoid is energized, supplying gas at a higher manifold pressure to the burners. The gas valve often contains separate manifold pressure adjustments for low and high fire.

Modulating gas valves are able to vary the input rate from as low as 40 percent of full input up to 100 percent of input in 5 percent increments, allowing the burner rate to be more closely matched to the actual heating load.

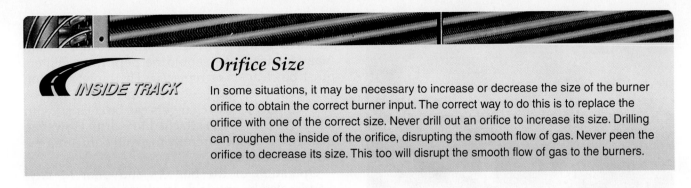

Orifice Size

In some situations, it may be necessary to increase or decrease the size of the burner orifice to obtain the correct burner input. The correct way to do this is to replace the orifice with one of the correct size. Never drill out an orifice to increase its size. Drilling can roughen the inside of the orifice, disrupting the smooth flow of gas. Never peen the orifice to decrease its size. This too will disrupt the smooth flow of gas to the burners.

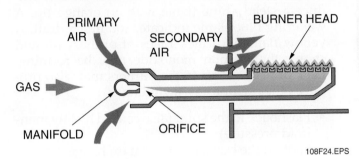

108F24.EPS

Figure 24 ◆ Combustion air supply.

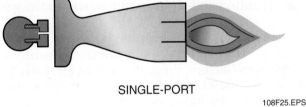

108F25.EPS

Figure 25 ◆ Gas burners.

4.4.0 Gas Burners

Gas burners can be either single-port (in-shot) or multi-port (*Figure 25*). Single-port burners, which are found in newer furnaces, direct a flame into an opening in the heat exchanger. Multi-port burners distribute flame through slots or holes along the length of the burner. Unlike single-port burners, multi-port burners are located inside the heat exchangers.

4.5.0 Gas Furnace Safety Switches

Gas furnaces are equipped with several switching devices that are designed to shut off the furnace if a hazardous condition is sensed. One of these is the high-temperature limit switch. *Figure 26* shows one example of a temperature limit switch. The switch is thermally actuated and is mounted near the heat exchangers. Electrically, it is wired in series with the gas valve. If the furnace overheats, the high-temperature limit switch will open, deenergizing the gas valve and shutting off the gas supply. These switches automatically reset when the temperature drops below the reset point. The high limit switch is often combined with the fan switch. These switches are called fan-limit switches.

Some downflow and horizontal furnaces are equipped with manual reset auxiliary limit switches because of the possibility of reverse air-flow occurring if the blower shuts down. Reverse airflow can cause the automatic high-temperature

108F26.EPS

Figure 26 ◆ Example of a high-temperature limit switch.

limit switch to reset, even though the conditions that caused it to trip are still present.

Flame rollout switches, such as the one shown in *Figure 27*, are placed near the opening in the heat exchangers. If there is insufficient combustion air, the burner flames can roll out of the

Figure 27 ◆ Flame rollout switch.

combustion chamber. The rollout switch senses this condition and shuts off the power to the gas valve.

A pressure switch such as the one shown in *Figure 28* is often used to ensure the presence of combustion airflow through the heat exchangers. If the pressure drops below a set point, indicating a failure of the induced draft fan or an airflow blockage, the switch will open and shut off the gas valve.

4.6.0 Maintenance

Gas furnaces require periodic cleaning, as well as maintenance and service on the gas burner assembly, pilot assembly, and automatic gas valve.

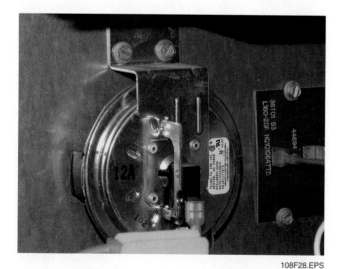

Figure 28 ◆ Combustion airflow pressure-sensing switch.

4.6.1 Maintenance Checks

At the beginning of each heating season, certain maintenance tasks should be performed. Filters must be cleaned or replaced. The blower motor, blower wheel, and heat exchangers must also be cleaned. The heat exchangers can be cleaned with a flexible wire brush attached to a drill. However, this must be done carefully to avoid damaging the heat exchangers.

Operation of the burners and pilot (if any) should also be checked. The burners should produce a steady blue flame with an orange tip. A wavering blue flame indicates too much draft. A yellow flame indicates a lack of primary air and shows that carbon monoxide may be forming. Lack of primary air to the burners may also produce other symptoms, including:

- Flashback to the spud (also caused by low manifold pressure)
- Soot on the burners and heat exchangers

Any excess of primary air will cause the flame to lift away from the burner. A flickering or distorted flame could be a sign of a crack in the heat exchanger. If there is any indication of a cracked heat exchanger, shut down the furnace and have the gas utility or other qualified agency inspect the furnace. A cracked heat exchanger can release carbon monoxide into the building, posing a serious risk for occupants.

Gas furnace maintenance and troubleshooting will be covered in detail in later modules.

4.6.2 Manifold Pressure

If the burners are not receiving the correct input, the heat exchangers may not warm up enough to prevent condensation and incomplete combustion may occur. The manufacturer's installation instructions supplied with the furnace will specify the correct manifold pressure that indicates input is adequate. You need to know the heat value and specific gravity of the gas, which you can get from the local utility. You also need to know the size of the orifice, which you can get by looking at the spud. An input gas pressure of 7 to 9 inches of water column (in. w.c.) is required, but 14 in. w.c. is maximum.

One way to check the manifold pressure is to connect a **manometer** to the pressure port on the gas valve (*Figure 29*). A manometer measures pressure by the amount of water that is pushed up the column. It is calibrated in in. w.c. The furnace manufacturer will state the correct manifold pressure to adjust to based on the factors described previously.

If the pressure needs to be adjusted, the pressure-adjustment screw on the gas valve is used.

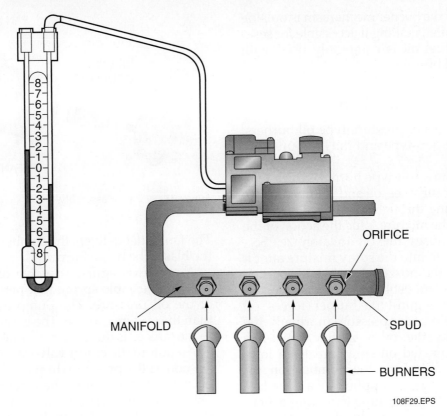

Figure 29 ◆ Adjusting manifold pressure.

5.0.0 ◆ OIL FURNACES

In a typical oil-fired furnace, the oil burner receives oil from a nearby storage tank. The oil burner shoots a spray of oil into the combustion chamber, where it is mixed with air supplied by a blower fan that is part of the oil burner. **Electrodes** located at the point where the oil spray enters the combustion chamber provide a high-voltage spark that ignites the oil.

The pressure-type burner (*Figure 30*) consists of a pump, an air tube and nozzle assembly, and a fan. Oil is pumped to the nozzle, which converts it to a fine mist. The oil mist is mixed with air in the air tube (also known as a blast tube) and ignited by an electric spark. The burner is not operated continuously, but is turned on or off in accordance with the demands of a room thermostat. The refractory firepot makes the unit more

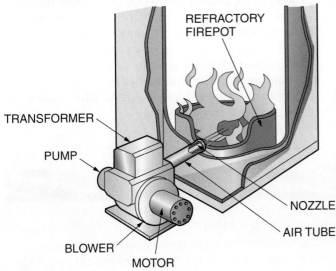

Figure 30 ◆ Pressure burner components.

efficient. Most of the burner mechanism is outside the furnace or boiler, making it accessible for servicing. Grade 2 fuel oil is commonly used with pressure-type oil burners.

5.1.0 Oil Burner Operation

There are two types of pressure-type oil burners, the low-pressure gun-type and the high-pressure gun-type.

In a low-pressure gun-type burner, which may be found in older furnaces, oil and primary air are mixed before going through a nozzle. A pressure of 1 to 15 psi on the mixture, plus the action of the nozzle orifice, causes the oil to atomize. Secondary air is drawn into the spray mixture after it is released from the nozzle. An electric spark is used to light the combustible mixture.

A high-pressure gun-type burner forces oil through the nozzle under pressure (normally 100 psi). This breaks the oil into fine, mist-like droplets. The atomized oil spray creates a low-pressure area into which the combustion air flows. Combustion air is supplied by a vane fan, creating turbulence and complete mixing action. The high-pressure, gun-type burner is the most popular domestic burner. It is simple in construction and efficient in operation. The parts are mass-produced, readily available, and relatively low in cost.

5.1.1 Power Assembly

The power assembly of the high-pressure burner unit consists of the motor, fan, and oil pump (*Figure 31*).

The motor drives the fan and oil pump. The fan forces air through a blast tube to provide combustion air for the atomized oil. The oil pump draws oil from the storage tank and delivers it to the nozzle. The fuel/air mixture is ignited by an electric spark that is formed between two electrodes in the nozzle assembly. The motor supplies power to operate the oil pump and fan. The fan delivers air for combustion. The inlet to the fan has an adjustable opening so that the air volume can be controlled manually.

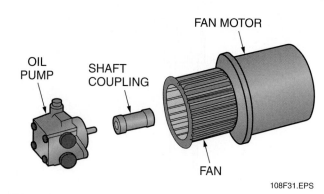

Figure 31 ◆ Power assembly.

The fan outlet delivers the combustion air through the blast tube of the burner.

A pressure-regulating valve on the oil pump has an adjustable spring that permits adjustment of the oil pressure. The pump delivers more oil than the system can use. The excess oil is returned to the tank or is dumped back into the supply line.

An automatic cutoff valve stops the flow of oil as soon as the pressure drops, thereby preventing oil from dripping into the combustion chamber. Oil pumps are designed for single-stage or two-stage operation. The single-stage pump is used where the oil supply is above the burner and the oil is fed to the pump by gravity. The two-stage unit is used where the storage tank is below the burner. The first stage draws the oil to the pump and the second stage provides the pressure required by the nozzle. Pump suction should not exceed 15 inches of vacuum.

5.1.2 Piping

Piping connections to the oil pump are of two types, single-pipe and two-pipe. In the single-pipe system, there is only one pipe from the storage tank to the burner. In the two-pipe system, two pipes are run from the tank to the burner. One carries supply oil; the other carries return oil. Compression fittings should never be used on oil system piping. The lines must be purged of air in order to avoid problems. A two-pipe system is easier to keep free of air.

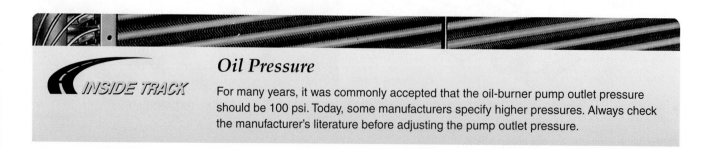

Oil Pressure

For many years, it was commonly accepted that the oil-burner pump outlet pressure should be 100 psi. Today, some manufacturers specify higher pressures. Always check the manufacturer's literature before adjusting the pump outlet pressure.

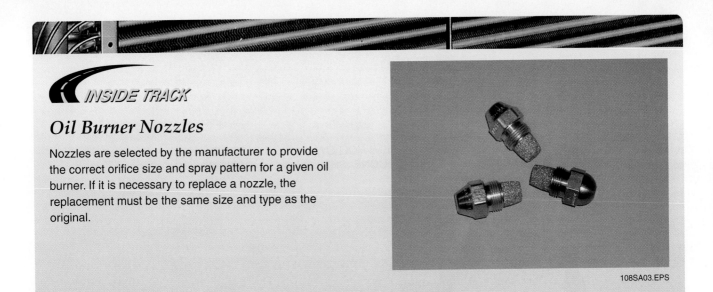
5.1.3 Nozzle

The nozzle assembly (*Figure 32*) consists of the oil feed line, the nozzle, ignition electrodes, and the transformer connections.

The nozzle prepares the oil for mixing with the air by atomizing the fuel. Oil passes through a strainer and enters slots that direct the oil to the swirl chamber. The swirl chamber gives the oil a rotary motion when it enters the nozzle orifice, shaping the spray pattern. The nozzle orifice increases the velocity of the oil. The oil leaves the nozzle in the form of a mist or spray and mixes with the air from the blast tube.

Due to the fine tolerances of the nozzle construction, dirty or defective nozzles are usually replaced. Nozzle spray patterns vary with the type of application. The shape of the spray and the angle between the sides of the spray can be varied. There are three spray shapes, as shown in *Figure 33*: hollow (H), semi-hollow (SH), and solid (S). The hollow and semi-hollow are most popular on domestic burners as they are more efficient when used with modern combustion chambers.

The angle of the spray must correspond to the type of combustion chamber (*Figure 34*). An angle of 70 degrees to 90 degrees is usually best for square or round chambers; an angle of 30 degrees to 60 degrees is best suited to long, narrow chambers.

5.1.4 Ignition System

The ignition system for high-pressure burners consists of a step-up transformer connected to two electrodes. The transformer supplies high voltage that causes a spark to jump between the two electrodes. The force of the air in the blast tube causes the arc to bend into the fuel-air mixture, igniting it.

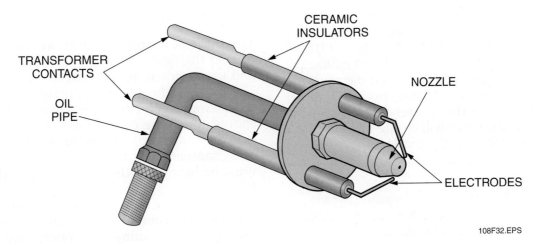

108F32.EPS

Figure 32 ◆ Nozzle assembly.

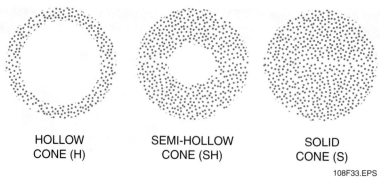

HOLLOW CONE (H)   SEMI-HOLLOW CONE (SH)   SOLID CONE (S)

108F33.EPS

Figure 33 ◆ Nozzle patterns.

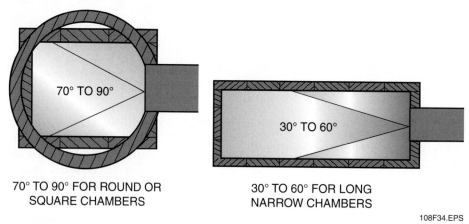

70° TO 90°

30° TO 60°

70° TO 90° FOR ROUND OR SQUARE CHAMBERS

30° TO 60° FOR LONG NARROW CHAMBERS

108F34.EPS

Figure 34 ◆ Spray angles.

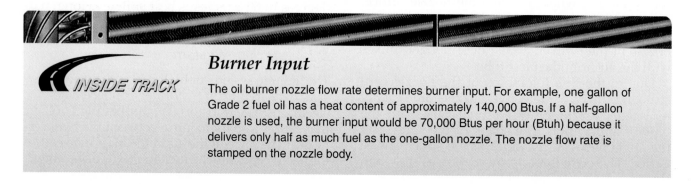

Burner Input

The oil burner nozzle flow rate determines burner input. For example, one gallon of Grade 2 fuel oil has a heat content of approximately 140,000 Btus. If a half-gallon nozzle is used, the burner input would be 70,000 Btus per hour (Btuh) because it delivers only half as much fuel as the one-gallon nozzle. The nozzle flow rate is stamped on the nozzle body.

Ceramic insulators surround the electrodes and serve to position them. The step-up transformer increases the voltage from 120V to about 10,000V, but reduces the amperage to about 20 milliamps. The low amperage reduces wear on the electrode tips. The correct position of the electrodes (*Figure 35*) is important. If the electrodes are not centered on the nozzle orifice, the flame will be one-sided and will cause carbon to form on the nozzle.

WARNING!

Due to the high voltage of the transformer and the line voltage delivered to it, caution must be used in checking the transformer.

5.2.0 Combustion Chamber

The purpose of the combustion chamber is to protect the heat exchanger from flame damage and to provide reflected heat to the burning oil. The reflected heat warms the tips of the flame, enhancing combustion. No part of the flame should touch the combustion chamber surface. If it does, incomplete combustion could result. The chamber must fit the flame and the nozzle must be located at the correct height above the floor.

Three types of materials are commonly used for combustion chamber construction: metal (stainless steel), insulating fire brick, and molded ceramic. In small furnaces such as those used in

Electrodes

Take care when handling electrodes. The ceramic insulators are fragile and easily cracked.

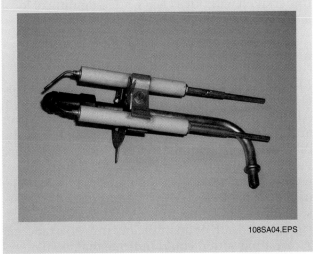

108SA04.EPS

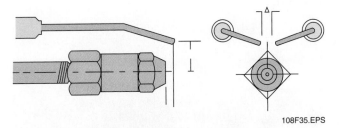

108F35.EPS

Figure 35 ◆ Electrode position.

residential applications, the combustion chamber is built into the furnace. For large commercial and industrial applications, the combustion chamber may be constructed on site. Such chambers can be made of fire brick or pre-cast ceramic (refractory) material. If pre-cast materials do not fit the job, a combustion chamber may be formed and poured on site using ceramic materials designed for this purpose.

A newer, moldable refractory material, called by the generic name of wet-pack, is used to build or re-line combustion chambers. This material is moist when it is purchased, and is shipped in sealed plastic bags. Because it is moist, it can be easily molded into the desired shape. Continued exposure to air causes the wet-pack to dry into a very hard, rigid structure.

5.3.0 Draft Regulator

A draft regulator is necessary to control heat and combustion. If too high a draft exists, it causes undue loss of heat through the chimney. If too little draft is available, incomplete combustion could result. The draft regulator should maintain a constant draft over the fire, usually 0.01 to 0.03 in. w.c. A draft regulator (*Figure 36*) consists of a small door in the side of the flue pipe. It is hinged near the center and controlled by adjustable weights.

5.4.0 Oil Safety Controls

The primary control unit, or primary (as it is known), is usually mounted on the burner assembly. The primary (*Figure 37*) provides both control and safety functions. As a control device, its job is to turn the burner on and off in response to commands from the thermostat. As a safety control, the primary shuts off power to the burner if the burner does not ignite or if the burner flame goes out. In modern furnaces, a cadmium sulfide (cad) cell (*Figure 38*) is used to sense the presence of a flame. The cad cell is photo-sensitive, and is usually installed in the burner assembly on a line of sight with the burner flame. If the cell does not sense a burner flame, it sends a signal to the primary to turn off the burner and stop the flow of oil.

In older installations, you may find a stack switch (*Figure 39*) used in place of a cad cell. The stack switch is installed in the flue vent (stack) ahead of the draft damper. It senses heat and shuts down the oil burner if heat is not detected in the stack.

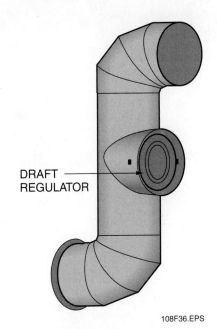

DRAFT REGULATOR

108F36.EPS

Figure 36 ◆ Draft regulator.

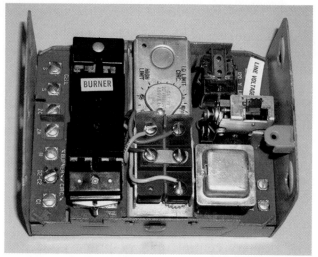

Figure 37 ◆ Primary control.

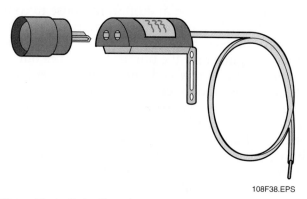

Figure 38 ◆ Cad cell.

Figure 39 ◆ Stack switch.

5.5.0 Oil Storage

Oil furnaces may be supplied from indoor or outdoor oil tanks. Outdoor tanks may be buried. Because many underground oil tanks have leaked, there are strict federal and usually local regulations governing placement, construction, and alarm systems for these tanks. Above-ground tanks over a certain size may require special containment arrangements. No installation should be undertaken without a clear understanding of local and national codes.

5.6.0 Oil Furnace Maintenance

Like the gas furnace, the oil furnace should be checked at the beginning of each heating season. Air filters must be cleaned or replaced, the blower and heat exchangers should be cleaned, and the oil filter (*Figure 40*) should be replaced. Oil burner operation should also be checked. If there is not enough combustion air, the flame may be orange or red instead of yellow. The presence of smoke, soot, or odors could indicate improper oil pressure, poor draft, or an improper mix of oil and air. A pulsating sound could indicate that the flame is touching the combustion chamber. In any of these cases, some cleaning, adjustment, or repair will be needed.

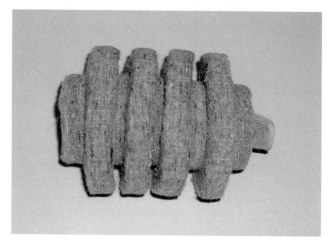

Figure 40 ◆ Oil filter.

6.0.0 ◆ ELECTRIC HEATING

The electric furnace (*Figure 41*) converts electrical energy directly into heat energy using resistance heaters.

Electric furnaces differ from fuel furnaces in that no combustion is required. Because no fuel is burned, there is no need for a chimney or vent to carry the products of combustion outdoors. This feature allows for greater operational safety and more installation flexibility. The return air from the conditioned space passes directly over the resistance heaters and into the supply air plenum. The amount of heat supplied by an electric furnace depends upon the number and size of the resistance heaters used in the application. To avoid overloading the electrical system on start-up, the elements are sequenced on in stages.

The major components of an electric furnace, excluding the controls, are the heating elements, the blower and motor assembly, and the furnace enclosure or cabinet. Accessories such as filters, a humidifier, and a cooling coil may be included.

6.1.0 Heating Elements

The function of the heating element is to provide the heat required for the conditioned space. *Figure 42* shows two types of heating elements. The heating element wires are made of nickel and chromium (Nichrome). The heating element wire

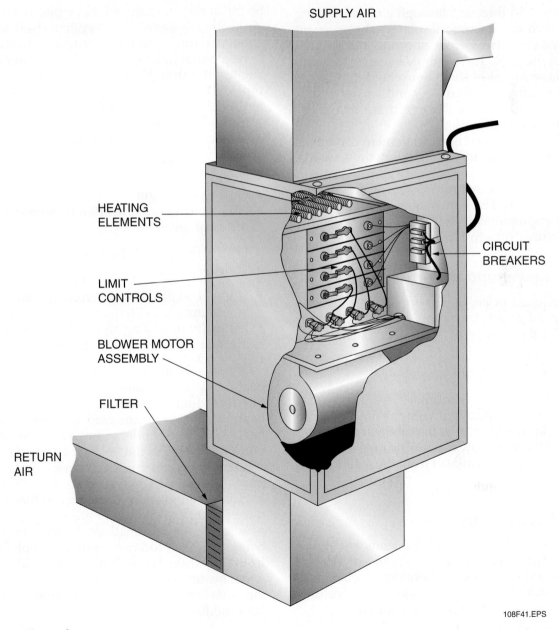

SUPPLY AIR

HEATING ELEMENTS

CIRCUIT BREAKERS

LIMIT CONTROLS

BLOWER MOTOR ASSEMBLY

FILTER

RETURN AIR

108F41.EPS

Figure 41 ◆ Electric furnace.

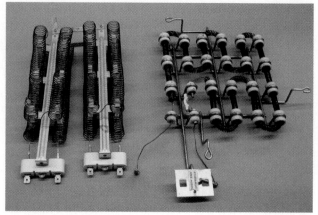

Figure 42 ◆ Electric heating element.

108F42.EPS

is spiraled and threaded through a metal holding rack, which has ceramic insulators that prevent the resistance wires from shorting out on the frame of the rack. The heating elements are similar to those found in small household appliances such as electric clothes dryers and toasters. Rod-type elements such as those you see in electric ovens are sometimes used.

The circuit for each heating element contains a fuse (or circuit breaker) and a safety limit switch. The fuse is a backup for the safety limit switch and is set to open at a temperature slightly higher than the limit switch. The limit switch is usually set to open at approximately 160°F and close when the temperature drops to 125°F.

6.2.0 Blower and Motor Assembly

The fans used in an electric furnace are usually multi-speed direct-drive. This provides larger air quantities if cooling is added to the system.

6.3.0 Furnace Enclosure

The casing of the electric furnace is similar to that of a gas or oil furnace but without the vent pipe connection. The interior of the cabinet is designed to permit the air to flow over the heating elements, which are usually insulated from the exterior casing by an air space.

6.4.0 Accessories

Filters, humidifiers, and cooling units are added to an electric furnace in much the same manner as with gas and oil furnaces. Therefore, electric heating can provide all of the climate control features found in other types of fuel-fired furnaces.

6.5.0 Power Supply

An electric furnace power supply is usually 208/240V, single-phase, 60 Hertz (Hz) alternating current. This type of heating unit is supplied by three wires: two hot conductors and one grounding conductor. The hot lines leading to the furnace contain fused disconnects. All the wiring should be enclosed in conduit with the proper connectors as specified by the *National Electrical Code®*. The *NEC®* also requires that the furnace unit be grounded. The supply ground is provided for this purpose.

7.0.0 ◆ HYDRONIC HEATING SYSTEMS

So far in this module, we have discussed forced-air heating systems. Those systems heat air that is circulated throughout the structure. Another type of residential heating system heats water and circulates it throughout the structure by way of pipes. This type of heating system is called a hydronic heating system (*Figure 43*).

The advantages of hydronic heating systems over forced-air systems include:

• Enhanced comfort
• Compact size
• Quieter operation
• Ability to maintain different temperatures in different areas

Disadvantages of hydronic heating systems include:

• Higher installed cost when compared to forced-air systems
• Difficulty in adding air conditioning and other air treatment options

A boiler is used to heat water in a hydronic heating system. Boilers can be gas- or oil-fired like forced-air furnaces, or they can use electric heating elements immersed in water to provide heat. Gas-fired boilers are available in high-efficiency versions that extract additional heat by condensing water from flue gases like condensing furnaces.

Boilers can produce hot water or they can produce steam. Steam boilers are rarely installed in new residential installations. Instead, most steam boilers installed today are put in as replacements. In this module, we will focus on residential hot water heating systems. All types of hydronic heating systems will be covered in more detail in a later module.

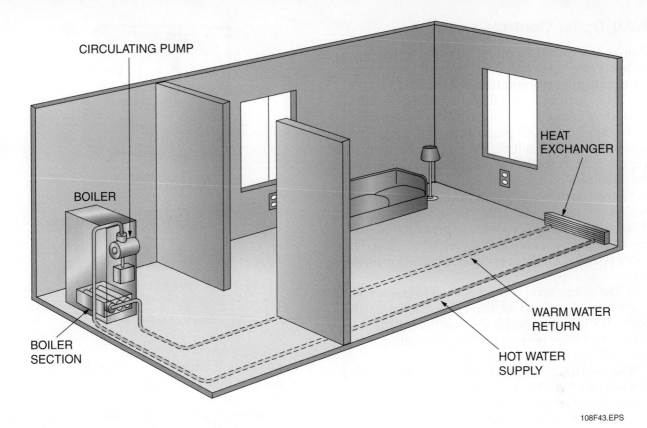

CIRCULATING PUMP

BOILER

BOILER
SECTION

HEAT
EXCHANGER

WARM WATER
RETURN

HOT WATER
SUPPLY

108F43.EPS

Figure 43 ◆ Residential hydronic heating system.

7.1.0 Major Boiler Components

The major components of a boiler include the burner and burner controls, the boiler sections where water is heated, and a pump to circulate the warmed water throughout the structure.

The gas- or oil-fired burner and controls are similar in design and function to the burners already discussed under forced-air furnaces. Controls used with boilers have similarities to and differences from the controls used in forced-air furnaces.

The boiler sections are usually made of cast iron, although steel is sometimes used. The burner flame passes over the outside of the boiler sections, warming the water within. Some hot water boilers contain a heat exchanger that allows the boiler to also provide domestic hot water.

Once the water is warmed to a pre-set temperature, the circulator pump is energized to move the water through the pipes to the terminals. There, the heat is given up. The circulator pump then returns the cooled water to the boiler, where the process repeats itself. The temperature/pressure gauge allows the condition of the water in the boiler to be monitored.

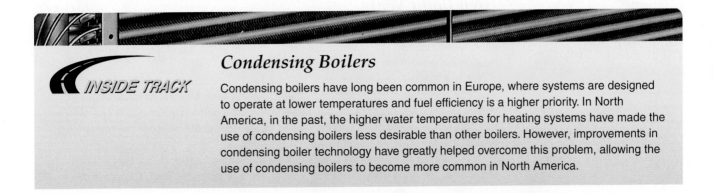

Condensing Boilers

Condensing boilers have long been common in Europe, where systems are designed to operate at lower temperatures and fuel efficiency is a higher priority. In North America, in the past, the higher water temperatures for heating systems have made the use of condensing boilers less desirable than other boilers. However, improvements in condensing boiler technology have greatly helped overcome this problem, allowing the use of condensing boilers to become more common in North America.

7.2.0 Boiler Controls

One of the most important controls found on a boiler is the aquastat. This temperature switch controls burner operation to maintain the water in the boiler within a specified temperature range. Aquastats come in a variety of types, depending on the application. Typical functions include preventing the water within the boiler from becoming too cold or too hot (limit switch) and turning the circulator pump on and off.

Water expands and increases in pressure when heated. If left unchecked, rising pressure in a boiler can cause it to explode. To prevent this, boilers are equipped with a pressure-relief valve. The valve opens and relieves pressure if it becomes too high.

A low-water level control is used to ensure that water is maintained at the correct level in the boiler. If the water level is too low, the control prevents the burner from operating, preventing damage to the boiler.

The centrifugal-type circulator pump is often factory-installed. However, many boiler manufacturers ship the pump loose and allow the installer to locate it in a convenient location.

7.3.0 Other Components of a Hydronic Heating System

There are a variety of other specialized components that are used in a hydronic heating system. In residential applications, the heat is usually delivered by finned baseboard radiators that use convection to transfer heat (*Figure 44*). Another method that has become very popular is the use of

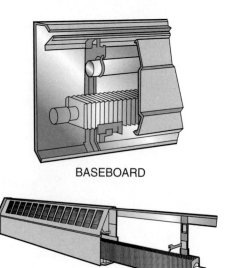

BASEBOARD

FINNED-TUBE

108F44.EPS

Figure 44 ◆ Hydronic heat radiators.

pipes or tubing embedded in the floor or ceiling that transfer heat by radiation.

One advantage of hydronic systems is that different rooms or zones within the same structure can be kept at different temperatures. This is accomplished by installing a zone control valve in the pipe that supplies hot water to each zone. Each zone valve is controlled by its own 24-volt room thermostat. The thermostat controls the opening and closing of the valve. This allows hot water to flow to the heat transfer device in the zone only when needed. Thermostats in other zones set at different temperatures open and close zone valves in their respective zones to maintain different temperatures.

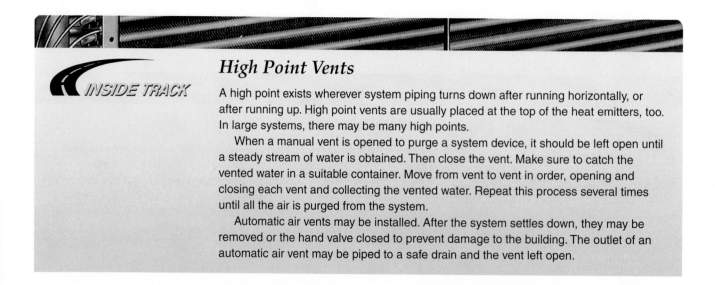

High Point Vents

INSIDE TRACK

A high point exists wherever system piping turns down after running horizontally, or after running up. High point vents are usually placed at the top of the heat emitters, too. In large systems, there may be many high points.

When a manual vent is opened to purge a system device, it should be left open until a steady stream of water is obtained. Then close the vent. Make sure to catch the vented water in a suitable container. Move from vent to vent in order, opening and closing each vent and collecting the vented water. Repeat this process several times until all the air is purged from the system.

Automatic air vents may be installed. After the system settles down, they may be removed or the hand valve closed to prevent damage to the building. The outlet of an automatic air vent may be piped to a safe drain and the vent left open.

How a Zone Valve Controls a Circulating Pump

Here's how a zone valve controls a circulating pump:

- A room thermostat in a calling zone energizes a motor-driven zone valve, causing it to open.
- The zone valve is equipped with normally open switch contacts that close when the valve is fully open.
- The closure of the switch contacts completes the path to energize the circulating pump.

A makeup water valve connected to city water or to a well system allows any water lost in the hydronic system to be automatically replaced. This valve may also contain a built-in pressure-reducing valve to prevent the hydronic system from being over-pressurized by city water pressure. The pressure-reducing valve may also be installed as a stand-alone component.

Water expands when heated. To allow for this normal expansion, hydronic heating systems include an expansion tank. This device contains an air space separated from the water by a flexible diaphragm. The expanding water causes the air to be compressed. This prevents pressure in the system from reaching unsafe levels.

Summary

Gas-fired and oil-fired warm air furnaces and boilers provide heat for the majority of buildings in the United States. Wood and coal are also used as heat sources, but their application is limited because they are less convenient and their combustion products are more damaging to the environment. Electric furnaces provide clean heat, but are usually much more expensive to operate than gas or oil. Oil and gas furnaces operate on the principle of transferring the heat generated by fuel combustion to air or water, which is then circulated through the conditioned space. Furnaces (whether gas or oil) contain a fuel control, a combustion chamber, and heat exchangers. They must also have a means of moving the heat through the conditioned space, as well as a system for venting flue gases to the outside.

Notes

1. A gas-fired warm air furnace heats by
   _____ .
   a. radiation
   b. convection
   c. conduction
   d. capillary action

2. If you burn your finger by touching a hot
   surface, it is an example of _____ .
   a. radiation
   b. convection
   c. conduction
   d. combustion

3. A temperature of 80°F is equal to _____ .
   a. 12°C
   b. 27°C
   c. 176°C
   d. 202°C

4. Carbon monoxide (CO) is produced when
   _____ .
   a. combustion is incomplete
   b. oil and oxygen are mixed
   c. there is water vapor in the flue gas
   d. there is excess air in the combustion
      chamber

5. A gas burner flame should be _____ .
   a. yellow with a blue tip
   b. completely blue
   c. completely yellow
   d. blue with an orange tip

6. An oil burner flame should be _____ .
   a. blue with a yellow tip
   b. red with a yellow tip
   c. yellow
   d. orange

7. Of the following gases, _____ has the highest heating value.
   a. propane
   b. butane
   c. natural gas
   d. manufactured gas

8. In a forced-air furnace, air _____ .
   a. flows through the inside of the heat
      exchangers
   b. is heated directly by the burners
   c. flows over the heat exchangers
   d. is circulated through the building and
      the furnace by convection

9. In a condensing furnace, the condensing
   heat exchanger _____ .
   a. extracts heat by condensing the condi-
      tioned air that passes over the heat
      exchanger
   b. is used only in the cooling mode
   c. extracts heat by condensing the flue gas
      that flows through the heat exchangers
   d. is used to heat the flue gases before they
      are vented

10. Humidifiers are used to _____ .
    a. reduce condensation
    b. add moisture to the air
    c. keep furniture dry
    d. lower the room temperature

11. Direct-venting of flue gases with PVC pipe
    may be used _____ .
    a. only for the highest efficiency condens-
       ing furnaces
    b. for fan-assisted and high-efficiency
       furnaces
    c. only for natural-draft furnaces
    d. for any furnace if a chimney is not
       available

12. A manometer may be used to measure
    _____ .
    a. how much water is in the flue gas
    b. the manifold pressure
    c. the flame rectification current
    d. the percentage of primary air used in
       combustion

13. In a typical oil burner, the oil is ignited by _____ .
    a. a pilot light
    b. a spark plug
    c. electrodes
    d. a hot surface igniter

14. The fan in a high-pressure oil burner is used to _____ .
    a. keep the furnace from overheating
    b. cool the oil
    c. atomize the oil
    d. provide combustion air

15. In an electric furnace, the heating element replaces which of the following item(s) found in a gas furnace?
    a. Burner assembly, heat exchangers, and induced-draft motor
    b. Burner assembly and blower
    c. Induced-draft motor and blower
    d. Condensing coil

Trade Terms Quiz

1. Air containing half of the moisture it is capable of holding has a 50 percent _____.

2. A hot water system uses a baseboard _____.

3. Furnaces with an electric spark ignition often use _____ for pilot safety.

4. A(n) _____ igniter contains a thermocouple.

5. The ultrasonic humidifier uses a(n) _____ crystal to atomize water.

6. When positioned correctly, the _____(s) are centered on the nozzle orifice.

7. The chemical reaction between fuel, heat, and oxygen is known as _____.

8. A furnace that uses a fan to draw in and exhaust combustion air is a(n) _____.

9. In a confined space or tightly sealed building there may not be enough _____ to provide makeup air to the furnace.

10. Oil-fired furnaces _____ fuel oil in order to burn it.

11. When the thermostat calls for heat from a gas furnace, a current flows through the _____, causing it to become extremely hot.

12. Older furnaces typically have a(n) _____, which always remains lit.

13. The current industry standard for defining furnace efficiency is _____.

14. Check the manifold pressure using a(n) _____.

15. If a safety pilot flame goes out, the _____ will signal the gas valve control circuit to turn off the gas.

16. Incomplete combustion can be caused by a lack of _____.

17. The size of the _____ determines the amount of gas that is delivered to the burner.

18. Most modern furnaces have two independent gas valves in series, which together are known as a(n) _____.

19. A(n) _____ shoots a spray of oil into the combustion chamber, where it is ignited by an electrode.

20. _____(s) can reach efficiencies of 95 percent and more.

21. Most existing _____(s) have an AFUE of less than 78 percent.

22. The gas burner will have a blue flame when approximately 50 percent of the _____ is mixed with gas before ignition.

23. A number on the _____ shows the orifice size.

Trade Terms

Annual Fuel Utilization
  Efficiency (AFUE)
Atomize
Combustion
Condensing furnace
Electrode

Flame rectification
Heat exchanger
Hot surface igniter
Induced-draft furnace
Infiltration
Manometer

Natural-draft furnace
Oil burner
Orifice
Piezoelectric
Primary air
Redundant gas valve

Relative humidity
Safety pilot
Secondary air
Standing pilot
Spud
Thermocouple

Troy Staton

Branch Manager, Upstate Service Division
W.B. Guimarin & Co., Inc.
Greenville, SC

How did you choose a career in HVAC?
After spending six years in the USAF as a fighter aircraft crew chief, I began to wonder what I would do after completing my tour of duty. By that time, I had begun teaching fighter aircraft maintenance programs on the F-15, and another instructor friend had decided to go into the HVAC service field. The idea sounded good to me—services that would always be needed, regardless of economic pressures.

What types of training have you been through?
My first training was a correspondence course offered by the National Radio Institute (NRI) from the back of a matchbook—yes, really! Most of that program was completed before I exited the military. My first job with a contractor led to an apprenticeship program in Virginia. It was a win-win situation, as the GI Bill paid a major portion of my wage on behalf of the employer, with the ratio changing every six months as I worked through the apprenticeship. After four years, the contractor was then entirely responsible for my wage. During the four years, though, he had extremely cheap but mechanically skilled labor at his disposal. I'll never forget that gentleman—Arthur E. Newsome of Newport News, VA, owner of Art Newsome, Inc. He hired me out of the Air Force strictly because another ex-military employee had performed so well for him in the past, so he took that chance with me, knowing I had never serviced a system before in my life.

Since then, I have attended many different manufacturers' programs, as well as programs at various tech schools. One of the best was offered by Frick Company—a two-week program in industrial refrigeration taught by one of the most treasured men in our trade's history, Milton Garland. He was an amazing man in his 90s at the time, and had been working for Frick since graduating from college—literally his entire life. Mr. Garland was one of the people responsible for the original piping codes in the U.S., as well as many other extraordinary accomplishments in our field.

What kinds of work have you done in your career?
The vast majority of my 27 years in the field have been spent with several contractors. After several years with Art Newsome, I worked for NASA for several years, primarily servicing computer room and other commercial systems in Langley, VA. I also spent a number of years with H.M. Webb and Associates, now known as Webb Technologies, servicing a variety of commercial cooling systems, but also a fantastic array of refrigeration systems using both halocarbon and ammonia refrigerants. The incredible variety of applications I was able to observe and service was a major contributor to my understanding of the refrigerant circuit. Throughout the years, though, I have always found myself in a position of leadership with almost every employer. I now manage a branch of the Service Division for W.B. Guimarin & Co, Inc., a company with a 104-year history in the trade.

What do you like about your job?

I truly enjoy the variety of applications and technical challenges that this field offers. I have been blessed with the opportunity to service and repair systems ranging from beer coolers at a local tavern to 500+ ton steam-powered absorption chillers. Thermodynamics has not changed, but the many ways we utilize those laws to our advantage never ceases to amaze. I have also always enjoyed doing a good job and providing a valuable service to people in need of a solution. I remember many homeowners in my residential days displayed tremendous gratitude for the service I performed and the way it was conducted. That, in turn, provided my enthusiasm for the next service call.

What factors have contributed most to your success?

Many things! I remember that, especially during my early years in the field and in search of knowledge and understanding, that I would take home catalogs from both manufacturers and parts houses to study. Although not textbooks, they still contained a wealth of knowledge. Most manufacturers still feel that providing a deep understanding of their product will result in sales. The endless variety of applications I was involved in with H.M. Webb was certainly a

benefit. Few people are exposed to the types of systems I serviced in those days. By and large, though, a hunger to truly understand the theory of operation for both individual components as well as systems has driven me to success.

What advice would you give to those who are new to the HVAC field?

Never let your integrity or your work ethic be considered any less than a top priority. It won't matter what you can do if your reputation is in shambles.

Although I hope all employers of people in our field will one day acknowledge the need for consistent training, don't wait for it! Study, study hard, and study anything you can get your hands on. You will be amazed at what you can learn from sources other than textbooks.

One thing has certainly held true throughout my experience: an air conditioning or refrigeration unit never signs a check for me! People do that, and you should never allow yourself to forget that we fix the problems of people, not machines. They are at the heart of our business, one way or another, and polishing the skills required to provide solid customer service will never be a waste of your time.

Trade Terms
Introduced in This Module

Annual Fuel Utilization Efficiency (AFUE): HVAC industry standard for defining furnace efficiency.

Atomize: The process by which a liquid is converted into a fine spray.

Combustion: The process by which a fuel is ignited in the presence of oxygen.

Condensing furnace: A furnace that contains a secondary heat exchanger that extracts latent heat by condensing exhaust (flue) gases.

Electrode: An electrical terminal that will conduct a current.

Flame rectification: The process by which a flame produces a sensible electrical current.

Heat exchanger: A device, usually metal, that is used to transfer heat from a warm surface or substance to a cooler surface or substance.

Hot surface igniter: A ceramic device that glows when an electrical current flows through it. Used to ignite gas in a gas furnace.

Induced-draft furnace: A furnace in which a motor-driven fan draws air from the surrounding area or from outdoors to support combustion.

Infiltration: Air that enters a building through doors, windows, and cracks in the construction.

Manometer: An instrument that measures air or gas pressure by the displacement of a column of liquid.

Natural-draft furnace: A furnace in which the natural flow of air from around the furnace provides the air to support combustion.

Oil burner: The main component of an oil-fired furnace. It combines oil and air and sprays the combination into the combustion chamber.

Orifice: A precisely drilled hole that controls the flow of gas to the burners.

Piezoelectric: The property of a quartz crystal that causes it to vibrate when a high-frequency voltage is applied to it.

Primary air: Air that is pulled or propelled into the combustion process along with the fuel.

Redundant gas valve: A gas control containing two gas valves in series. If one fails, the other is available to shut off the gas when needed.

Relative humidity: The amount of moisture in the air in relation to the capacity of the air to hold moisture.

Safety pilot: A pilot light with a flame-sensing element.

Secondary air: Air that is added to the mix of fuel and primary air during combustion.

Standing pilot: A gas pilot that is on continuously.

Spud: A threaded metal device that screws into the gas manifold. It contains the orifice that meters gas to the burners.

Thermocouple: A device made up of two unlike metals that generates electricity when there is a difference in temperature from one end to the other.

Additional Resources

This module is intended to present thorough resources for task training. The following reference works are suggested for further study. These are optional materials for continued education rather than for task training.

Fundamentals of Gas Heating, Latest Edition. Tyler, TX: The Trane Company.

General Training—Heating (GTH), Latest Edition. Syracuse, NY: Carrier Corporation.

Heating, Ventilating, and Air Conditioning Fundamentals, Latest Edition. Upper Saddle River, NJ: Prentice Hall.

NCCER makes every effort to keep these textbooks up-to-date and free of technical errors. We appreciate your help in this process. If you have an idea for improving this textbook, or if you find an error, a typographical mistake, or an inaccuracy in NCCER's Contren® textbooks, please write us, using this form or a photocopy. Be sure to include the exact module number, page number, a detailed description, and the correction, if applicable. Your input will be brought to the attention of the Technical Review Committee. Thank you for your assistance.

Instructors – If you found that additional materials were necessary in order to teach this module effectively, please let us know so that we may include them in the Equipment/Materials list in the Annotated Instructor's Guide.

Write: Product Development and Revision
   National Center for Construction Education and Research
   3600 NW 43rd St., Bldg. G, Gainesville, FL 32606

Fax: 352-334-0932

E-mail: curriculum@nccer.org

Craft   Module Name
_____

Copyright Date   Module Number   Page Number(s)
_____

Description
_____
_____
_____
_____

(Optional) Correction
_____
_____
_____

(Optional) Your Name and Address
_____
_____
_____

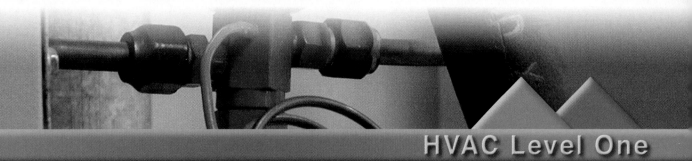

Air Distribution Systems
03109-07

03109-07
Air Distribution Systems

Topics to be presented in this module include:

Overview

The vast majority of heating and cooling systems use ductwork to deliver conditioned air to the spaces being cooled or heated. The ductwork may be made of sheet metal, fiberglass ductboard, fabric, or flex duct. The performance of an HVAC system is closely linked to the quality of the air system. The ductwork must be of the proper size and type, and must be correctly installed and sealed in order for the system to work properly. For those reasons, duct installation is one of the most important duties of the HVAC installer.

Objectives

When you have completed this module, you will be able to do the following:

1. Describe the airflow and pressures in a basic forced-air distribution system.
2. Explain the differences between propeller and centrifugal fans and blowers.
3. Identify the various types of duct systems and explain why and where each type is used.
4. Demonstrate or explain the installation of metal, fiberboard, and flexible duct.
5. Demonstrate or explain the installation of fittings and transitions used in duct systems.
6. Demonstrate or explain the use and installation of diffusers, registers, and grilles used in duct systems.
7. Demonstrate or explain the use and installation of dampers used in duct systems.
8. Demonstrate or explain the use and installation of insulation and vapor barriers used in duct systems.
9. Identify the instruments used to make measurements in air systems and explain the use of each instrument.
10. Make basic temperature, air pressure, and velocity measurements in an air distribution system.

Trade Terms

Air handler
Angle bracket
ASHRAE
Atmospheric pressure
Channel
Cubic feet per minute (cfm)
Dew point
Dry-bulb temperature
Free air delivery
National Fire Protection Association (NFPA)
Plenum

Psychrometrics
Relative humidity (RH)
Revolutions per minute (rpm)
Sheet metal saddle
SMACNA
Static pressure (s.p.)
Total pressure
Velocity
Velocity pressure
Venturi
Volume
Wet-bulb temperature

Required Trainee Materials

1. Paper and pencil
2. Appropriate personal protective equipment

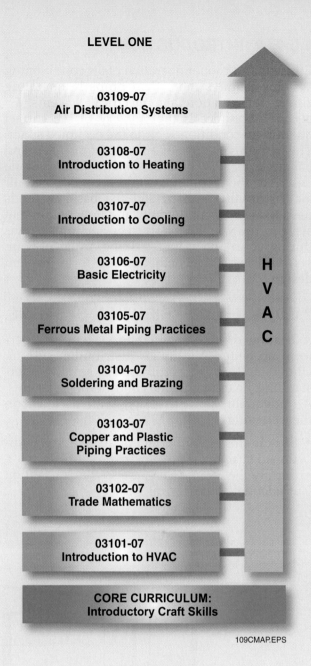

LEVEL ONE

03109-07
Air Distribution Systems

03108-07
Introduction to Heating

03107-07
Introduction to Cooling

03106-07
Basic Electricity

03105-07
Ferrous Metal Piping Practices

03104-07
Soldering and Brazing

03103-07
Copper and Plastic Piping Practices

03102-07
Trade Mathematics

03101-07
Introduction to HVAC

CORE CURRICULUM:
Introductory Craft Skills

H V A C

109CMAP.EPS

Prerequisites

Before you begin this module, it is recommended that you successfully complete *Core Curriculum*; and *HVAC Level One*, Modules 03101-07 through 03108-07.

This course map shows all of the modules in the first level of the *HVAC* curriculum. The suggested training order begins at the bottom and proceeds up. Skill levels increase as you advance on the course map. The local Training Program Sponsor may adjust the training order.

1.0.0 ◆ INTRODUCTION

Efficient and proper operation of air conditioning systems requires more than properly operating closed refrigeration and electrical systems. Of equal importance is the delivery of the correct quantity of conditioned air to the occupied space. This requires the use of a properly installed and balanced air distribution system. This module describes the components that form a forced-air air distribution system and the basic methods used to measure air quantity and flow within that system. You will learn how to balance air systems during your Level Three training. In Level Four, you will learn the methods used in air system design.

2.0.0 ◆ AIR DISTRIBUTION SYSTEMS

A heating or air conditioning system will perform no better than its air distribution system. Understanding air distribution is key to understanding heating and air conditioning systems. All adequate air distribution systems must do the following:

- Supply the right quantity of air to each conditioned space.
- Supply the air in each space so that air motion is adequate but not drafty.

- Condition the air to maintain the proper comfort zones for people, or to maintain the proper conditions needed for a commercial or manufacturing process.
- Provide for the return of air from all conditioned areas to the **air handler.**
- Operate efficiently without excessive power consumption or noise.
- Operate with minimum maintenance.

Most air distribution systems are forced-air systems. The major components that make up a forced-air system are the air handler, air supply system, return air system, and the grilles and registers that allow the circulated air to enter the conditioned space and then return to the conditioning equipment. The air handler is the device that moves the air in a forced-air system. In a split system, it could be a furnace or air handler that contains the blower fan, cooling coil, metering device, air filter, and related housing. The conditioning equipment can be a cooling coil or furnace. *Figure 1* shows the basic components of a forced-air system. The operation of forced-air distribution systems is basically the same for all systems. What changes from system to system is the type of conditioning equipment, size and style of the components, and

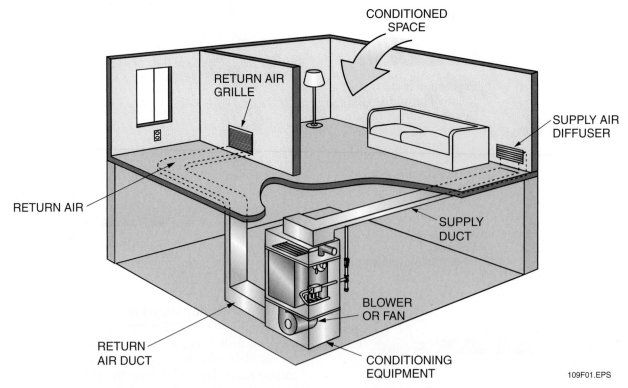

Figure 1 ◆ Basic forced-air distribution system.

109F01.EPS

Duct System Design

INSIDE TRACK

the installed locations of the components. Additional devices called accessories may be used in some systems to gain the desired temperature, humidity, air movement, or air cleanliness. This section describes the basics of a forced-air distribution system.

2.1.0 Airflow and Pressures in the Distribution System Ductwork

Air can be moved by creating a pressure above **atmospheric pressure** (positive pressure) or a pressure below atmospheric pressure (negative pressure). All blowers (or fans) produce both conditions. The air inlet to a blower is below atmospheric pressure, while the exhaust of a blower is above atmospheric pressure.

When a blower is inserted into a duct system, the airflow through the system, except within the blower itself, is the natural flow from a higher to a lower pressure area (*Figure 2*). Normal atmospheric pressure exists in the conditioned space. At the return air grille, the air pressure is slightly lower than atmospheric pressure; therefore, air moves into the duct. The pressure decreases to its lowest point at the blower input. Through the action of the blower, the air pressure is increased to its highest level at the blower discharge. From there, the air resumes its normal natural flow from the higher pressure area at the blower discharge to the lower pressure area of the conditioned space.

The amount of pressure difference needed to move air through a duct system depends on the

velocity, the **volume**, the cross-section area of the duct, and the length of the duct. Velocity is how fast the air is moving and is usually measured in feet per minute (fpm). Volume is a measure, in cubic feet, of the amount of air that flows past a point in one minute. Air velocity and volume can be measured in air distribution systems using anemometers and velometers. You will learn more about these instruments later in this module. Volume in **cubic feet per minute (cfm)** can be calculated by multiplying the velocity of air (in fpm), times the area it is moving through (in square feet) as follows:

$$\text{cfm} = \text{area} \times \text{velocity}$$

Other variations of the formula are:

$$\text{Velocity} = \text{cfm} \div \text{area}$$

$$\text{Area} = \text{cfm} \div \text{velocity}$$

For example, assume that you need to learn if a given duct will support the required air volume. The duct is 14" × 12" and has an air velocity of 900 feet per minute. Because air volume is measured in cubic feet per minute, you must convert the duct dimensions from inches to feet before using the formula. Divide each dimension by 12 to convert from inches to feet.

$$14" \times 12"$$

$$(14 \div 12) \times (12 \div 12)$$

$$1.166 \times 1 = 1.166$$

$$\text{Cfm} = \text{area} \times \text{velocity}$$

$$\text{Cfm} = 1.166 \times 900 = 1{,}050 \text{ cfm}$$

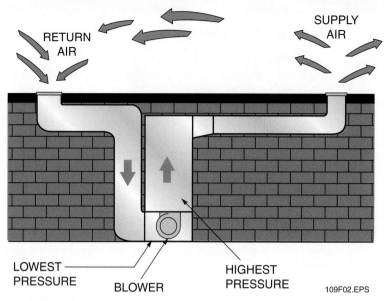

AIR FLOWS FROM HIGHER TO LOWER PRESSURE AREA

RETURN AIR

SUPPLY AIR

LOWEST PRESSURE

BLOWER

HIGHEST PRESSURE

109F02.EPS

Figure 2 ◆ Pressures in an air distribution system.

The inside surface of the duct offers resistance to the flow of air. The velocity of airflow within a duct is not uniform. It varies from zero at the duct walls to a maximum in the center of the duct. This variation can be caused by joints or elbows in the duct, screws or fasteners protruding into the airstream, and various duct lining materials.

In addition to the friction loss in the ductwork, the blower must provide the additional pressure needed to overcome the friction or pressure loss caused by duct fittings (*Figure 3*). Fittings such as elbows, takeoffs, or boots that change the direction of airflow, or change its velocity, also add friction and decrease the quantity of air a duct can carry. The quantity and size of the registers and grilles also affects the airflow in a system. Friction losses have a great impact on the size of the blower and ductwork used in a system. These losses are referred to as static pressure drop, static pressure loss, or friction loss.

The three pressures that exist in a duct system are **static pressure (s.p.), velocity pressure,** and **total pressure.**

Static pressure is the pressure exerted uniformly in all directions within a duct system. In a supply air duct, it is the bursting or exploding pressure that acts on all surfaces of the duct. As shown in *Figure 4*, static pressure can be applied to

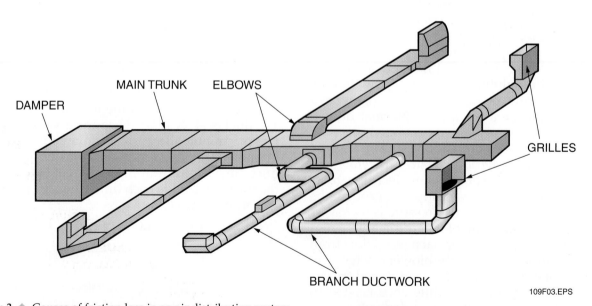

DAMPER

MAIN TRUNK   ELBOWS

GRILLES

BRANCH DUCTWORK

109F03.EPS

Figure 3 ◆ Causes of friction loss in an air distribution system.

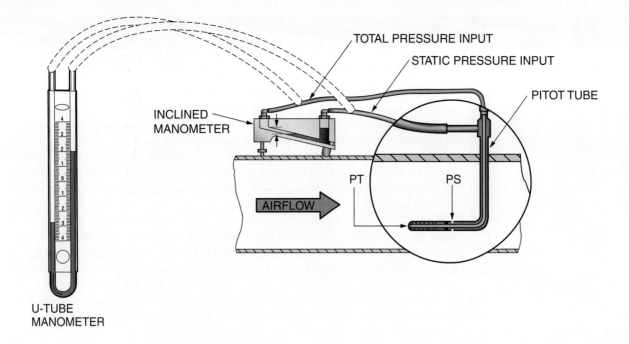

PITOT TUBE SENSES TOTAL AND STATIC PRESSURES. MANOMETER MEASURES
VELOCITY PRESSURE (DIFFERENCE BETWEEN TOTAL AND STATIC PRESSURES).

PT = TOTAL PRESSURE
PS = STATIC PRESSURE

109F04.EPS

Figure 4 ◆ Static, total, and velocity pressures.

a pressure gauge (manometer) for measurement
via the static pressure openings of a pitot tube or
static pressure tip connected to the manometer.
You will learn more about the manometer, pitot
tube, and static pressure tips later in this module.

Velocity pressure is the pressure in a duct system caused by the movement of the air. It acts in
the direction of airflow only. It is the difference
between the total pressure and the static pressure.
Figure 4 shows a manometer connected to give a
reading of velocity pressure. As shown, the pitot
tube must have its opening pointing in the direction of airflow.

Total pressure is the sum of the static and the
velocity pressures in a duct system (*Figure 4*). It is
the pressure produced by the fan or blower.

Due to the low pressures inside a duct system, a
manometer is used to measure duct static, velocity,
and total pressures in inches of water column (in.
w.c.). Inches of water column is the height, in inches,
to which the pressure will lift a column of water.
The atmosphere exerts a pressure of 14.7 psi at sea
level with 70°F dry air. This atmospheric pressure
level of 14.7 psi will support a column of water 33.9',
or 406.8" high (*Figure 5*). Therefore, for every one
pound per square inch of pressure, a column of
water will rise 27.68" (406.8 ÷ 14.7), or about 2.3'.

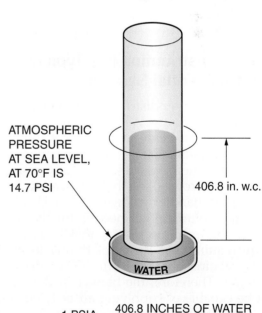

$$1 \text{ PSIA} = \frac{406.8 \text{ INCHES OF WATER}}{14.696 \text{ PSIA}}$$

$$1 \text{ PSIA} = 27.68 \text{ INCHES OF WATER}$$

109F05.EPS

Figure 5 ◆ Comparison of atmospheric pressure to inches
of water.

Absolute Pressure vs. Gauge Pressure

Every square inch of the Earth's surface at sea level at 70°F has 14.7 pounds of air pressure pushing down on it. We have learned that the atmospheric pressure of 14.7 pounds per square inch (psi) at sea level will support a column of water 406.9" high when measured with a manometer. When measured with a mercury-tube barometer like the ones used by meteorologists, the same atmospheric pressure of 14.7 psi will cause the mercury in the tube to rise to a height of 29.92". The values of 14.7 psi and 29.92" of mercury (in. Hg) at sea level, at 70°F, are standards that are used frequently in HVAC work.

The absolute pressure scale is based on the barometer measurements just described. On this scale, pressures are expressed in pounds per square inch (psi) or pounds per square inch absolute (psia) starting from 0 psi. Another scale, called the gauge pressure scale, is frequently used to define air pressure levels. Gauge pressure scales use atmospheric pressures as their starting point. Positive pressures above zero (14.7 psi) are expressed in pounds per square inch gauge (psig). Negative pressures (those below 0 psig) are expressed in inches of mercury vacuum (in. Hg vacuum). Gauge pressures can easily be converted to absolute pressures by adding 14.7 to the gauge pressure value. Absolute pressures can be converted to gauge pressures by subtracting 14.7. Conversion between absolute and gauge pressure scales is often necessary when making calculations concerned with air pressure relationships.

In special low-pressure applications such as blower door testing, pascals (Pa) are used as a unit of measure for air pressure. The normal pressure of the atmosphere at sea level is 101.3 kPa, usually rounded to 100 kPa. To convert psi to kPa, multiply psi by 6.895.

2.2.0 Air Distribution in a Typical Residential System

Figure 6 shows an air distribution system for a typical single-story house. We will discuss the airflow in this system in detail to demonstrate the concepts and pressure relationships you have learned so far, and some new ones. Generally, more airflow is needed for cooling than for heating. Therefore, the air handler (blower) must be able to supply the volume of air needed for the cooling mode. In this example, assume that 3 tons of cooling are required. As a rule of thumb in HVAC, cooling requires about 400 cfm ±50 of air per ton of cooling. Therefore, the blower in this system must be capable of supplying air at 1,200 cfm (3 tons × 400 cfm) or more. Assume that the blower develops a static pressure of 1.0 in. w.c., and that the supply and return ducts have external static pressures of 0.5 in. w.c. and –0.5 in. w.c., respectively. As shown, the system has 11 air supply outlets, each requiring 100 cfm, and two smaller outlets, each requiring 50 cfm. The return air is taken into the system through two centrally located grilles.

While studying this system, consider the entire house as part of the system. The supply air leaves the supply registers and sweeps the walls of the house. Then, it travels through the conditioned spaces within the house as it flows toward the return air grilles. The air is at room temperature at this time. The duct system begins at the two return-air grilles. Relative to the atmospheric pressure of the rooms, there is a slightly negative pressure at the grilles. As shown, the pressure on the blower side of the return-air grille filters is about –0.03 in. w.c., which is lower than the pressure in the rooms. This results in the higher room pressures pushing the air through the return-air filters and into the return duct. As the air flows down the return duct towards the blower, the pressure continues to decrease as a result of friction losses in the duct. At the inlet to the blower, the air pressure is at its lowest point in the system. For our example, it is at –0.5 in. w.c., which is well below the room pressure (0.03 in. w.c.). The return air is forced through the blower, and at the blower output, is increased to its highest level in the duct system. For our example, it is 0.50 in. w.c., which is well above the room pressure. The difference in static pressure between the input and output of the blower is 1.0 in. w.c.

The air at the blower output is pushed through the furnace heat exchanger and the cooling coil,

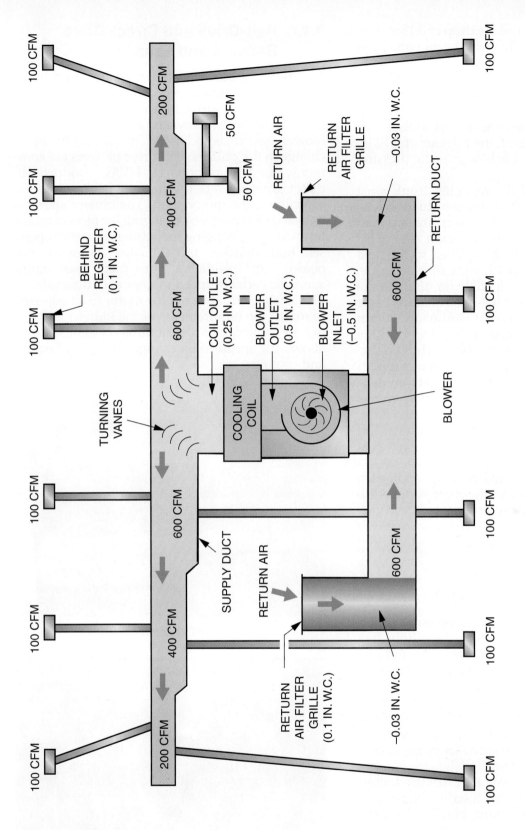

Figure 6 ◆ Typical residential air distribution system.

109F06.EPS

where it encounters a pressure drop of 0.15 in. w.c. At the input to the supply duct, the air enters at a pressure of 0.35 in. w.c. (0.50 in. w.c. – 0.15 in. w.c.). After the air enters the supply duct, it undergoes a slight pressure drop at the tee where the duct is split into two reducing trunks, one to feed each end of the house. It then encounters a discharge loss into the room through the diffuser of 0.10 in. w.c. for a total supply side blower static pressure of 0.50 in. w.c.

Each first section of the reducing trunk must handle 600 cfm of air. Two branch duct outlets, each with an air capacity of 100 cfm, are supplied from the first trunk section on each side. This reduces the quantity of air supplied to the next sections of the trunk to 400 cfm for each side. These sections each supply 200 cfm of air to the conditioned space. This reduces the quantity of air supplied to the last sections of the trunk to 200 cfm for each side, allowing another reduction in the trunk size for each of these sections. The last section of trunk on each side of the system supplies 200 cfm of air to the remaining outlets on each side. In this example, smaller reducing trunks were used to save the cost of materials. Also, reducing the duct size as air was distributed off the trunk keeps the pressure in the duct system at the desired level all along the duct.

Normally, dampers would be installed in each branch to balance the quantity of air supplied to each room. The system in our example will furnish 100 cfm to each outlet, but if a room does not need that much air, the dampers can be adjusted to reduce the quantity.

3.0.0 ◆ FANS AND BLOWERS

The blower or fan provides the pressure difference necessary to force the air into the supply duct system, through the grilles and registers, and into the conditioned space. It must overcome the pressure loss involved in the return of the air as it flows into the return air grilles and through the return ductwork system back to the air handler. In addition, the blower must also overcome the resistance of any other components in the system through which the air passes.

The terms fan and blower are often used interchangeably, but they usually describe the application. For example, the word blower describes applications where the device must work against the resistance of a duct system, like a forced-air system. On the other hand, the word fan describes applications where high quantities of air are needed with little resistance to airflow, such as a ventilation fan.

3.1.0 Belt-Drive and Direct-Drive Blowers and Fans

Two types of blowers are commonly used in air distribution systems: belt-drive and direct-drive (*Figure 7*). In belt-drive blowers, the blower motor is connected to the blower by a belt and pulley. The blower speed is adjusted mechanically by a change in the pulleys. Belt-drive blowers are commonly used in commercial HVAC products. In direct-drive blowers, the blower wheel is mounted directly on the motor shaft. The blower speed is adjusted electrically by changing the blower motor terminals, or changing the settings of motor speed selection switches on a related motor control board. Most residential equipment uses multi-speed or variable-speed motors with direct-drive. This enables the speed of the motor to be adjusted to match the requirements of the individual heating or cooling air distribution system. It also allows the speed to be changed between heating and cooling seasons.

BELT-DRIVE

DIRECT-DRIVE

109F07.EPS

Figure 7 ◆ Belt-drive and direct-drive blowers.

3.2.0 Centrifugal Blowers

Centrifugal blowers (*Figure 8*) are used with forced-air systems because they are designed to work against the resistance of the duct system. They can be used in very large systems that are considered high-pressure systems. High-pressure systems are those that have pressures of 3.5 in. w.c. or greater. In centrifugal blowers, airflow is at right angles (perpendicular) to the shaft on which the wheel is mounted. The wheel is mounted in a scroll-type housing, which is necessary to develop the needed pressures. Centrifugal blowers are identified by the wheel blade position with respect to the direction of rotation. The types of centrifugal blowers include the following:

- Forward-curved
- Backward-inclined
- Radial

3.2.1   Forward-Curved Centrifugal Blowers

Forward-curved centrifugal blowers are normally used in residential and light commercial heating and air conditioning systems. They are also used in light-duty exhaust systems where maximum air delivery and low noise are required. Typically, these blowers are capable of producing pressures up to 3.0 in. w.c. As shown in *Figure 9*, the tips of the blades in a forward-curved blower are inclined in the direction of rotation.

3.2.2   Backward-Inclined Centrifugal Blowers

The blades of the backward-inclined centrifugal blower are inclined away from the direction of rotation. Typically, these blowers are used in commercial and industrial heating and cooling systems that require heavy-duty blower construction and stable air delivery. They are also used extensively as ventilators. Backward-inclined blowers operate at higher efficiencies than forward-curved blowers. They also operate at higher speeds, and therefore tend to be noisier. Typically, these blowers are capable of producing pressures up to 6 in. w.c.

Smaller wheels are usually supplied with flat blades, while larger wheels are supplied with airfoil blades to improve efficiency. Blowers using the airfoil blades generally run more quietly than other types of centrifugal blowers. Also, they do not pulsate within their operating range because air flows through the wheel with less turbulence. *Figure 10* shows a backward-inclined (back-curved) fan wheel and an airfoil fan wheel.

3.2.3   Radial Blowers

Radial blower wheels have straight blades that are, to a large extent, self cleaning. This makes radial blowers more suitable for use in air systems that have large amounts of particles or grease in the air. They can also be used in other applications such as pneumatic conveying systems involved with material handling.

The wheels of radial blowers are simple in construction with narrow blades. They can withstand the high speeds needed to operate at higher static

Figure 8 ◆ Centrifugal blower.

109F08.EPS

Figure 9 ◆ Forward-curved centrifugal blower wheel.

109F09.EPS

pressures (up to 12 in. w.c.). For static pressures above 5.6 in. w.c., the fan rotation speed requires that the wheels and blades be welded. *Figure 11* shows some examples of radial blower wheels.

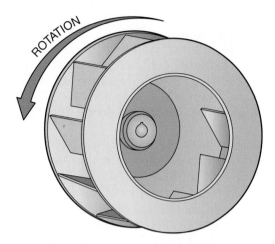

BACKWARD-CURVED FAN WHEEL

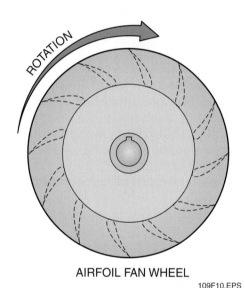

AIRFOIL FAN WHEEL

109F10.EPS

Figure 10 ◆ Backward-inclined centrifugal blower wheels.

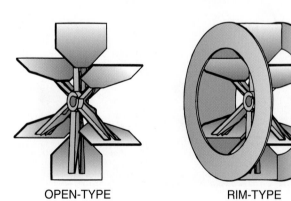

OPEN-TYPE   RIM-TYPE

Figure 11 ◆ Radial centrifugal blower wheels.

3.3.0 Fans

Fans are typically used in applications where high quantities of air are needed with little resistance to airflow, such as with air ventilation and exhaust fans. Although there are many variations in fans, most are either propeller fans (axial) or duct fans (tube-axial).

3.3.1 Propeller Fans

Propeller or axial fans (*Figure 12*) produce an airflow which is parallel or axial to the shaft on which the propeller is mounted. These fans have good efficiency and near **free air delivery** and are commonly used as condenser fans in HVAC applications. Free air delivery is the condition that exists when there are no effective restrictions to airflow (no static pressure) at the inlet or outlet of an air-moving device (fan). Propeller fans are usually mounted in a **venturi** to cause the air to flow in a straight line from one side of the fan to the

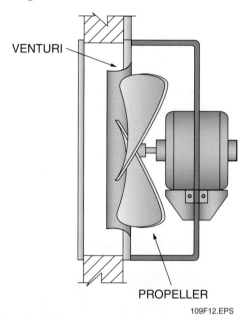

VENTURI

PROPELLER

109F12.EPS

Figure 12 ◆ Propeller-type fan.

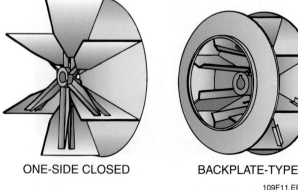

ONE-SIDE CLOSED   BACKPLATE-TYPE

109F11.EPS

other. A venturi is a ring or panel surrounding the blades on a propeller fan. To achieve the best performance from a propeller-type fan, the blade must be properly set in the venturi opening. If the setting is other than that specified by the manufacturer, performance will drop off and the fan might be noisy. Propeller fans make more noise than centrifugal blowers or fans, so they are normally used where noise is not a factor.

3.3.2 Duct Fans

In duct fans, airflow is also parallel or axial to the shaft on which the wheel is mounted. However, duct fans have the propeller housed in a cylindrical duct or tube. This design allows duct fans to operate at higher static pressures than propeller fans. Duct fans are commonly used in spray booths and other ducted exhaust systems. A fan is considered ducted if the duct length is more than the distance between the inlet to and the outlet from the fan blades. The two types of ducted fans commonly used are tube-axial fans and vane-axial fans. A tube-axial fan discharges air in a helical or screw-like motion. A vane-axial fan has vanes on the discharge side of the propeller, which cause the air to discharge in a straight line. This reduces the amount of turbulence, thereby improving the fan efficiency and pressure capabilities. *Figure 13* shows both tube-axial and vane-axial versions of duct fans.

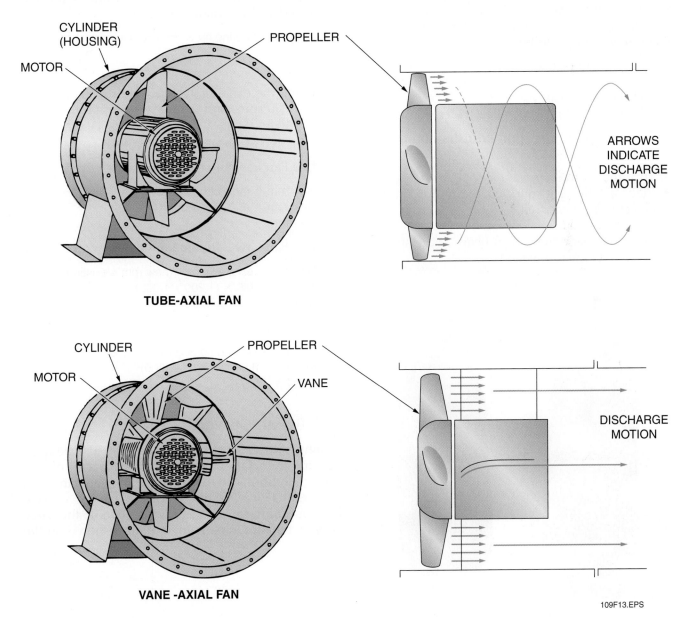

TUBE-AXIAL FAN

VANE -AXIAL FAN

109F13.EPS

Figure 13 ◆ Tube-axial and vane-axial duct fans.

3.4.0 Fan Laws

The performance of all fans and blowers is governed by three rules commonly known as the Fan Laws. Cubic feet per minute (cfm), **revolutions per minute (rpm)**, static pressure (s.p.), and horsepower (hp) are all related. For example, when the cfm changes, the rpm, s.p., and hp will also change. The speed at which the shaft of an air-moving device is rotating is measured in rpm. The easiest way to determine the fan rpm is to measure it directly with a tachometer (*Figure 14*).

There are two types of tachometers: contact and non-contact. Use of the non-contact type is safer and more convenient when the motor is located in a hard-to-reach place. Some manufacturers make a combination contact/non-contact model tachometer that can be used to make rpm measurements either by the contact or no-contact method. To measure motor rpm with the contact-type tachometer, do the following:

- Turn the motor on.
- Contact the end of the motor shaft with the tachometer sensor tip.
- Allow the reading to stabilize, then read the rpm.

To measure motor rpm with the non-contact tachometer:

- Turn the motor off.
- Place a reflecting mark on the motor shaft or object to be measured.
- Turn the motor on.
- Point the tachometer light beam at the shaft or object, then read the rpm.

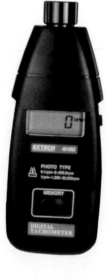

109F14.EPS

Figure 14 ◆ Tachometer.

Fan Laws 1, 2, and 3 are as follows:

- *Fan Law 1* states that the amount of air delivered by a fan varies directly with the speed of the fan. Stated mathematically:

  New cfm = (new rpm × existing cfm) ÷ existing rpm

  or

  New rpm = (new cfm × existing rpm) ÷ existing cfm

- *Fan Law 2* states that the static pressure (resistance) of a system varies directly with the square of the ratio of the fan speeds. Stated mathematically:

  New s.p. = existing s.p. × (new rpm ÷ existing rpm)^2

- *Fan Law 3* states that the horsepower varies directly with the cube of the ratio of the fan speeds. Stated mathematically:

  New hp = existing hp × (new rpm ÷ existing rpm)^3

Example

Assume the existing system conditions are 5,000 cfm, 1,000 rpm, and 0.5 in. w.c., with a fan hp of 0.5. With an increase in airflow to 6,000 cfm, what are the new rpm, s.p., and hp?

- Use Fan Law 1 to calculate the new rpm:

  New rpm = (new cfm × existing rpm) ÷ existing cfm
  New rpm = (6,000 × 1,000) ÷ 5,000
  New rpm = 1,200

- Use Fan Law 2 to calculate the new s.p.:

  New s.p. = existing s.p. × (new rpm ÷ existing rpm)^2
  New s.p. = 0.5 × (1,200 ÷ 1,000)^2
  New s.p. = 0.72 in. w.c.

- Use Fan Law 3 to calculate the new hp:

  New hp = existing hp × (new rpm ÷ existing rpm)^3
  New hp = 0.5 × (1,200 ÷ 1,000)^3
  New hp = 0.864

3.5.0 Fan Curve Charts

Manufacturer's fan curve charts can also be used to find the relationships that exist for a set of system conditions involving s.p., blower/fan rpm, and cfm. *Figure 15* shows a typical fan curve chart. If you know the values for any two of the three characteristics (s.p., rpm, and cfm) shown on the

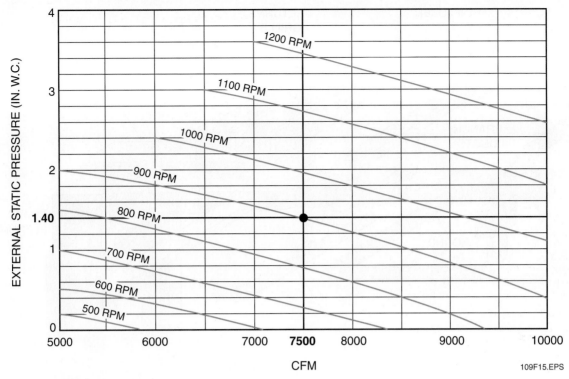

Figure 15 ◆ Typical fan curve chart.

109F15.EPS

chart, you can easily find the value for the other characteristic. For example, assume that the s.p. is 1.4 in. w.c. and the blower is running at 900 rpm, as shown on the chart in *Figure 15*. To find the cfm, locate the intersection point of the 1.4 in. w.c. static pressure line and the 900 rpm curve. From this point, drop down vertically to the cfm scale, then read the value of 7,500 cfm.

3.6.0 System Air Handler Blowers

The blower used in the system air handler must provide the pressure required to send the air to each conditioned space through the supply ductwork system. It must also overcome the pressure loss involved in pushing the air through the diffusers into each room. Additional pressure loss occurs as the air moves into the return air grilles and through the return ductwork system back to the air handler. The air handler must also overcome the added resistance encountered by the system air as it passes through any other system components. The total pressure loss from the duct system external to the air handler is called the external static pressure loss.

Internal resistances of the air handler include pressure loss from the filter, losses in the blower itself, and losses in other components of the air handler. These losses are accounted for by the manufacturer of the air handling unit. Therefore, in the field, you will only be concerned with the external static pressure losses resulting from the system ductwork and its components.

4.0.0 ◆ AIR DISTRIBUTION DUCT SYSTEMS

Air distribution systems consist of the supply duct system and the return duct system. The supply duct system receives air from the output of the system air handler, then distributes the air to the terminal units, and through the registers or diffusers into the conditioned space. The return duct system collects and routes air contained in the conditioned space for return to the input of the system air handler. Air distribution systems used in commercial and industrial structures vary depending on the structure and its intended use. Since air distribution systems used for residential applications are more uniform, they will be used as the basis for discussion in the rest of this section. Except for the system layout and size of the parts, the principles of operation and types of parts used in all duct systems are basically the same.

Figure 16 shows a typical residential distribution duct system that consists of the supply and return air **plenums**, supply and return trunk ducts, branch ducts, supply diffusers, and return grilles. The supply and return trunk ducts attach to their respective plenums. A plenum is a sealed chamber at the inlet or outlet of an air handler.

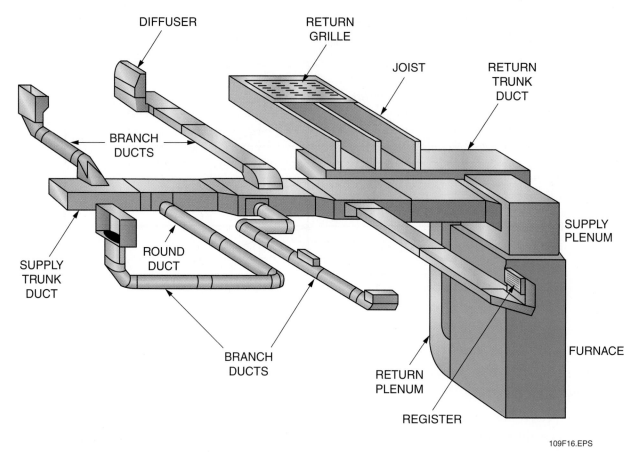

Figure 16 ◆ Typical residential air distribution duct system.

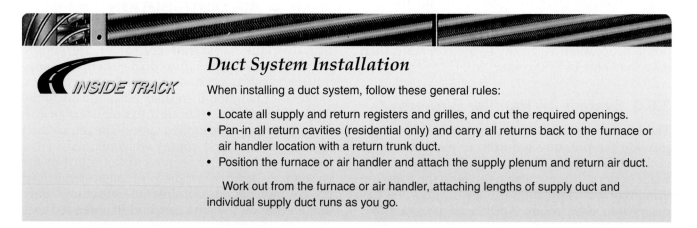

Duct System Installation

When installing a duct system, follow these general rules:

- Locate all supply and return registers and grilles, and cut the required openings.
- Pan-in all return cavities (residential only) and carry all returns back to the furnace or air handler location with a return trunk duct.
- Position the furnace or air handler and attach the supply plenum and return air duct.

Work out from the furnace or air handler, attaching lengths of supply duct and individual supply duct runs as you go.

A good practice is to line the return air plenums with an acoustical duct liner to reduce fan noise, especially if the return grille is close to the furnace. All duct connections to a furnace must extend outside the furnace closet. Return air must not be taken from the furnace room or closet. Also, the height of the return air duct must be high enough to allow the air filter(s) to be removed and replaced. All return air must pass through the air filter after it enters the return air plenum.

Before getting into the details of duct systems, it is helpful to have a basic understanding of duct system design. The main factors that affect the form of a duct system are climate and building construction.

4.1.0 Duct Systems Used in Cold Climates

The type of duct system used in a building is mainly determined by the climate. In cold climates, most buildings use perimeter duct systems. Perimeter systems have floor or baseboard supply diffusers along the perimeter of the building walls.

Use of floor or baseboard supply diffusers provides a good trade-off for heating and cooling performance.

In winter, the warm air supplied by the furnace blankets the outside walls and windows. This compensates for the cold downdrafts that tend to develop at the outside walls, windows, and doors. The return air grilles are located on the interior partition walls, at or near the floor. Central returns may be used, or for better performance, individual returns can be installed in each room. Location of the return grilles on the interior walls near floor level helps to remove any cool air from the floor, where it tends to collect or stratify.

Figure 17 shows the room airflow during the heating and cooling modes of system operation. During the heating mode, the heated air blankets the outside walls and windows. Because it is warmer and lighter than the room air, it spreads across the ceiling and down the inside wall. Room air is drawn (induced) into the flow of warm air and mixes with it. A resulting stratified zone of cool air is formed that tends to collect near the floor. This is mitigated by the use of a low sidewall return.

During the cooling mode, cold supply air travels up the outside wall and windows and strikes the ceiling. Because it is cooler and heavier than the room air, it travels a short distance along the ceiling, then drops back down into the room as shown. The cold air induces the room air fairly well, leaving only a small stratified layer of warm air near the ceiling. High sidewall returns would minimize this problem, but would result in a loss of heating performance.

Perimeter systems can have various layouts. Four common layouts are as follows:

- Loop perimeter
- Radial perimeter
- Extended plenum
- Reducing extended plenum

4.1.1 Loop Perimeter Duct System

Loop perimeter duct systems are common in structures built on concrete slabs (*Figure 18A*). They are easily used with air handlers that are centrally located. The perimeter loop is a continuous round duct of constant size imbedded in the slab. It runs close to the outer walls with the outlets located next to the wall. The perimeter loop is fed by several branches from the plenum. When the furnace fan is running, there is warm air in the whole loop, which helps to keep the slab at a more even temperature. Heat loss to the outside is minimized by the use of insulation around the slab. The loop has constant pressure around the system and provides the same pressure to all outlets.

4.1.2 Radial Perimeter Duct System

In a radial perimeter system (*Figure 18B*) each outlet is fed from a central supply plenum through a separate branch duct. This system is most often used in small homes or with additions containing less than 1,200 square feet of space. The ductwork can be installed in a concrete slab or through a crawl space under the floor. It can also be run in an attic. Ductwork running through unconditioned or vented crawl spaces and attics should be insulated. It should also include a proper vapor barrier if used for cooling. This type of system is the most economical. However, it usually has a poor airflow performance because of static pressure losses at the furnace plenum. These losses are due to duct fittings at the plenum that are poor or excessive in number.

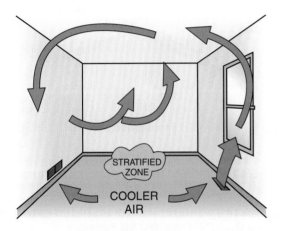

HEATING MODE

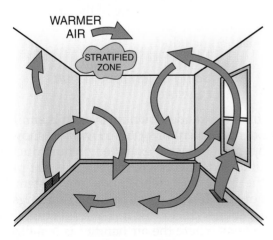

COOLING MODE

109F17.EPS

Figure 17 ◆ Room air distribution patterns for a perimeter duct system.

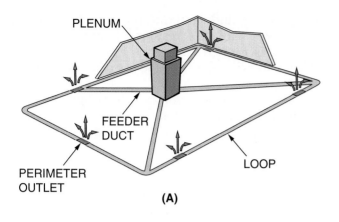

PLENUM

FEEDER DUCT

PERIMETER OUTLET

LOOP

(A)

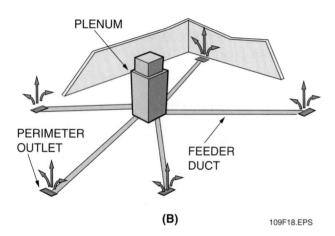

PLENUM

PERIMETER OUTLET

FEEDER DUCT

(B)

109F18.EPS

Figure 18 ◆ Perimeter duct systems.

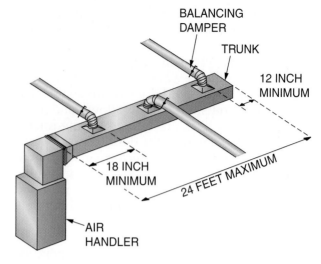

BALANCING DAMPER

TRUNK

12 INCH MINIMUM

18 INCH MINIMUM

24 FEET MAXIMUM

AIR HANDLER

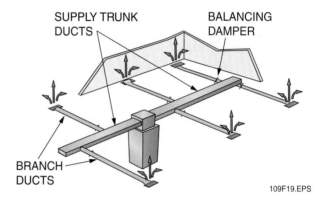

SUPPLY TRUNK DUCTS

BALANCING DAMPER

BRANCH DUCTS

109F19.EPS

Figure 19 ◆ Extended plenum duct system.

4.1.3 Extended Plenum Duct System

The extended plenum duct system (*Figure 19*) uses rectangular trunk ducts as the main supply and return ducts. The supply and return trunk ducts are a constant size over the whole length. This is the reason it is called an extended plenum system. Separate branch ducts run from the trunk duct to each supply outlet. The extended plenum works best when the air handler is located in the center of the main duct; however, it can be run in one direction. The trunk ducts are normally installed near the center line of the building, and their dimensions are constant over the entire length. The branch ducts are normally round, but can be rectangular. An air volume damper is usually installed in each branch duct near the trunk. This allows the airflow to be balanced with all supply air outlets fully open.

An extended plenum duct system is recommended for use in many applications. It requires accurate design, but gives the best performance for both the equipment and customer by allowing proper and equal air distribution throughout the system. The following are some recommended

practices for laying out an extended plenum duct system:

- The supply and return ducts should extend no more than 24' from the air handler.
- The first branch duct should be at least 18" from the beginning of the main duct. This helps to achieve the best balancing of the branch ducts.
- The main trunk should extend at least 12" from the last branch duct.

4.1.4 Reducing Extended Plenum Duct System

A reducing extended plenum system is similar to the extended plenum system. *Figure 20* shows an example of a reducing extended plenum duct system. If an extended plenum system is not practical, the reducing extended plenum system is a good option. It works well in larger buildings that require longer duct runs. It is also a better choice for systems where the air handler is installed on one end of the main trunk duct rather than in the middle. When properly designed, the same pressure drop is maintained from one end of the duct system to the other. This allows each branch duct

Duct Systems in Concrete Slabs

Duct systems in concrete slabs must be installed to prevent heat loss through the slab into the ground. Rigid foam boards are placed between the ducts and the ground before the concrete is poured. In areas with high water tables, a vapor barrier must be installed to prevent ground water from infiltrating the ducts. Only PVC and fiberglass-reinforced plastic duct can be used for this purpose.

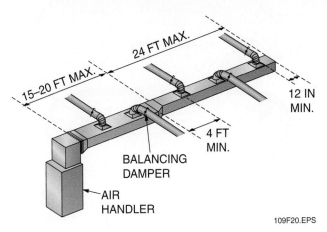

109F20.EPS

Figure 20 ◆ Reducing extended plenum duct system.

to have about the same pressure pushing the air into its takeoff from the trunk duct. The following are some recommended practices for laying out a reducing extended plenum duct system:

- The first main duct section should be no longer than 20'.
- The length of each reducing section should not exceed 24'.
- The first branch duct connection down from a single-taper transition should be at least 4' from the beginning of the transition fitting. This distance allows the air turbulence caused by the fitting to die down before the air is sent into the next branch duct. If the distance is less than 4', the branch ducts near the transition can be hard to balance and may cause the system to be noisy.
- The trunk duct should extend at least 12" from the last branch duct.

4.2.0 Duct Systems Used in Warm Climates

In warm climates, buildings should have duct systems that favor cooling over heating. Perimeter systems like those used in cold climates can work reasonably well in some warm areas. However, their use is normally limited to buildings

constructed over a basement or crawl space. Since cold floors and downdrafts from the outside walls, etc., are not normally a problem in warm climates, the air supply outlets do not need to be located at the building perimeter. In warm climates, supply air openings can be mounted high on the interior walls or in the ceiling to intensify cooling. Return openings are usually wall mounted near the baseboards on interior walls.

Figure 21 shows the room airflow with high sidewall outlets. In the cooling mode, cool air moves across the ceiling and wraps around the far wall. The room air mixes well with the supply air and almost no stratification occurs. In the heating mode, the pulling effect of the low sidewall return air grille tends to draw some of the warm air down from the ceiling and prevents a stratified layer of cold air from building up near the floor.

Ceiling diffusers are one of the best air supply methods used for cooling, but they are the poorest for heating. In the cooling mode, supply air from the diffuser mixes well with the room air (*Figure 22*). Air motion in the room is good with no stagnant areas. In the heating mode, the ceiling diffuser can perform poorly if the return grille is also ceiling mounted. The warm air clings to the ceiling with almost none of it reaching the occupied space in the room. Using a return grille mounted near the floor on an inside wall can help prevent this.

Duct systems used in warm climates can have various layouts. Four typical layouts include:

- Overhead trunk
- Overhead radial
- Attic extended plenum
- Attic radial

4.2.1 Overhead Trunk and Overhead Radial Duct Systems

The overhead trunk and overhead radial duct systems are typically used in buildings on concrete slabs. They are often used in buildings with no attic space, or where the entire air system and the air handler are installed on one floor, such as in

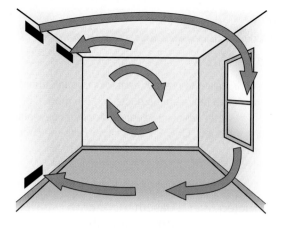

HEATING MODE

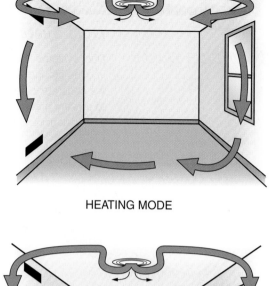

HEATING MODE

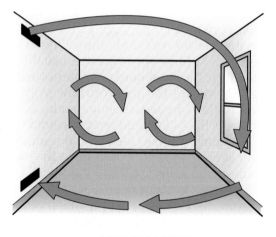

COOLING MODE

109F21.EPS

Figure 21 ◆ Room air distribution patterns for high sidewall outlets.

COOLING MODE

109F22.EPS

Figure 22 ◆ Room air distribution patterns for ceiling outlets.

apartments. Because the equipment is installed in the conditioned space, duct insulation and vapor barriers are normally not required.

In the overhead trunk duct system (*Figure 23*), the air handler and ductwork are usually placed in a drop ceiling area, such as a hallway or closet. The rectangular trunk ducts run from the side of the supply air plenum straight to each high sidewall outlet located in the rooms. If the trunk duct length extends more than 24' from the air handler, a reducing plenum should be used. Overhead trunk systems most often use a central return.

In the overhead radial duct system (*Figure 24*), separate branch ducts are run from a common supply air plenum to each high sidewall outlet. The branch ducts are round metal or flex duct installed within a soffit or drop ceiling. Overhead radial systems most often use a central return.

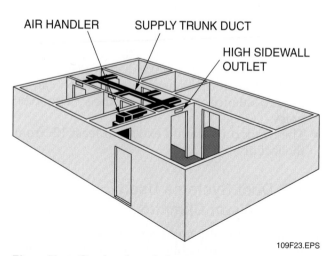

AIR HANDLER   SUPPLY TRUNK DUCT

HIGH SIDEWALL OUTLET

109F23.EPS

Figure 23 ◆ Overhead trunk duct system.

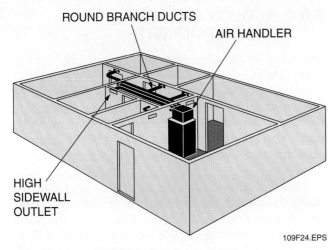

ROUND BRANCH DUCTS

AIR HANDLER

HIGH
SIDEWALL
OUTLET

109F24.EPS

Figure 24 ◆ Overhead radial duct system.

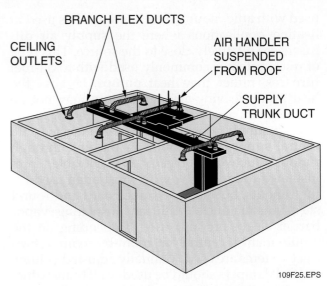

BRANCH FLEX DUCTS

CEILING
OUTLETS

AIR HANDLER
SUSPENDED
FROM ROOF

SUPPLY
TRUNK DUCT

109F25.EPS

Figure 25 ◆ Attic extended plenum duct system.

4.2.2 Attic Extended Plenum and Attic Radial Duct Systems

Attic extended plenum and attic radial duct systems are installed in buildings with attic areas. If a building has an attic, it can be used to house both the air handler and air duct system. This allows the equipment to be installed out of sight. Attic installations eliminate the need for drop ceilings or soffit enclosures in the building. As a result, the amount of living or usable space in the building is increased. The rooms in the building are quieter, because the equipment and its noise have been removed from the rooms. Normally, the air handler is hung from the roof supports with vibration isolators to keep equipment noises from being transmitted through the ceiling.

In the extended plenum duct system (*Figure 25*), the trunk ductwork can be made from insulated fiberglass ductboard or insulated sheet metal with a vapor barrier. If the trunk duct length extends more than 24' from the air handler, a reducing plenum trunk should be used. The branch ducts can be pre-insulated round flex duct, fiberglass ductboard, or round sheet metal covered with an insulating sleeve that has a vapor barrier outside. It should be pointed out that excessive use of flexible duct will result in poor performance and shorter equipment life. Excessive use of flexible duct is considered to be anything more than a 6' section connected at each grille, register, or diffuser. Flexible duct should be used mainly for noise and vibration attenuation. Attic extended plenum systems typically use a central return system. However, because there is full access to the room ceilings, room-by-room returns are also used.

The attic radial duct system, also known as a spider system or plenum system, is commonly

Installing Overhead Supply Duct

INSIDE TRACK

In some locations, a common method of installing an overhead supply duct is to run it concentrically inside the return air plenum. To do this, a chase for the return trunk is framed up around the supply trunk after it and the individual supply ducts are run. Returns are then tied into the framed-in area and the chase is fully enclosed, usually with gypsum drywall.

In operation, return air flows back to the furnace or air handler in the space surrounding the supply duct. To prevent leakage between the supply and return air sides, all supply duct joints should be tight and leak free.

used with attic-mounted air handlers. It is used in heating applications where the supply air diffusers are relatively close to the source. This type of duct system is commonly used with fossil-fuel furnaces, rather than heat pumps, because the heat pump leaving-air temperature may not be high enough for efficient system operation.

Each supply room air outlet is connected to a central supply air plenum on the unit through runout ductwork (*Figure 26*). Flexible, pre-insulated round ductwork is most often used for the runouts, but fiberglass ductboard and round sheet metal ducts with an insulating/vapor barrier sleeve are also used. Depending on the requirements, central or room-by-room return duct systems are used. Manually adjusted volume control dampers should be used in all branch duct runs to facilitate balancing the system airflow. Balancing air volume at the supply air diffusers is poor practice and should be avoided.

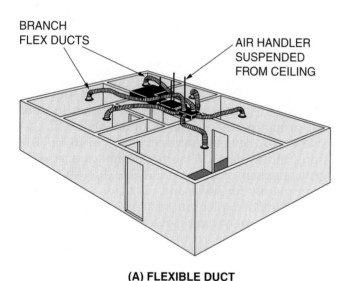

(A) FLEXIBLE DUCT

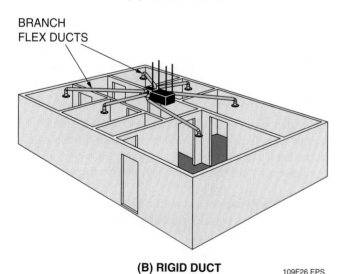

(B) RIGID DUCT

109F26.EPS

Figure 26 ◆ Attic radial duct system.

5.0.0 ◆ DUCT SYSTEM COMPONENTS

Building code requirements pertaining to the installation of air distribution systems are not standard across the nation. Almost all localities have minimum standards or codes that determine the type of materials and methods that must be used. The HVAC technician should become familiar with and follow the local codes that apply to each job.

The selection and size of trunk and branch ducts used in an air distribution system are based on the air volume (cfm) needed to satisfy the heating and/or cooling requirements for the building. For new buildings, this information can usually be found in the design specifications. The method normally used to find the required air volume for an existing building is to survey the structure. Armed with known values for air volume, you can use friction charts, duct sizing charts, and/or duct size calculators to find the correct duct sizes for the job. After the duct sizes are known, the

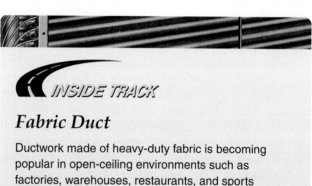

INSIDE TRACK

Fabric Duct

Ductwork made of heavy-duty fabric is becoming popular in open-ceiling environments such as factories, warehouses, restaurants, and sports complexes. This ductwork is available in round and half-round configurations. The latter is ideal for surface-mount installations. Duct sections can be zippered together to obtain the required length. Fittings such as tees and elbows are available.

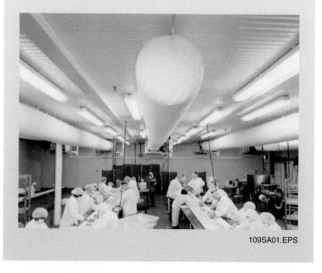

109SA01.EPS

selection of the parts used for the layout is mainly determined by the construction of the building.

The tables in the appendix provide typical sizes for residential duct system components based on various airflow requirements. This section describes these components and overviews some of the application considerations. The basic components of a duct system include the following:

- Main trunk and branch ducts
- Fittings and transitions
- Air diffusers, registers, and grilles
- Dampers
- Insulation and vapor barriers

5.1.0 Duct Materials

There is a wide variety of materials used in HVAC system duct designs. *Table 1* shows a few of the common materials used in duct systems, and where the material would be applied. As the table shows, the most common material used is galvanized steel. The other materials are used for more specific applications such as kitchen exhausts, moist air, and so on. Airshafts, typically found in multi-story buildings, are usually made of concrete or gypsum board.

5.2.0 Main Trunk and Branch Ducts

Duct systems can be installed in basements, crawl spaces, attics, and within concrete floors or slabs. Ducts can be made from metal, fiberglass ductboard, ceramic, fabric, or plastic materials. Galvanized sheet metal or fiberboard ducts are typically used for heating/cooling air distribution. When installed in a concrete slab, ducts are usually made of metal, plastic, or ceramic. Spiral metal and flexible ducts are also in common use. Where weight is a factor, aluminum duct is sometimes used.

5.2.1 Galvanized Steel Trunk and Branch Ducts

Galvanized steel or sheet metal duct can be round, square, or rectangular (*Figure 27*). All three types are often used in the same duct system. Popular sizes of round, square, and rectangular steel ducts, along with an assortment of standard fittings, can be obtained from HVAC supply houses. For large commercial jobs involving customized ductwork, the ducts and fittings are often made separately in a metal shop or fabricated at the job site.

Because sheet metal duct is rigid, the layout must be well planned, and all the pieces cut precisely, or the system will not fit together.

The thickness of galvanized steel and other metal duct is expressed in terms of gauge thickness. When a duct is made of 28-gauge sheet metal, this means that the thickness of the duct walls is $\frac{1}{28}$". Likewise, a sheet metal duct made out of 24-gauge metal has a wall thickness of $\frac{1}{24}$", etc. Larger ducts are made from thicker metal and are more rigid than smaller ducts. This prevents them

Table 1   Duct Materials and Their Applications

Material	Applications
Galvanized steel	Widely used for most HVAC systems
Aluminum	For systems with high moisture-laden air, or special exhaust systems
Stainless steel	For kitchen exhaust, fume exhaust, or high moisture-laden air
Concrete	Underground ducts and air shafts
Rigid fibrous glass	Interior, low-pressure HVAC systems
Gypsum board	Ceiling plenums, corridor ducts, and air shafts

109T01.EPS

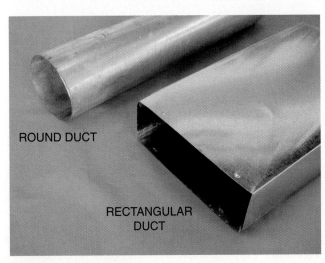

ROUND DUCT

RECTANGULAR DUCT

Figure 27 ◆ Metal ducts.

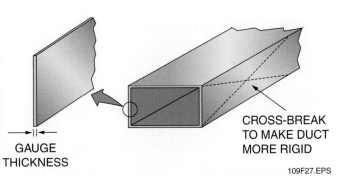

GAUGE THICKNESS

CROSS-BREAK TO MAKE DUCT MORE RIGID

109F27.EPS

from swelling and making popping noises when the system blower starts and stops. Also, lines or ridges, normally called cross-breaks, are used on large sheet metal panels or ducts to make them more rigid (*Figure 27*).

The aspect ratio of a duct is often used to classify a duct size and estimate its cost. Aspect ratio is the ratio of the duct's width to its height. For example, if a duct is 18" wide and 6" high, the aspect ratio is 18:6, or 3:1. *Table 2* shows some typical gauge thicknesses used for rectangular and round metal ducts. It also shows a tabulation of duct aspect ratios.

Sections of square or rectangular duct are assembled using any one of several fasteners that are available. Typically, S-fasteners and drive clips and/or snap-lock fasteners are used (*Figure 28*). In addition to the connection methods shown, there are numerous other types of connection devices for metal duct. *Figure 29* shows several examples.

Round duct sections are normally fastened together with self-tapping sheet metal screws. These fasteners make a nearly airtight connection. When further sealing is needed, the joint can be taped using special duct tape or sealed with a flexible duct sealing compound which can be applied with a paint brush or caulking gun. Leaking joints cut down on the amount of air available for delivery to the outlets at the end of long runs.

Table 2 Typical Metal Duct Gauge Thickness and Aspect Ratios

| Rectangular Duct Width in Inches | Commercial | | Residential |
	Sheet Metal Galvanized	Aluminum	Sheet Metal Galvanized
UP TO 12	26	0.020	28
13–23	24	0.025	26
24–30	24	0.025	24
31–42	22	0.032	–
43–54	22	0.032	–
55–60	20	0.040	–
61–84	20	0.040	–
85–96	18	0.050	–
OVER 96	18	0.050	–

RECTANGULAR DUCT

Round Duct Diameter	Commercial Sheet Steel Galvanized Gauge	Residential Sheet Steel Galvanized Gauge
UP TO 12	26	30
13–18	24	28
19–28	22	–
27–36	20	–
35–52	18	–

ROUND DUCT

Duct Class (Aspect Ratio)	Width in Inches	Perimeter
1	6–8	24–72
2	12–24	36–72
3	26–40	70–106
4	24–88	60–220
5	48–90	116–216
6	90–145	210–336

ASPECT RATIO

109T02.EPS

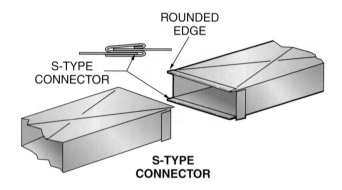

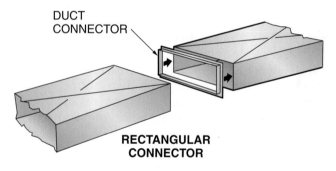

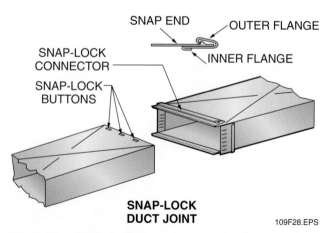

109F28.EPS

Figure 28 ◆ Typical square and rectangular sheet metal duct fasteners.

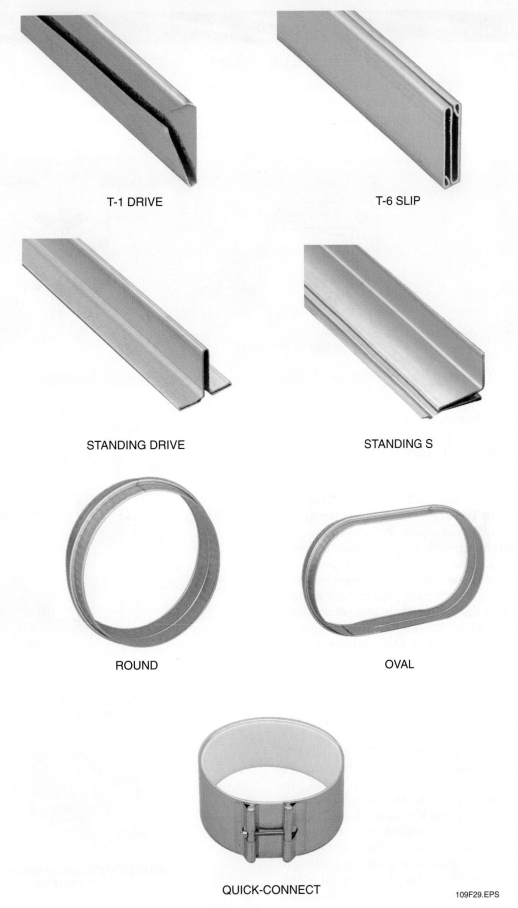

T-1 DRIVE

T-6 SLIP

STANDING DRIVE

STANDING S

ROUND

OVAL

QUICK-CONNECT

109F29.EPS

Figure 29 ◆ Metal duct connectors.

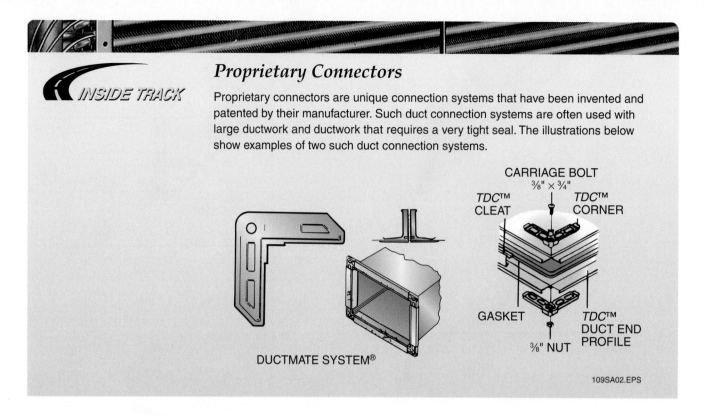
A ductwork system must be well supported so that it does not move. If it is not properly supported, movement can occur when the fan starts, causing a rush of air through the system. Sheet metal ductwork can also move as a result of expansion and contraction as it heats and cools. This type of movement can be contained by using flexible or fabric joints at different points in the system.

Sheet metal ductwork systems also transmit vibrations from the air handling equipment. Transmission of these vibrations to the duct system can be prevented by using flexible connectors or fabric joints at the main supply and return ductwork connections to the air handler. *Figure 30* shows some typical devices for controlling ductwork noise and vibration.

5.2.2 Fiberglass Duct

Fiberglass ductboard can be used instead of metal duct. It has more friction losses than metal duct, but is quieter because the ductboard absorbs blower and air noises better. Fiberglass duct is available in flat sheets for fabrication, or as prefabricated round duct sections. Fiberglass duct is normally 1", 1½", or 2" thick with an aluminum foil backing. This backing is reinforced with fiber to make it strong. The inside surface of the ductboard is coated with plastic or a similar coating to prevent the erosion of the duct fibers into the supply air. Fiberglass particles released into the air can be harmful to health.

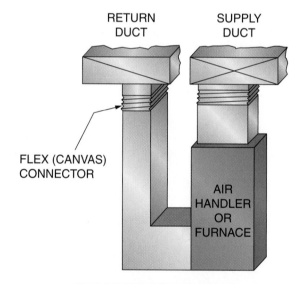

RETURN DUCT   SUPPLY DUCT

FLEX (CANVAS) CONNECTOR

AIR HANDLER OR FURNACE

VIBRATION/NOISE CONTROL AT AIR HANDLER

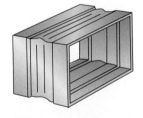

VIBRATION/NOISE/MOVEMENT CONTROL IN DUCT RUNS

109F30.EPS

Figure 30 ◆ Ductwork vibration and noise control devices.

Health Issues

At one time, there were concerns about ductboard on three fronts. The first was the potential for the porous material to absorb moisture from the air and allow the growth of mold and fungus spores. This could potentially cause severe health problems for building occupants. The second concern was the difficulty in cleaning ductboard because of the potential to damage the rather fragile material and release fiberglass particles into the airstream. The third concern was the potential for the glass fibers to loosen from the duct over time and be circulated with the conditioned air. All of these concerns have been largely eliminated by the use of biocides in the fiberglass to prevent growth of microorganisms and the application of an interior coating to contain the fiberglass and make the inside of the duct easier to clean.

The ductwork is made from sheets of fiberglass ductboard using special machines or knives. These devices cut away the fiberglass to form the edges, and the reinforced foil backing is left intact to support the connections. When two pieces are fastened together, an overlap of foil is left so that one piece can be stapled to the other using special staples (*Figure 31*). The joint is then taped to make it airtight.

Once the ductboard has been assembled, it must be sealed using the appropriate closure system, in accordance with applicable codes and job specifications. The duct tape you can buy at the local discount store is not acceptable for this use. Only closure systems that comply with *UL 181A* are suitable for use with rigid fiberglass duct systems. These closure systems include the following:

- Pressure-sensitive aluminum foil tapes listed in *UL 181A, Part I (P)*
- Heat-activated aluminum foil or scrim tapes listed in *UL 181A, Part II (H)*. This is the preferred method.
- Mastic and glass fabric tape closure systems listed in *UL 181A, Part III (M)*.

Round fiberglass duct is also easy to install because it can be cut to size with a knife. Fiberglass ductwork systems must be properly supported or they will sag over long runs. Special hangers designed not to cut the outside cover of the ductboard must be used.

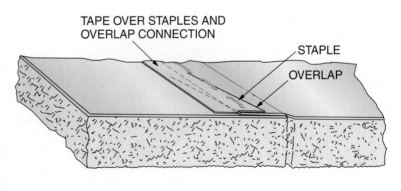

109F31.EPS

Figure 31 ◆ Fiberglass ductboard.

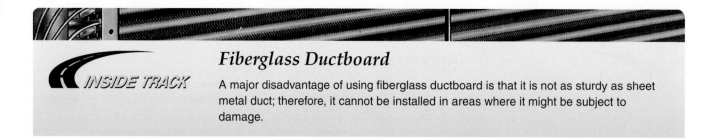

Fiberglass Ductboard

A major disadvantage of using fiberglass ductboard is that it is not as sturdy as sheet metal duct; therefore, it cannot be installed in areas where it might be subject to damage.

5.2.3 Flexible Duct

Flexible round duct (*Figure 32*) comes in sizes up to 24" in diameter. It is available with a reinforced aluminum foil backing for use in conditioned areas. It also comes wrapped with insulation protected by a vapor barrier made of fiber-reinforced vinyl or foil backing for use in unconditioned areas.

Flexible duct is typically used in spaces where obstructions make the use of rigid duct difficult or impossible. Flexible duct is easy to route around corners and other bends. Duct runs should be kept as short and as straight as possible. Gradual bends should be used, since tight turns can greatly reduce the airflow and may even cause the duct to collapse. If a connection to a ceiling diffuser needs an elbow, it is better to use an insulated metal elbow at the input to the diffuser than to bend flexible duct tightly to form the connection. This is because diffuser performance is disrupted far less by the metal elbow.

Long runs of flexible duct are not recommended unless the friction loss is taken into account. Even when properly installed, most flex

ducts cause at least two to four times as much resistance to airflow as the same size diameter sheet metal duct. To avoid sags in the run, flexible duct should be amply supported with 1" wide or wider bands to keep the duct from collapsing and reducing the inside dimension. Some flexible duct comes with built-in eyelet holes for hanging.

As mentioned earlier, any duct system that uses excessive flexible duct will result in poor performance and shorter equipment life. Excessive use of flexible duct is considered to be anything more than a 6' section connected at each grille, register, or diffuser. Flexible duct should be used mainly for noise and vibration attenuation.

5.3.0 Fittings and Transitions

Fittings in ducts, such as elbows, takeoffs, or boots change the direction of airflow or change its velocity. Transitions are typically used to change from one size duct to another. They are also used to change from one duct shape to another. *Figure 33* shows examples of duct fittings.

Air moving in a duct has inertia that makes it want to continue flowing in a straight line. Each fitting in a duct run adds friction and decreases the quantity of air the duct can carry. It takes energy to overcome the resistance (friction) inherent in a fitting. Because of this, the number of fittings used in a duct system, and their types, must be carefully selected in order to minimize the total amount of friction added to the system.

Fittings and/or transitions are rated by equivalent feet of length. This means that a specific fitting produces a pressure drop equal to a certain number of feet of straight duct length of the same size. Therefore, adding fittings has the same effect on the pressure loss of a duct as increasing its overall length. This is why duct runs should be made as straight as possible to each room. Also, the use of unnecessary fittings, or ones not best suited for the job, must be avoided. For each standard type of fitting, the pressure drop is known and has been converted to the equivalent

109F32.EPS

Figure 32 ◆ Flexible duct.

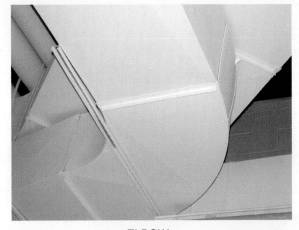
ELBOW

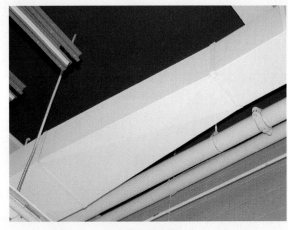
REDUCER

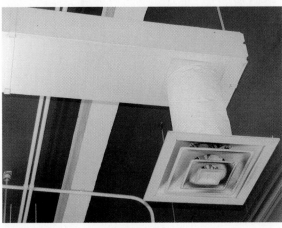
TAKEOFF

TEE

109F33.EPS

Figure 33 ◆ Duct fittings.

feet of duct length. This information is available in a set of charts that show the standard types of fittings and/or transitions and the value for the equivalent feet of length used for each one. These charts are available in **ASHRAE** and **SMACNA** publications.

The total equivalent feet of length for a duct run is calculated by adding all the equivalent lengths for fittings in the run to the actual length of straight duct used. In the example shown in *Figure 34*, an elbow with an equivalent length of 30' is added to a duct with 100' of straight length. The resulting pressure drop is the same as that of a straight duct 130' long.

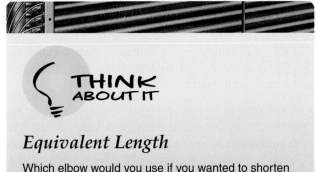

THINK ABOUT IT

Equivalent Length

Which elbow would you use if you wanted to shorten the equivalent length of the run in the example shown in *Figure 34* from 130' to 110'?

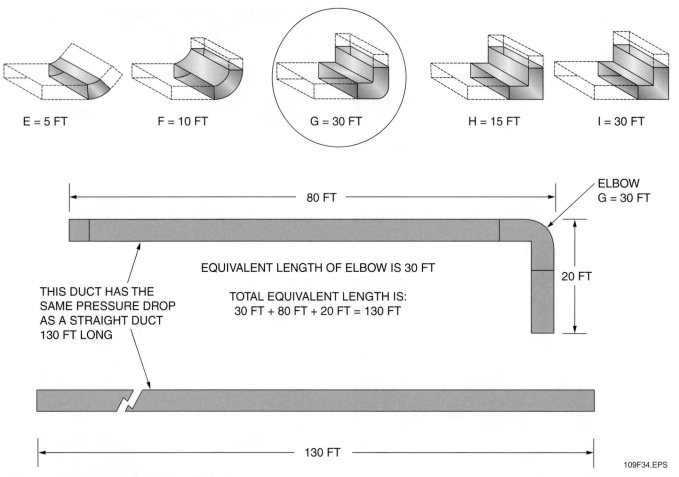

Figure 34 ◆ Example of equivalent length.

5.4.0 Air Diffusers, Registers, and Grilles

Air outlets distribute the supply air into the conditioned space. When properly selected, they act to blend the supply air with the room air so that the room is comfortable without excess noise or drafts. The terms diffuser, register, and grille are used to describe different kinds of outlets.

- A diffuser is an outlet that discharges supply air into a room in a widespread fan-shaped pattern.
- A register is an outlet that discharges supply air into a room in a concentrated non-spreading stream. Many have one-way and two-way adjustable air stream deflectors.
- A grille is the louvered covering of an opening created for the passage of air into a room. It controls the distance, height, and spread of airflow, as well as the amount of air delivered to the space. Grilles have many different designs. Some are fixed and can direct air in one direction only. Others are adjustable and can be set to send air in different directions. Grilles with no adjustments are typically used as covers for the return air duct.

Figure 35 shows several registers and diffusers in common use. As shown, the floor register is relatively long and narrow. It gives excellent performance for both heating and cooling when used in perimeter duct systems. A floor register is usually installed parallel to the room's outside wall at about 6" to 8" away from the wall. Floor registers are fed from below and discharge air upward. Typically, fixed vanes are installed crosswise at various angles to spread the air stream so that it blankets the outside wall and windows with supply air. Floor registers normally have a built-in shutoff damper.

Low sidewall registers are excellent for heating when used in perimeter systems. They also work well for cooling, if designed to discharge air upward. If not so designed, a plastic air scoop accessory may be added to the exterior to redirect the supply air upward for cooling. Low sidewall registers are fed from the back and mounted flush with the wall just above the baseboard trim. Air is discharged outward and slightly upward. Some are made to discharge air in two or three directions. Low sidewall registers normally include a built-in shutoff damper.

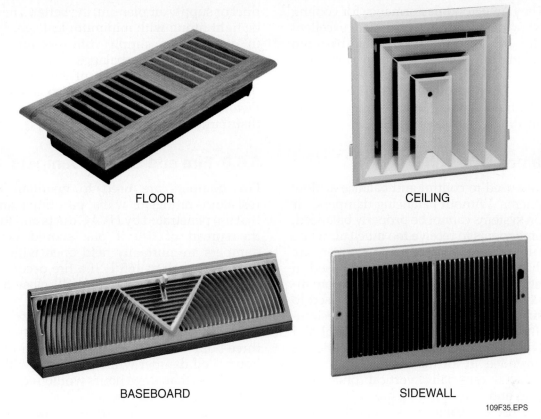

FLOOR

CEILING

BASEBOARD

SIDEWALL

109F35.EPS

Figure 35 ◆ Air registers and diffusers.

High sidewall registers provide poor heating performance in cold climates because they leave a cold layer of air in the lower half of the room, especially with slab floors. However, they are adequate for heating in warm climates. High sidewall registers provide good performance in cooling when used with central returns, and when used with room-by-room returns, the cooling performance is even better. High sidewall registers are mounted flush on the room's inside wall and fed from behind. The top edge is usually mounted 6" to 12" down from the ceiling. They typically discharge air horizontally toward the outside wall. Some are made to discharge air in two or three directions. High sidewall registers normally include a built-in shutoff damper.

Baseboard diffusers are long and narrow. They are mounted on the floor with their back mounted snugly against the wall. Supply air is fed from below and discharged upward close to the wall so as to blanket a wide area of the outside wall and windows. Baseboard diffusers normally have a built-in shutoff damper. They perform well in both heating and cooling when used with perimeter duct systems.

Ceiling diffusers can be round or rectangular in shape. They mount flush to the ceiling and are fed supply air from above. Ceiling diffusers are made that distribute supply air equally in all directions. Other styles are made that distribute air in one, two, three, or four directions. Ceiling diffusers are made both with and without a built-in shutoff

INSIDE TRACK

Registers and Grilles

Not all registers and grilles are created equal. Inexpensive grilles and registers can add significant static pressure losses to the duct system, contributing to poor system performance. Always check the register or grille manufacturer's specification sheets to make sure the grilles and registers selected for a job will enhance and not hinder system performance.

damper. They give good performance for cooling when used with a central return, and excellent performance when used with room-by-room returns. Ceiling diffusers give poor performance for heating in cold climates because they leave a cold layer of air in the lower half of the room, especially with slab floors. However, they are adequate for heating in warm climates.

5.5.0 Dampers

Dampers are used to control and balance airflow in duct systems. Without balancing dampers, air distribution systems cannot be properly balanced, causing some rooms to receive too much air while others do not receive enough. Some dampers are made with manual adjustments. Others, used in zoned heating or cooling systems, are automatically controlled. Sometimes dampers are used to mix two airflows, such as with fresh and recirculated air. By code requirements, commercial and industrial buildings normally have automatic fire dampers installed in all the vertical duct runs, since all ducts, especially vertical ones, will spread fumes and flames from a fire.

A damper used to balance a system should be installed in an accessible place in each branch supply duct. The closer the dampers are to the main

duct or supply air plenum, the better. They should be tight fitting with minimum leakage. The built-in dampers on supply diffusers and registers should not be used to balance an air system. When partially closed, they disrupt the performance of the diffuser or register and also make it noisy. *Figure 36* shows three types of dampers used in air distribution systems.

5.6.0 Fire and Smoke Dampers

Fire dampers are used to maintain the fire-resistance ratings of walls, partitions, and floors that are penetrated by HVAC ducts and to prevent the spread of fire if one should occur. The dampers are normally held open with a fusible link set to melt at 165°F. If a fire occurs, the link will melt and the dampers will close automatically by spring action or gravity.

All fire dampers have a resistance rating of either 1½ or 3 hours. Building partitions with a three-hour fire rating require the use of a three-hour rated damper while partitions with fire ratings of less than three hours would require the use of 1½-hour rated dampers.

In addition to the fire-resistance rating, fire dampers have a static or dynamic air closure rating. Static-rated dampers can only be used in

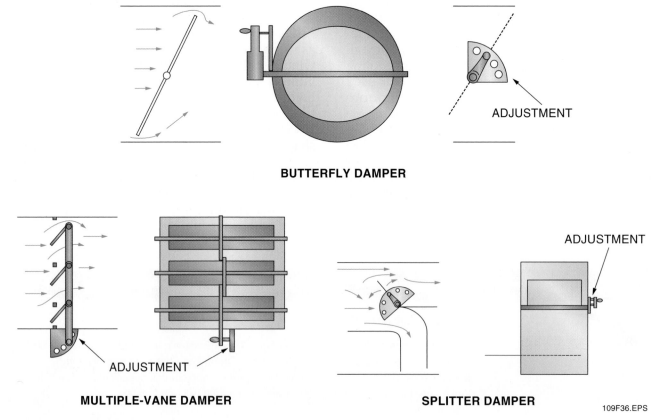

BUTTERFLY DAMPER

MULTIPLE-VANE DAMPER

SPLITTER DAMPER

109F36.EPS

Figure 36 ◆ Typical dampers.

HVAC systems where the HVAC equipment will be automatically shut down in case of fire. In that case, no air would be flowing within the ducts. Dynamic-rated dampers are designed to close even if the HVAC system remains running and air is moving through the ducts.

Smoke can be equally as deadly as fire, so controlling the spread of smoke in a building is critical. Smoke dampers in the ducts of HVAC systems perform this critical task. Smoke dampers can be passive in their function where they operate to simply shut off and isolate a section of duct. They also can be part of an engineered smoke control system that allows the building duct system to direct air and/or smoke in such a way as to prevent the spread of fire and smoke or to move the smoke to an area where it poses no problem.

Some smoke dampers are operated electrically or pneumatically and are controlled by a smoke or heat detector, fire alarm, or automated building control system. Others are controlled by a fusible link. Smoke dampers are rated for leakage. Class 1 has the lowest leakage rating with classes 2, 3, and 4 having higher leakage ratings. Smoke dampers are rated for temperature and have a velocity and pressure rating which indicates how they will operate against specific airflow and pressure differential conditions within the duct.

Combination fire and smoke dampers (*Figure 37*) are also available. These dampers must conform to the rating agency requirements for both fire and smoke dampers.

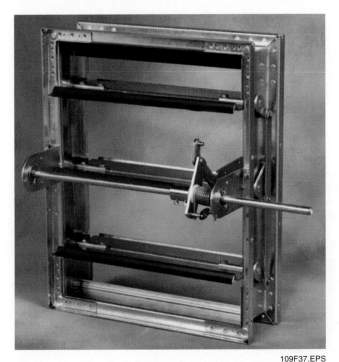

109F37.EPS

Figure 37 ◆ Combination fire and smoke damper.

5.7.0 Insulation and Vapor Barriers

When ductwork passes through an unconditioned space, heat transfer may take place between the air in the duct and the air in the unconditioned space. If the heat exchange adds or removes very much heat from the conditioned air, insulation should be applied to the ductwork. A difference of 15°F from the inside to the outside of the duct is considered the maximum difference allowed before insulation is required. Many installations use preinsulated ductboard for the main supply and return ducts. In this case, the insulation and vapor barrier is the duct itself.

Metal duct can be insulated in two ways: on the outside or on the inside. Insulation inside the duct is installed by the duct manufacturer. It is either glued or fastened to tabs mounted on the inside duct wall. Insulation and a vapor barrier can also be wrapped around the outside of the ductwork after it has been installed. The insulation is usually a foil or vinyl-backed fiberglass. It comes in several thicknesses, with 2" being typical. The backing creates a moisture vapor barrier. If the duct operates below the dew-point temperature of the outside air, use of a vapor barrier is important in order to prevent the moisture in the air outside the ductwork from condensing on the duct. Once installed, all joints must be properly sealed with approved duct tape. To avoid condensation damage, any punctures, seams, and slits in the vapor barrier must also be sealed.

Ductwork systems with outside insulation have a lower pressure loss, and are therefore more efficient than systems made from ductboard or sheet metal and lined on the inside. Another advantage is that the cost for the metal duct is cheaper because the physical size of the duct can be smaller. A duct with 1" of internal insulation must have a width and height dimension that is 2" larger in order to deliver the same amount of air.

A disadvantage of using duct with outside insulation is that it takes longer to install and there is a greater chance of damaging the insulation during installation. They also tend to be noisier than a lined system.

Special care must be taken when installing duct systems in crawl spaces, attics, and similar unconditioned areas. When an enclosed crawl space itself is properly insulated and protected by an adequate vapor barrier, the ductwork may be installed without any additional insulation or vapor barrier. If a crawl space is ventilated rather than enclosed, or if an existing vapor barrier is questionable, the duct system should be insulated and must include an adequate vapor barrier. Sheet metal main trunk ductwork with inside insulation and taped joints can be used with no extra vapor sealing. Sheet metal ductwork with outside insulation can also be used, provided it is covered with a vapor barrier.

When branch ductwork is constructed of sheet metal, an external sleeve is normally installed over each branch duct. The insulating sleeve must have a plastic or foil layer outside the insulation to act as a vapor barrier. All vapor barrier joints must be sealed with approved tape, otherwise the moisture from the air will go through the insulation and condense on the duct, causing the duct to drip and damage the insulation. *Figure 38* shows a typical residential installation for a ventilated crawl space.

In attic installations, the ductwork must be insulated in order to maintain proper cooling/heating in the conditioned rooms of the building. *ASHRAE Standard 90-80* specifies the minimum acceptable R-value of insulation that must be used. *Figure 39* shows an example of *ASHRAE Standard 90-80* used to calculate the R-value for insulation related to the cooling mode. The R-value must also be calculated for the heating mode in the same way. The amount of insulation actually used for the system is determined by the mode with the greatest need for insulation. Since attic systems are more common in warm climates, it is the cooling mode that usually determines the amount of insulation required.

Uninsulated sheet metal ductwork provides no resistance to heat transmission. This means that the entire resistance for the main supply duct must come from insulation either inside or outside of the duct. Typically, 1" of insulation is equal to an R-value of 4. Branch runs are usually made using

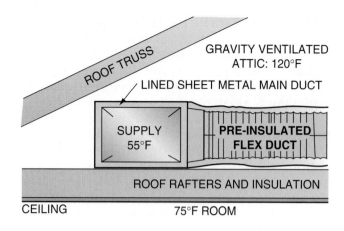

$$\text{TD DUCT} = 120 - 55 \qquad R = TD/15 \qquad 1" = \text{ABOUT}$$
$$= 65°F \qquad\qquad = 65/15 \qquad\quad R\text{-}4$$
$$= 4.3$$

$$R = TD/15$$

WHERE:

R = THERMAL RESISTANCE TO HEAT FLOW OF THE DUCT WITH INSULATION (R-VALUE)

TD = TEMPERATURE DIFFERENCE BETWEEN INSIDE AND OUTSIDE DUCT

109F39.EPS

Figure 39 ◆ Example of R-value calculation using *ASHRAE Standard 90-80*.

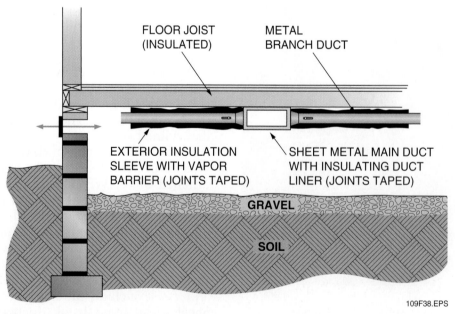

109F38.EPS

Figure 38 ◆ Typical duct installation in a ventilated crawl space.

pre-insulated, flexible duct, but other material combinations can also be used. Even though return duct systems have less difference in air temperature than supply ductwork, they should also be insulated. Any duct system that runs through an unconditioned space and has a temperature difference greater than 15°F should be insulated.

6.0.0 ◆ DUCT HANGERS AND SUPPORTS

HVAC ductwork must be properly supported. Each type of duct material has its own special support requirements. An overview of these requirements is provided in this section. An HVAC duct support system is composed of three elements:

- The upper attachment to the structure
- The hanger
- The lower attachment to the duct

When planning the installation of hangers and supports, you must consider such things as proper spacing, materials, and installation practices. The following sections present some construction code specifications for hanging straight ducts, fiberglass duct hangers, fiberglass risers, riser supports, and flexible ducts.

6.1.0 Hanger Spacing: Straight Duct

In straight duct sections, the joints are the weakest points, so support must be provided. Standards published by the Sheet Metal and Air Conditioning Contractors National Association (SMACNA) specify the allowable loads and maximum spacing for hangers. These specifications vary according to hanger and duct type and size as well as the type of installation.

6.2.0 Riser Supports

Rectangular risers (vertically running ducts) must be supported with angle irons or **channels** secured to the sides of the ducts with welds, bolts, sheet metal screws, or blind rivets. Riser supports are required at one- or two-story height intervals (every 12 to 24 feet). Some risers are supported from the floor (*Figure 40*). Reinforcing for the riser support is located below the duct joint. It may be necessary to install vibration isolators to eliminate duct-borne noise. These isolators keep vibration noise from transferring through the riser supports to surrounding walls, floors, or ceilings, which might act like sounding boards and amplify the noise.

Some risers are supported from the wall (*Figure 41*) and may be held in place by a band or

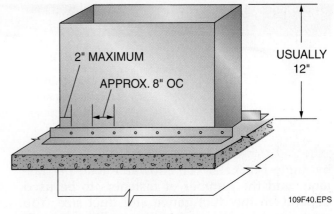

Figure 40 ◆ Floor-supported risers.

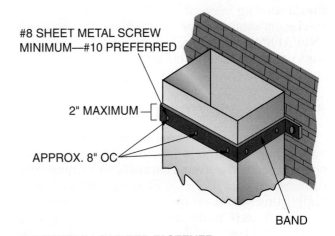

ALLOWABLE LOAD PER FASTENER:
25 LBS – 26 to 28 Gauge
35 LBS – 20, 22, 24 Gauge
50 LBS – 16, 18 Gauge

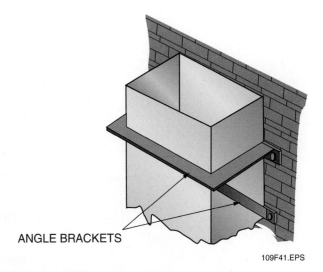

Figure 41 ◆ Wall-supported risers.

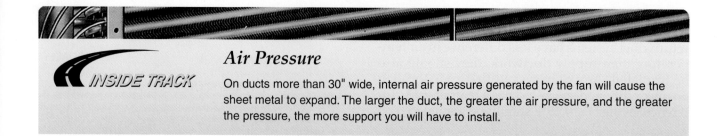

Air Pressure

INSIDE TRACK

On ducts more than 30" wide, internal air pressure generated by the fan will cause the sheet metal to expand. The larger the duct, the greater the air pressure, and the greater the pressure, the more support you will have to install.

by **angle brackets**. The allowable load per fastener and the number of fasteners to be used depend on the duct gauge and duct size. You must locate ducts against the wall or a maximum of 2" away from the wall.

Fiberglass duct risers 8' long or longer require the use of special support (*Figure 42*). This support does not reinforce the duct for deflection (deformation of the duct from loads placed upon it). Vertical riser supports should be installed at maximum spacing intervals of 12'.

The **National Fire Protection Association (NFPA)** limits riser height for vertical risers in air duct systems serving buildings with more than two stories.

6.3.0 Fiberglass Duct Hangers

Trapeze hangers (*Figure 43*) are recommended for supporting fiberglass ducts. These hangers should be made from channels with minimum dimensions of 3" × 1½" × 24-gauge metal. The 1" supporting metal strap should be 22 gauge or heavier. Rods ¼" in diameter may be used in place

of hanger straps; when local codes permit, 12-gauge hanger wire may be substituted.

The maximum on-center (OC) spacing for trapeze hangers on fiberglass ducts is shown in *Table 3* and *Figure 44*.

In addition, reinforcement of pressurized fiberglass air duct systems is often necessary to reduce sagging of larger panels and to eliminate excess stress.

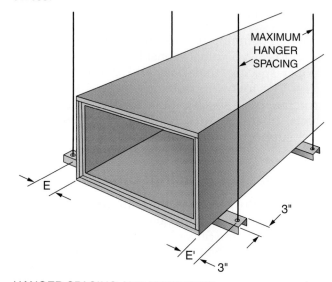

HANGER SPACING AND EXTENSION
3" Wide Channels

STANDARD 3" WIDE HANGERS
Hanger extension is defined as the sum of the distances between the hanging wires and the duct walls (both sides).

TOTAL HANGER EXTENSION (E + E')

CHANNEL SELECTION		
If Total Extension Is Not Greater Than:	Minimum Channel Gauge	Minimum Channel Profile
6"	24	3" × 1½"
18"	22	3" × 2"
30"	18	3" × 2"

CHANNEL PER HANGER SCHEDULE

#10 PLATED SHEET METAL SCREWS 8" OC MINIMUM 1 PER SIDE

2½" SQUARE WASHERS OR 2" WIDE SHEET METAL SLEEVE INSIDE DUCT

1" × 1" × ⅛" WALL SUPPORT ANGLE

109F42.EPS

Figure 42 ◆ Fiberglass duct riser support.

109F43.EPS

Figure 43 ◆ Trapeze hangers for fiberglass ducts.

Table 3 Maximum On-Center (OC) Spacing for Trapeze Hangers on Fiberglass Ducts

Duct Size (Width)	Maximum OC Spacing
49" or greater	4' OC
48" or less and less than 12" high	6' OC
48" or less and from 24" high	8' OC
24" wide or less and more than 24" high	8' OC
More than 24" wide and more than 24" high	6' OC

109T03.EPS

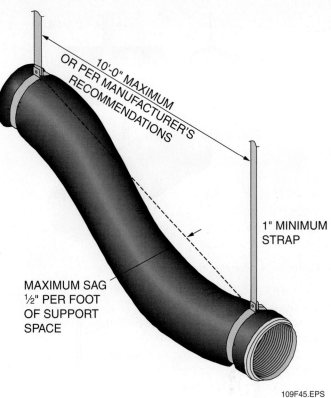

Figure 45 ◆ Flexible duct support.

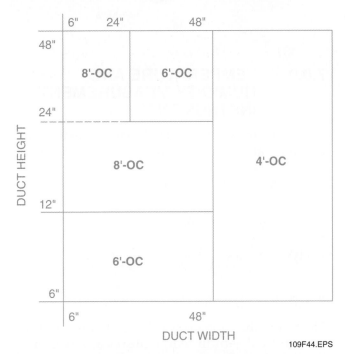

Figure 44 ◆ Maximum on-center spacing for trapeze hangers.

6.4.0 Flexible Duct and Round Duct

When supporting flexible duct, you should never place supports more than 10 feet apart (*Figure 45*) or as recommended by the manufacturer. The most sag that is allowed is ½" per foot between supports. Note that connections to other ducts or equipment are also considered to be flexible duct supports.

You must ensure that the straps used to wrap around and hang the duct do not pinch the duct diameter (*Figure 46*). Unless you use a **sheet metal saddle** underneath the duct, the strapping material must never be less than 1" wide. To avoid tearing any vapor barrier, especially during installation, never support the entire weight of the duct on one hanger. You must always avoid contact between flexible duct and sharp metal edges. You may sometimes be permitted to repair damage to the vapor barrier with the proper kind of tape. But if the internal core is punctured, you must replace that section of flexible duct or else cut it completely and treat it as a connection.

6.5.0 Rectangular Duct Hangers

Rectangular duct hangers include strap hangers (*Figure 47*) and trapeze hangers (*Figure 48*). Duct up to 60" wide must be supported with strap hangers. Duct more than 60" wide must be supported with trapeze hangers.

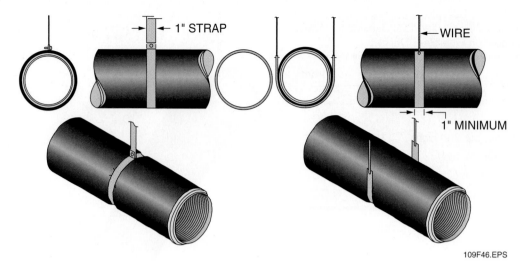

1" STRAP

WIRE

1" MINIMUM

109F46.EPS

Figure 46 ◆ Hanging flexible duct.

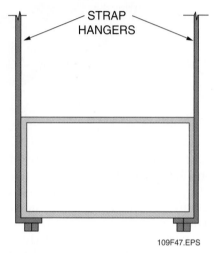

STRAP HANGERS

109F47.EPS

Figure 47 ◆ Strap hangers.

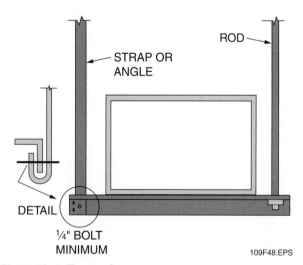

ROD

STRAP OR ANGLE

DETAIL

¼" BOLT MINIMUM

109F48.EPS

Figure 48 ◆ Trapeze hanger.

7.0.0 ◆ TEMPERATURE AND HUMIDITY MEASUREMENT INSTRUMENTS

Several instruments are used to measure the temperature and humidity of the conditioned air when working on air distribution systems. The following are three of the most common instruments:

* Electronic thermometers
* Psychrometers
* Hygrometers

7.1.0 Electronic Thermometers

Electronic thermometers (*Figure 49*) are used for measuring **dry-bulb temperatures** and **wet-bulb temperatures** in air distribution systems and other HVAC equipment.

Electronic thermometers display the temperature on either an analog meter or a digital readout. Digital thermometers are the most commonly used instrument for HVAC field service work. They use either a thermocouple or thermistor-type temperature probe, or both, to sense the heat and generate the temperature reading. Often, several different probes are used with the same instrument to allow measurement of a wide range of temperatures. Many electronic thermometers have two or more probes so that measurements can be made at several locations within the equipment at the same time. Most electronic thermometers of this type can calculate and display the difference in temperature between the different locations being measured.

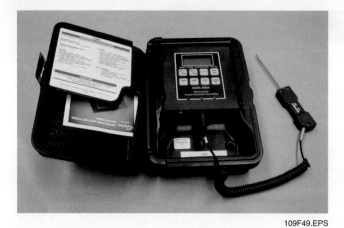

Figure 49 ◆ Electronic thermometer.

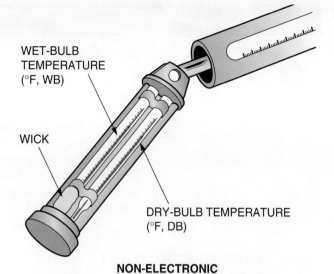

WET-BULB TEMPERATURE (°F, WB)

WICK

DRY-BULB TEMPERATURE (°F, DB)

NON-ELECTRONIC

Many digital multimeters (DMMs) can also be used to measure temperature. This feature requires the use of thermocouple and/or thermistor-temperature probe accessories that convert the DMM into an electronic thermometer. Some DMMs can use a non-contact infrared probe to measure temperature.

Electronic thermometers are precise measuring instruments. Be sure to read and follow the manufacturer's instructions for operating electronic thermometers. Also, be sure to follow the manufacturer's instructions for calibration of the instrument.

7.2.0 Psychrometers

Psychrometers are used to measure temperature. They have two thermometers, one to measure the dry-bulb temperature and the other to measure the wet-bulb temperature. A sling psychrometer (*Figure 50*) is often used to measure temperatures when working on air distribution systems. The sensing bulb of the wet-bulb thermometer is covered with a wick that is saturated with distilled water before taking a reading. To make sure the wet-bulb temperature is accurate, the sling psychrometer must be spun rapidly in the air that is being tested. This is necessary so that evaporation occurs at the wet-bulb thermometer, giving it a lower temperature reading. The measured wet-bulb and dry-bulb temperatures can be used to find the percent of **relative humidity (RH)** in the air. This is done using either a built-in chart on the psychrometer or a separate **psychrometric** chart. Relative humidity is the ratio of the amount of moisture present in a given sample of air to the amount it can hold at saturation. Relative humidity is expressed as a percentage.

DIGITAL

Figure 50 ◆ Psychrometers.

Squeeze-bulb and battery-operated aspirating psychrometers are also available. These are used in confined spaces where the use of the sling psychrometer is restricted. The squeeze-bulb aspirating psychrometer is operated by rapidly squeezing the bulb to draw air over the thermometers. In the battery-operated psychrometer, a fan draws air over the thermometers.

Psychrometrics

Our atmosphere consists of a mixture of air (mostly nitrogen and oxygen) and water vapor. The study of air and its properties is called psychrometrics. In 1911, Dr. Willis Carrier presented his Rational Psychrometric Formula to the American Society of Mechanical Engineers. This formula led to the development of the psychrometric chart like the one shown here. It gives a graphical representation of the interrelationships that exist for all the properties of air. A psychrometric chart is typically used by HVAC engineers, designers, and technicians to predict the values for the various properties of air when designing an HVAC system, or before adjusting or modifying an existing HVAC system.

PSYCHROMETRIC CHART
Normal Temperatures
Barometric Pressure
29.92 Inches of Mercury

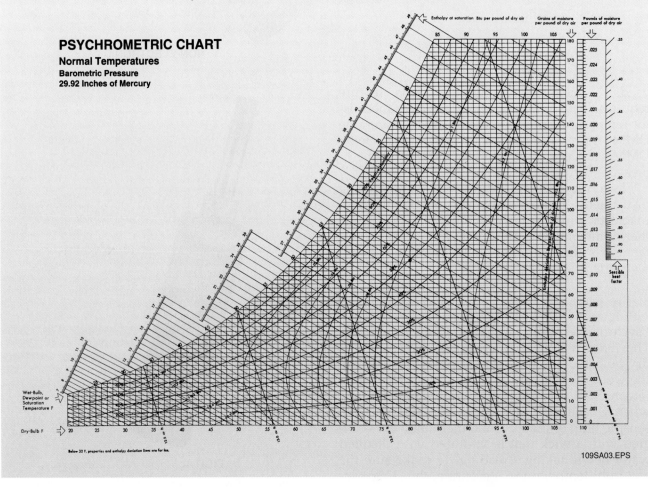

109SA03.EPS

7.3.0 Hygrometers

Hygrometers or relative humidity meters are used to measure and give direct readings of relative humidity. Many varieties of hygrometers are available, including both electronic and dial types.

For field service work, the electronic hygrometer is the most commonly used. Electronic hygrometers (*Figure 51*) have temperature/humidity sensing probes used to measure the conditioned air. The measured relative humidity and/or temperature readings are usually displayed on a digital readout.

Some hygrometers are also capable of measuring and giving a direct reading of the **dew point**. Dew point is the temperature at which water vapor in the air becomes saturated and starts to condense into water droplets. Some hygrometers use two or more temperature probes to calculate the differential temperature measured between the probes. Still others are available that can display temperature and relative humidity readings simultaneously. Be sure to follow the manufacturer's instructions for hygrometer usage and calibration.

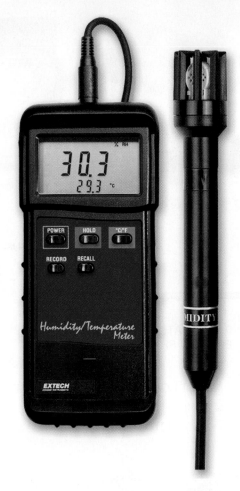

109F51.EPS

Figure 51 ◆ Electronic hygrometers.

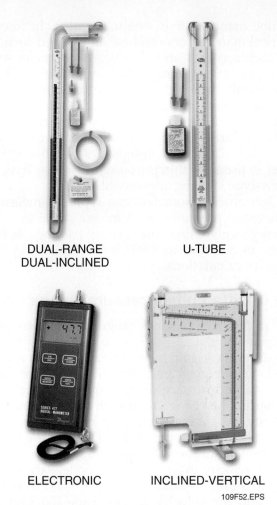

DUAL-RANGE
DUAL-INCLINED

U-TUBE

ELECTRONIC

INCLINED-VERTICAL

109F52.EPS

Figure 52 ◆ Manometers.

8.0.0 ◆ AIR DISTRIBUTION SYSTEM MEASUREMENT INSTRUMENTS

Several instruments and accessories are used to measure the static, velocity, and total pressures in an air distribution system. Among these are three common devices:

- Manometers
- Differential pressure gauges
- Pitot tubes and static pressure tips

8.1.0 Manometers

Manometers are used to measure the low-level static, velocity, and total air pressures found in air distribution duct systems. Manometers used for air distribution servicing are calibrated in inches of water column (in. w.c.). Manometers can use water or oil as the measuring fluid. Popular ones use an oil which has a specific gravity of 0.826 as the measuring fluid. The manufacturer of the gauge specifies the type of oil to be used, so substitution for the

specified oil is not recommended. Manometers come in many types, including U-tube, inclined, and combined U-inclined. Electronic manometers are also widely used. Pitot tubes or static pressure tips, described later in this section, are almost always used with manometers when measuring pressures in duct systems. *Figure 52* shows four types of manometers.

Manometers work on the principle that air pressure is indicated by the difference in the level of a column of liquid in the two sides of the instrument. If there is a pressure difference, the column of liquid will move until the liquid level in the low-pressure side is high enough so that its weight and the low air pressure being measured will equal the higher pressure in the other tube.

Individual U-tube and inclined manometers are available in many pressure ranges. Inclined manometers are usually calibrated in the lower pressure ranges and are more sensitive than U-tube manometers. U-inclined manometers combine both the sensitivity of the inclined manometer with the high-range capability of the

U-tube manometer in one instrument. Inclined-vertical manometers combine an inclined section for high accuracy and a vertical manometer section for extended range. They also have an additional scale that indicates air velocity in feet per minute (fpm). To get accurate readings with inclined and vertical-inclined manometers, the inclined portion of the scale must be at the exact angle for which it is designed. A built-in spirit level is used for this purpose. Most also have a screw-type leveling adjustment.

Electronic manometers can typically measure differential pressures of –1 in. w.c. to 10 in. w.c. Many can give direct air velocity readings in the range of 300 fpm to 9,990 fpm, eliminating the need for calculations.

8.2.0 Differential Pressure Gauge

The differential pressure gauge, also known as a magnahelic gauge, provides a direct reading of pressure. These gauges are typically used to measure fan and blower pressures, filter resistance, air velocity, and furnace draft. Some are capable of measuring just pressure or both pressure and air velocity. Single-scale pressure models are calibrated in either in. w.c. or psi. Dual-scale gauges are normally calibrated for pressure in in. w.c. and for air velocity in fpm. Several models are available covering pressures from 0.0 in. w.c. to 10 in. w.c., and air velocity ranges from 300 fpm to 12,500 fpm. Normally these gauges are installed in the equipment, but portable models are available. Pitot tubes and/or static pressure tips are normally used with portable models to make air pressure and velocity measurements in air distribution system ductwork. *Figure 53* shows a portable differential pressure gauge.

8.3.0 Pitot Tubes and Static Pressure Tips

The pitot tube and static pressure tips (*Figure 54*) are probes used with manometers and pressure gauges when making measurements inside the ductwork of an air distribution system. The standard pitot tube used for measurements in ducts 8" and larger has a ⁵⁄₁₆" outer tube with eight equally spaced 0.04" diameter holes used to sense static pressure. For measurements in ducts smaller than 8", use of pocket size pitot tubes with a ⅛" outer tube and four equally spaced 0.04" diameter holes are recommended.

The pitot tube consists of an impact tube, which receives the total pressure input, fastened concentrically inside a larger tube, which receives static pressure input from the radial sensing holes

109F53.EPS

Figure 53 ◆ Portable differential pressure gauge.

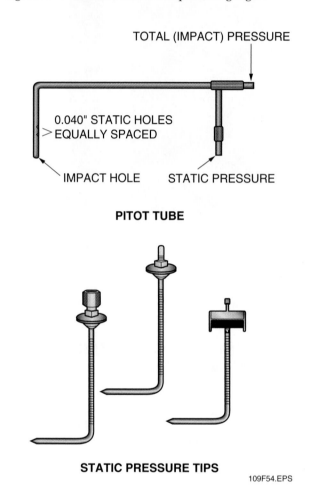

PITOT TUBE

STATIC PRESSURE TIPS

109F54.EPS

Figure 54 ◆ Pitot tube and static pressure tips.

around the tip. The air space between the inner and outer tubes permits transfer of pressure from the sensing holes to the static pressure connection at the opposite end of the pitot, and then through the connecting tubing to the low- or negative-pressure side of the manometer.

When the total pressure tube is connected to the high-pressure side of the manometer, velocity pressure is indicated directly. To be sure of accurate velocity pressure readings, the pitot tube tip must be pointed directly into the duct air stream.

Pitot tubes come in various lengths ranging from 6" to 60", with graduation marks at every inch to show the depth of insertion in the duct.

Static pressure tips, like pitot tubes, are used with manometers and differential pressure gauges to measure static pressure in a duct system. They are typically L-shaped with four radially drilled 0.04" sensing holes.

9.0.0 ◆ AIR VELOCITY MEASUREMENT INSTRUMENTS

Velometers (*Figure 55*) are used to measure the velocity of airflow. Measurement of air velocity is done to check the operation of an air distribution system. It is also done when balancing system airflow.

Most velometers give direct readings of air velocity in fpm. Some can provide direct readings in cfm. Velometers with analog scales and digital readouts are in common use.

Some velometers use a rotating vane (propeller) or balanced swing vane to sense the air movement. When the rotating vane velometer is positioned to make a measurement, the vane rotates at a rate determined by the velocity of the air stream. This rotation is converted into an equivalent velocity reading for display. In the swinging vane velometer, the air stream causes the balanced vane to tilt at different angles in response to the measured air velocity. The position of the vane is converted into an equivalent velocity reading for display.

Another type of velometer, also called a hot wire anemometer, gives direct readings of air velocity in fpm. This instrument uses a sensing probe which contains a small resistance heater element. When the probe is held perpendicular to the air stream being measured, the temperature of the heater element changes due to changes in the airflow. This causes its resistance to change, which alters the amount of current flow being applied to the meter circuitry. There, it automatically calculates the air velocity for display on the meter.

Some velometers use probes that have a sensitively balanced vane or a small resistance heater element that, when placed in the air stream, produces a measurement of airflow for display on the velometer meter scale. Depending on the sensing probe or attachment used, velometers can measure air velocities in several ranges within the overall range of 0 to 10,000 fpm. Some electronic velometers that use a microprocessor can automatically average up to 250 individual readings taken across a duct area to provide an average air velocity. Certain velometers also include an optional micro-printer to record the readings.

Special velometers, known as air volume balancers, can be used when balancing air distribution systems. This type of velometer is held directly against the diffuser or register to get a direct reading of air velocity. Another type of velometer called a flow hood is frequently used to get direct velocity readings in cfm when measuring the output of large air diffusers in commercial systems. Air velocity can also be measured using a manometer, and then applying the formula for cfm.

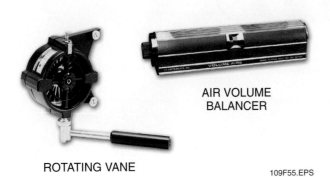

AIR VOLUME BALANCER

ROTATING VANE

109F55.EPS

Figure 55 ◆ Velometers.

Summary

Proper measurement and control of air flow is necessary so that the human body can feel comfortable in a room environment. It is also critical to many commercial and manufacturing processes. As an HVAC technician, you will need to evaluate the air in a conditioned space, then make knowledgeable decisions pertaining to the service or adjustments that may be needed for the related HVAC equipment. A heating or air conditioning system will perform no better than its air distribution system. All adequate air distribution systems must perform the following functions:

- Supply the right quantity of air to each conditioned space.
- Supply the air in each space so that air motion is adequate but not drafty.
- Condition the air to maintain the proper comfort zones for humans, or the proper conditions needed for a commercial or manufacturing process.

- Provide for the return of air from all conditioned areas to the air handler.
- Operate efficiently without excessive power consumption or noise.
- Operate with minimum maintenance.

The layout and equipment used for air distribution systems is normally determined by the climate in which the system must operate and the construction of the building in which the equipment is installed. Ideally, the system air handler will be located in an area that allows the duct length to be as short as possible and the number and types of fittings as few as possible.

Ductwork must be properly sealed and supported. Improper support can result in duct failure and improper sealing can cause inefficient, noisy operation.

Notes

1. Within an air distribution system, the highest pressure level is found at the _____ .
   a. conditioned space
   b. input to the return duct
   c. input to the blower
   d. output of the blower

2. The static pressure, velocity pressure, and total pressure measured in a duct system are measured in _____ .
   a. psi
   b. inches of mercury
   c. inches of water column
   d. inches of mercury vacuum

3. An air distribution system for a light commercial building is designed to work with a 5-ton cooling system. About how many cfm of airflow should the system blower be capable of supplying?
   a. 1,200 cfm
   b. 1,650 cfm
   c. 2,000 cfm
   d. 2,350 cfm

4. The type of blower or fan most often used in residential heating and air conditioning systems is the _____ .
   a. forward-curved centrifugal blower
   b. backward-inclined centrifugal blower
   c. duct fan
   d. radial centrifugal blower

5. An existing air distribution system has an airflow of 6,000 cfm created by a blower operating at a speed of 1,200 rpm. If you wanted to change the airflow to 5,000 cfm, what is the required new blower speed?
   a. 833 rpm
   b. 1,000 rpm
   c. 1,150 rpm
   d. 1,300 rpm

6. External static pressure loss is the _____ .
   a. total pressure loss in the air handler and its component parts
   b. total pressure loss of an air distribution system, excluding the air handler
   c. total pressure drop across the air handler
   d. difference between the total pressure and the static pressure

7. Perimeter duct systems are used in _____ .
   a. warm climates
   b. cold climates
   c. primarily in cold climates, but also homes in warm climates if they are constructed over a basement or crawl space
   d. locations where the outside air temperature difference between the heating and cooling seasons does not exceed 70°F

8. The duct material that has the least friction loss is _____ .
   a. sheet metal
   b. fiberglass ductboard
   c. flexible duct
   d. internally insulated sheet metal

9. Each fitting and transition used in a duct system has _____ .
   a. the same friction loss per foot as the same size straight duct
   b. less friction loss per foot as the same size straight duct
   c. a friction loss equal to a predetermined length of the same size straight duct
   d. approximately the same friction loss

10. High sidewall registers work best in _____ .
    a. cold climates for heating
    b. both heating and cooling when used with a central return
    c. warm climates for heating
    d. cooling when used with room-by-room returns

11. Dampers built into supply diffusers and registers are used to _____ .
  a. add moisture to the air
  b. reduce duct noise
  c. control and balance airflow in a duct system
  d. vent stale air to the outdoors

12. A fire damper will close when _____ .
  a. the fire has burned for 3 hours
  b. the fusible link melts
  c. the HVAC equipment shuts off
  d. smoke is detected

13. Air duct systems in attics must be insulated with an insulation having an R-value based on the _____ .
  a. cooling mode
  b. heating mode
  c. greater value as needed by the cooling or heating mode
  d. insulation manufacturer's recommendations

14. Rectangular risers must be supported with _____ .
  a. vibration isolators
  b. temporary supports
  c. studs or pins
  d. angle irons or channels

15. An instrument that can be used to measure pressure losses in duct systems is the _____ .
  a. psychrometer
  b. velometer
  c. differential pressure gauge
  d. hygrometer

Trade Terms Quiz

1. The volume of air flowing past a point in one minute is measured in _____.

2. The device that moves the air across the heat exchanger in a forced-air system is called a(n) _____.

3. A(n) _____ is an L-shaped metal supporting member used to support vertical risers.

4. The weight of our atmosphere pressing on all things on the surface of the earth is called _____.

5. A U-shaped piece of structural steel used as a supporting device is called a(n) _____.

6. The temperature measured using a standard thermometer is called _____.

7. The study of air and its properties is called _____.

8. The speed at which the shaft of an air-moving device is rotating is measured in _____.

9. _____ is the sum of the static pressure and the velocity pressure in an air duct.

10. The ring or panel surrounding the blades of a propeller fan that increases the fan's performance is called a(n) _____.

11. The temperature at which water vapor in the air becomes saturated and starts to condense into water droplets is called the _____.

12. An association that develops and publishes standards about fire prevention is called the _____.

13. _____ is the amount of air, in cubic feet, flowing past a point in one minute.

14. A sling-like device that cradles insulated duct from underneath is called a(n) _____.

15. _____ is the ratio of the amount of moisture present in a given sample of air to the amount it can hold at saturation.

16. The pressure exerted uniformly in all directions within a duct system is called _____.

17. _____ is how fast air is moving, usually measured in feet per minute.

18. _____ is the pressure in a duct due to the movement of the air.

19. The condition that exists when there are no effective restrictions to airflow at the inlet or outlet of an air-moving device is called _____.

20. The sealed chamber at the inlet or outlet of an air handler is called a(n) _____.

21. Temperature taken with a thermometer that has a wick wrapped around its sensing bulb that has been saturated with clean, distilled water before taking a reading is called a(n) _____.

22. Standards published by _____ specify allowable loads for duct hangers.

23. Standards published by _____ are used to calculate appropriate insulation for ductwork.

Trade Terms

Air handler
Angle bracket
ASHRAE
Atmospheric pressure
Channel
Cubic feet per minute (cfm)

Dew point
Dry-bulb temperature
Free air delivery
National Fire Protection Association (NFPA)
Plenum
Psychrometrics

Relative humidity (RH)
Revolutions per minute (rpm)
Sheet metal saddle
SMACNA
Static pressure (s.p.)
Total pressure

Velocity
Velocity pressure
Venturi
Volume
Wet-bulb temperature

Wayne Culp

Commercial Service and Training Coordinator
Total Comfort Service Center, Inc.
Columbia, South Carolina

How did you choose a career in HVAC?
I was looking for a career that was both challenging and would last throughout my working years.

What kinds of work have you done in your career?
I've worked in automobile and mobile refrigeration and commercial refrigeration and kitchen equipment. I've been a residential service technician, a commercial service technician, and a commercial service supervisor.

Tell us about your present job.
I am service coordinator for a large commercial service company. As part of this job I am also the service training coordinator for three sites statewide in South Carolina. I have seven craft instructors and three assessment sites to coordinate the use of the NCCER curriculum for our continuing education. I'm chairperson of the SkillsUSA S.C. State HVAC-R competition. As continuing education instructor at Midlands Technical College, I've taught heat pump courses and journeyman and master HVAC-R prep courses.

What factors have contributed most to your success?
First, a love for this career field, and second, a desire to grow my knowledge and talents to share with others.

What types of training have you been through?
I studied HVAC technologies at Midlands Technical College. Bell & Gossett provided training in large and built-up hydronic systems. The Trane factory provided Commercial II Technician training. I've also taken the Amerisure 10-hour OSHA training and have South Carolina State Master Mechanic Certification.

What advice would you give to those who are new to the HVAC field?
You'll be most successful if you determine to make this a career path and not just a job. Be willing to continue learning as technological change is rapid in HVAC. Try to work for companies that are willing to challenge you to grow through continued education. A company that is unwilling to invest in your training is not willing to invest in its own future.

Trade Terms
Introduced in This Module

Air handler: The device that moves the air across the heat exchanger in a forced-air system. In a split system, it normally contains the blower fan, cooling coil, metering device, air filter, and related housing.

Angle bracket: An L-shaped metal supporting member used to support vertical risers. Also called angle iron.

ASHRAE: American Society of Heating, Refrigeration, and Air Conditioning Engineers.

Atmospheric pressure: The pressure exerted on all things on Earth's surface as the result of the weight of our atmosphere.

Channel: A U-shaped piece of structural steel used as a supporting device.

Cubic feet per minute (cfm): A measure of the amount or volume of air in cubic feet flowing past a point in one minute. Cubic feet per minute can be calculated by multiplying the velocity of air, in feet per minute (fpm), times the area it is moving through, in square feet (cfm = fpm × area).

Dew point: The temperature at which water vapor in the air becomes saturated and starts to condense into water droplets.

Dry-bulb temperature: The temperature measured using a standard thermometer. It represents a measure of the sensible heat of the air or surface being tested.

Free air delivery: The condition that exists when there are no effective restrictions to airflow (no static pressure) at the inlet or outlet of an air-moving device.

National Fire Protection Association (NFPA): An association that develops and publishes standards about fire prevention.

Plenum: A sealed chamber at the inlet or outlet of an air handler. The duct attaches to the plenum.

Psychrometrics: The study of air and its properties.

Relative humidity (RH): The ratio of the amount of moisture present in a given sample of air to the amount it can hold at saturation. Relative humidity is expressed as a percentage.

Revolutions per minute (rpm): The speed at which the shaft of an air-moving device is rotating.

Sheet metal saddle: A sling-like device that cradles insulated duct from underneath.

SMACNA: Sheet Metal and Air Conditioning Contractors National Association.

Static pressure (s.p.): The pressure exerted uniformly in all directions within a duct system, as measured in in. w.c.

Total pressure: The sum of the static pressure and the velocity pressure in an air duct. It is the pressure produced by the fan or blower.

Velocity: How fast air is moving. The rate of airflow usually measured in feet per minute.

Velocity pressure: The pressure in a duct due to the movement of the air. It is the difference between the total pressure and the static pressure in w.c.

Venturi: A ring or panel surrounding the blades on a propeller fan; used to improve fan performance.

Volume: The amount of air in cubic feet flowing past a point in one minute (cfm).

Wet-bulb temperature: Temperature taken with a thermometer that has a wick wrapped around its sensing bulb that is saturated with clean, distilled water before taking a reading. The reading from a wet-bulb thermometer, through evaporation of the distilled water, takes into account the moisture content of the air. It reflects the total heat content of the air.

Typical Air Distribution System Duct and Supply Outlet Data

SUPPLY OUTLETS

Floor Outlets - Perimeter

CFM	SIZE (IN)	APPROX. SPREAD (FT)	FACE VELOCITY (FPM)	FREE AREA (SQ IN)
70	2¼ × 10	9	535	18.6
80	2¼ × 12	10	565	21.1
100	2¼ × 14	11	610	23.6
110	4 × 10	10	500	32.4
135	4 × 12	13	500	39.0
175	4 × 14	14	555	45.5

Low Sidewall - Perimeter

CFM	SIZE (IN)	APPROX. SPREAD (FT)	FACE VELOCITY (FPM)	FREE AREA (SQ IN)
80	10 × 6	13	430	26.7
100	12 × 6	10	440	32.6
120	14 × 6	8	450	38.4

Baseboard

CFM	SIZE (FT)	APPROX. SPREAD (FT)	OUTLET VELOCITY (FPM)	FREE AREA (SQ IN)
80	2	7.5	430	26.6

High Sidewall

CFM	SIZE (IN)	HORIZ. THROW (FT)	FACE VELOCITY (FPM)	FREE AREA (SQ IN)
80	10 × 4	8	390	29.0
125	10 × 6	10	415	43.3
150	12 × 6	10	410	52.7
165	14 × 6	9.5	375	62.1

Round Ceiling Outlets

CFM	SIZE (IN)	HORIZ. THROW (FT)	OUTLET VELOCITY (FPM)	FREE AREA (SQ IN)
45	6	3	500	12.2
105	8	5	580	26.1
185	10	7	580	43.8
285	12	8.5	575	65.7
425	14	10.5	600	91.9

Square Ceiling Outlets

CFM	SIZE (IN)	HORIZ. THROW (FT)	OUTLET VELOCITY (FPM)	FREE AREA (SQ IN)
50	6 × 6	3.5	450	15.4
135	8 × 8	5	550	35.1
250	10 × 10	6	620	58.1
325	12 × 12	7	550	85.1

Tables based on Lima Register Co. Catalog Data, @ 0.028 inch w.g. drop across outlet.

EQUIVALENT RECTANGULAR DUCT SIZES

CFM	DUCT SIZES						CFM
	Supply			Return			
200	10 × 6	8 × 8		12 × 6	10 × 8		200
300	12 × 6	10 × 8	10 × 10	16 × 6	12 × 8	10 × 10	300
400	16 × 6	12 × 8	11 × 10	20 × 6	14 × 8	11 × 10	400
500	18 × 6	14 × 8	12 × 10	24 × 6	16 × 8	12 × 10	500
600	22 × 6	16 × 8	12 × 12	20 × 8	16 × 10	12 × 12	600
700	18 × 8	14 × 10	13 × 12	22 × 8	18 × 10	14 × 12	700
800	20 × 8	16 × 10	14 × 12	24 × 8	19 × 10	15 × 12	800
900	22 × 8	16 × 10	16 × 12	26 × 8	20 × 10	17 × 12	900
1,000	24 × 8	18 × 10	18 × 12	30 × 8	22 × 10	18 × 12	1,000
1,200	28 × 8	22 × 10	20 × 12	36 × 8	26 × 10	20 × 12	1,200
1,400	32 × 8	24 × 10		30 × 10	24 × 12		1,400
1,600	28 × 10	22 × 12		34 × 10	26 × 12		1,600
1,800	30 × 10	24 × 12		37 × 10	30 × 12		1,800
2,000	32 × 10	26 × 12		42 × 10	32 × 12		2,000

SUPPLY sizes based on friction rate of 0.08 in w.g. per 100 ft equivalent length metal duct.

RETURN sizes based on friction rate of 0.05 in w.g. per 100 ft equivalent length metal duct.

RETURN AIR GRILLES

CFM	SIZE (IN)	FREE AREA (SQ IN)
100	10 × 6	36.4
125	12 × 6	44.4
170	12 × 8	61.0
145	14 × 6	52.4
200	14 × 8	72.0
245	24 × 6	89.6
335	24 × 8	122.0
310	30 × 6	110.8
425	30 × 8	152.0

VERTICAL STACKS

SUPPLY CFM	STACK SIZE (IN)	RETURN CFM
100	3¼ × 10	75
125	3¼ × 12	90
150	3¼ × 14	110

2¼" stacks = 55% of 3¼" stack capacity.

PANNED JOIST (16 IN OC)

RETURN CFM	NOMINAL JOIST DEPTH (IN)	ACTUAL JOIST DEPTH (IN)
260	6	5½
375	8	7½
525	10	9½

109A01.EPS

Additional Resources

This module is intended to be a thorough resource for task training. The following reference work is suggested for further study. This is optional material for continued education rather than for task training.

Residential Air System Design, 1993. Syracuse, NY: Carrier Corporation.

NCCER makes every effort to keep these textbooks up-to-date and free of technical errors. We appreciate your help in this process. If you have an idea for improving this textbook, or if you find an error, a typographical mistake, or an inaccuracy in NCCER's Contren® textbooks, please write us, using this form or a photocopy. Be sure to include the exact module number, page number, a detailed description, and the correction, if applicable. Your input will be brought to the attention of the Technical Review Committee. Thank you for your assistance.

Instructors – If you found that additional materials were necessary in order to teach this module effectively, please let us know so that we may include them in the Equipment/Materials list in the Annotated Instructor's Guide.

Write: Product Development and Revision
National Center for Construction Education and Research
3600 NW 43rd St., Bldg. G, Gainesville, FL 32606

Fax: 352-334-0932

E-mail: curriculum@nccer.org

Craft   Module Name

Copyright Date   Module Number   Page Number(s)

Description

(Optional) Correction

(Optional) Your Name and Address

Glossary

Absolute pressure: The total pressure that exists in a system. Absolute pressure is expressed in pounds per square inch absolute (psia). Absolute pressure = gauge pressure + atmospheric pressure 14.7 psi at sea level at 70°F.

Acceleration: The rate of change of velocity; also, the process by which a body at rest becomes a body in motion.

Air handler: The device that moves the air across the heat exchanger in a forced-air system. In a split system, it normally contains the blower fan, cooling coil, metering device, air filter, and related housing.

Alloy: Any substance made up of two or more metals.

Alternating current (AC): An electrical current that changes direction on a cyclical basis.

Ammeter: A test instrument used to measure current flow.

Ampere (amp): The unit of measurement for current flow. The magnitude is determined by the number of electrons passing a point at a given time.

Analog meter: A meter that uses a needle to indicate a value on a scale.

Angle bracket: An L-shaped metal supporting member used to support vertical risers. Also called angle iron.

Annealed copper refrigeration (ACR) tubing: Copper tubing made especially for refrigeration and HVAC work. It is especially clean and is usually charged with dry nitrogen. The ends are sealed to prevent contamination.

Annealing: A process in which a material is heated, then cooled to strengthen it.

Annual Fuel Utilization Efficiency (AFUE): HVAC industry standard for defining furnace efficiency.

Area: The amount of surface in a given plane or two-dimensional shape.

ASHRAE: American Society of Heating, Refrigeration, and Air Conditioning Engineers.

Atmospheric pressure: The standard pressure exerted on all things on the Earth's surface. Atmospheric pressure is normally expressed as 14.7 pounds per square inch (psi) or 29.92 inches of mercury at sea level at 70°F.

Atomize: The process by which a liquid is converted into a fine spray.

Barometric pressure: The actual atmospheric pressure at a given place and time.

Black iron pipe: Carbon steel pipe that gets its black coloring from the carbon in the steel.

Brazing: A method of joining metals with a non-ferrous filler metal using heat above 800°F but below the melting point of the base metals being joined. Also known as hard soldering.

British thermal unit (Btu): The amount of heat needed to raise the temperature of one pound of water one degree Fahrenheit.

Bushing: A pipe fitting with male threads on the outside and female threads on the inside. It is most often used to connect the male end of a pipe to a fitting of a larger size.

Cap: A female pipe fitting that is closed at one end. It is used to close off the end of a piece of pipe.

Capillary action: The movement of a liquid along the surface of a solid in a kind of spreading action.

Chain vise: A device used to clamp pipe and other round metal objects. It has one stationary metal jaw and a chain that fits over the pipe and is clamped to secure the pipe.

Chain wrench: An adjustable tool for holding and turning large pipe up to 4" in diameter. A flexible chain replaces the usual wrench jaws.

Channel: A U-shaped piece of structural steel used as a supporting device.

Chiller: A high-volume cooling unit. The chiller acts as an evaporator.

Chlorofluorocarbon (CFC) refrigerant: A class of refrigerants that contains chlorine, fluorine, and carbon. CFC refrigerants have a very adverse effect on the environment.

Clamp-on ammeter: A current meter in which jaws placed around a conductor sense the magnitude of the current flow through the conductor.

Coefficient: A multiplier (e.g., the numeral 2 as in the expression 2b).

Cold: A relative term for temperature. Cold means having less heat energy than another object against which it is being compared.

Combustion: The process by which a fuel is ignited in the presence of oxygen.

Compression joint: A method of connection in which tightening of a threaded nut compresses a compression ring to seal the joint.

Compressor: In a refrigeration system, the mechanical device that converts low-pressure, low-temperature refrigerant gas into high-temperature, high-pressure refrigerant gas.

Condenser: A heat exchanger that transfers heat from the refrigerant flowing inside it to the air or water flowing over it.

Condensing furnace: A furnace that contains a secondary heat exchanger that extracts latent heat by condensing exhaust (flue) gases.

Conduction: A means of heat transfer in which heat is moved from one material to another by means of direct contact.

Conductor: A material that readily conducts electricity; also, the wire that connects components in an electrical circuit.

Constant: An element in an equation with a fixed value.

Contactor: A control device consisting of a coil and one or more sets of contacts used as a switching device in high-voltage circuits.

Continuity: A continuous current path. Absence of continuity indicates an open circuit.

Convection: The transfer of heat by the flow of liquid or gas caused by a temperature differential.

Coupling: A pipe fitting containing female threads on both ends. Couplings are used to join two pipes in a straight run or to join a pipe and fixture.

Cross: A pipe fitting with four female openings at right angles to one another.

Cubic feet per minute (cfm): A measure of the amount or volume of air in cubic feet flowing past a point in one minute. Cubic feet per minute can be calculated by multiplying the velocity of air, in feet per minute (fpm), times the area it is moving through, in square feet (cfm = fpm × area).

Current: The rate at which electrons flow in a circuit, measured in amperes.

Dew point: The temperature at which water vapor in the air becomes saturated and starts to condense into water droplets.

Die: A tool insert used to cut external threads by hand or machine.

Digital meter: A meter that provides a direct numerical reading of the value measured.

Direct current (DC): An electric current that flows in one direction. A battery is a common source of DC voltage.

Dry-bulb temperature: The temperature measured using a standard thermometer. It represents a measure of the sensible heat of the air or surface being tested.

Elbow: An angled pipe fitting having two openings. It is used to change the direction of a run of pipe.

Electrode: An electrical terminal that will conduct a current.

Electromagnet: A coil of wire wrapped around a soft iron core. When a current flows through the coil, magnetism is created.

Enthalpy: The total heat content (sensible and latent) of a refrigerant or other substance.

Evaporator: A heat exchanger that transfers heat from the air flowing over it to the cooler refrigerant flowing through it.

Expansion device: Also known as the liquid metering device or metering device. Provides a pressure drop that converts the high-temperature, high-pressure liquid refrigerant from the condenser into the low-temperature, low-pressure liquid refrigerant entering the evaporator.

Exponent: A small figure or symbol placed above and to the right of another figure or symbol to show how many times the latter is to be multiplied by itself (e.g., $b^3 = b \times b \times b$).

Flame rectification: The process by which a flame produces a sensible electrical current.

Flange: A flat plate attached to a pipe or fitting and used as a means of attaching pipe, fittings, or valves to the piping system.

Flare fitting: A fitting in which one end of each tube to be joined is flared outward using a special tool. The flared tube ends mate with the threaded flare fitting and are secured to the fitting with flare nuts.

Flashback arrestor: A valve that prevents the flame from traveling back from the tip and into the hoses.

Floodback: Refrigerant returning to the compressor in the liquid state.

Fluorocarbons: Halocarbons in which at least one or more of the hydrogen atoms has been replaced with fluorine.

Flux: A chemical substance that prevents oxides from forming on the surface of metals as they are heated for soldering, brazing, or welding.

Force: A push or pull on a surface. Force is considered to be the weight of an object or fluid. This is a common approximation.

Free air delivery: The condition that exists when there are no effective restrictions to airflow (no static pressure) at the inlet or outlet of an air-moving device.

Galvanized pipe: Carbon steel pipe that has been coated with zinc to prevent rust.

Gauge pressure: The pressure measured on a gauge, expressed as pounds per square inch gauge (psig) or inches of mercury vacuum (in Hg vac.). Also pressure measurements that are made in comparison to atmospheric pressure.

Grooved pipe: A piping method for connecting piping systems. The use of grooved piping eliminates the need for threading, flanging, or welding when making connections. Connections are made with gaskets and couplings installed using a wrench and lubricant.

HACR (heating, air conditioning, and refrigeration) circuit breaker: A circuit breaker with a built-in trip delay used specifically in HVAC circuits because of the power surge that occurs with compressor startup.

Halocarbon refrigerants: Short for halogenated hydrocarbons, a class of refrigerants that includes most of the refrigerants used in residential and small commercial air conditioning systems.

Halocarbons: Hydrocarbons, like methane and ethane, that have most or all of their hydrogen atoms replaced with the elements fluorine, chlorine, bromine, astatine, or iodine.

Halogens: Substances containing chlorine, fluorine, bromine, astatine, or iodine.

Heat: A form of energy. It causes molecules to be in motion and raises the temperature of a substance. Other forms of energy like electricity, light, and magnetism deteriorate into heat.

Heat content: The amount of heat energy contained in a substance. Measured in Btus.

Heat exchanger: A device, usually metal, that is used to transfer heat from a warm surface or substance to a cooler surface or substance.

Heat transfer: The transfer of heat from a warmer substance to a cooler substance.

Hot surface igniter: A ceramic device that glows when an electrical current flows through it. Used to ignite gas in a gas furnace.

HVAC plan: Added to the mechanical plans for complex jobs that require separate heating, ventilating, and air-conditioning systems.

Hydrocarbons: Compounds containing only hydrogen and carbon atoms in various combinations.

Hydrochlorofluorocarbon (HCFC) refrigerant: A class of refrigerants that contains hydrogen, chlorine, fluorine, and carbon. Although not as high in chlorine as chlorofluorocarbon (CFC) refrigerants, HCFCs are still considered hazardous to the environment.

Induced-draft furnace: A furnace in which a motor-driven fan draws air from the surrounding area or from outdoors to support combustion.

Induction: To generate a current in a conductor by placing it in a magnetic field and moving the conductor or magnetic field.

Infiltration: Air that enters a building through doors, windows, and cracks in the construction.

In-line ammeter: A current-reading meter that is connected in series with the circuit under test.

Inside diameter (ID): The distance between the inner walls of a pipe. Used as the standard measure for tubing used in heating and plumbing applications.

Insulation: A substance that retards the flow of heat.

Insulator: A device that inhibits the flow of current (opposite of conductor); also, material that resists heat transfer by conduction.

International Building Code: A series of model construction codes. These codes set standards that apply across the country. This is an ongoing process led by the International Code Council (ICC).

Ladder diagram: A simplified schematic diagram in which the load lines are arranged like the rungs of a ladder between vertical lines representing the voltage source.

Latent heat: The heat energy absorbed or rejected when a substance is changing state (solid to liquid, liquid to gas, or vice versa) but maintaining its measured temperature.

Latent heat of condensation: The heat given up or removed from a gas in changing back to a liquid state (steam to water).

Latent heat of fusion: The heat gained or lost in changing to or from a solid (ice to water or water to ice).

Latent heat of vaporization: The heat gained in changing from a liquid to a gas (water to steam).

Length: The distance from one point to another; typically refers to a measurement of the long side of an object or surface.

Line duty: A protective device connected in series with the supply voltage.

Liter: A standard unit of volume in the metric system. It is equal to one cubic decimeter.

Load: A device that converts electrical energy into another form of energy (heat, mechanical motion, light, etc.). Motors are the most common loads in HVAC systems.

Manometer: An instrument that measures air or gas pressure by the displacement of a column of liquid.

Mass: The quantity of matter present.

Mechanical refrigeration: The use of machinery to provide cooling.

Multimeter: A test instrument capable of reading voltage, current, and resistance.

National Fire Protection Association (NFPA): An association that develops and publishes standards about fire prevention.

Natural-draft furnace: A furnace in which the natural flow of air from around the furnace provides the air to support combustion.

Newton (N): The amount of force required to accelerate one kilogram at a rate of one meter per second.

Nipple: A short length of pipe that is used to join fittings. It is usually less than 12" long and has male threads on both ends.

Nominal size: The approximate dimension(s) by which standard material is identified.

Nonferrous: A group of metals and metal alloys that contain no iron.

Noxious: Harmful to health.

Ohm: The unit of measurement for electrical resistance.

Oil burner: The main component of an oil-fired furnace. It combines oil and air and sprays the combination into the combustion chamber.

Orifice: A precisely drilled hole that controls the flow of gas to the burners.

Outside diameter (OD): The distance between the outer walls of a pipe. Used as the standard measure for ACR tubing.

Oxidation: The process by which the oxygen in the air combines with metal to produce tarnish and rust.

Piezoelectric: The property of a quartz crystal that causes it to vibrate when a high-frequency voltage is applied to it.

Pilot duty: A protective device that opens the motor control circuit, which then shuts off the motor.

Pipe dope: A putty-like pipe joint material used for sealing threaded pipe joints.

Piping: A generic term used to designate thick-wall pipe that can be threaded and joined with threaded fittings.

Plenum: A sealed chamber at the inlet or outlet of an air handler. The duct attaches to the plenum.

Plug: A pipe fitting with external threads and head that is used for closing the opening in another fitting.

Power: The amount of energy (measured in watts) consumed by an electrical load. Power = voltage × current.

Pressure: Force per unit of area.

Pressurestat: A pressure-sensitive switch used to protect compressors.

Primary air: Air that is pulled or propelled into the combustion process along with the fuel.

Psychrometrics: The study of air and its properties.

Purging: Releasing compressed gas to the atmosphere through some part or parts, such as a hose or pipeline, for the purpose of removing contaminants.

Radiation: The movement of heat in the form of invisible rays or waves, similar to light.

Reamer: A tool used to remove the burr from the inside of a pipe that has been cut with a pipe cutter.

Reclamation: The remanufacturing of used refrigerant to bring it up to the standards required of new refrigerant.

Recovery: The removal and temporary storage of refrigerant in containers approved for that purpose.

Rectifier: A device that converts AC voltage to DC voltage.

Recycling: Circulating recovered refrigerant through filtering devices that remove moisture, acid, and other contaminants.

Redundant gas valve: A gas control containing two gas valves in series. If one fails, the other is available to shut off the gas when needed.

Refrigerant: A liquid or gas that picks up heat by evaporating at a low temperature and pressure, and gives up heat by condensing at a higher temperature and pressure.

Refrigeration: The transfer of heat from a space or object where it is not wanted to a space or object where it is not objectionable.

Refrigeration cycle: The process by which a circulating refrigerant absorbs heat from one location and transfers it to another location.

Relative humidity (RH): The ratio of the amount of moisture present in a given sample of air to the amount it can hold at saturation. Relative humidity is expressed as a percentage.

Relay: A magnetically operated device consisting of a coil and one or more sets of contacts. Used in low-current circuits.

Resistance: The opposition to the flow of electrons (i.e., load).

Revolutions per minute (rpm): The speed at which the shaft of an air-moving device is rotating.

Safety pilot: A pilot light with a flame-sensing element.

Secondary air: Air that is added to the mix of fuel and primary air during combustion.

Sensible heat: Heat that can be measured by a thermometer or sensed by touch. The energy of molecular motion.

Sheet metal saddle: A sling-like device that cradles insulated duct from underneath.

Short circuit: A situation in which a conductor bypasses the load, causing a very high current flow.

Slow-blow fuse: A fuse with a built-in time delay.

Slug: A large amount of liquid refrigerant and/or oil entering a compressor cylinder.

SMACNA: Sheet Metal and Air Conditioning Contractors National Association.

Solder: A fusible alloy used to join metals.

Soldering: A method of joining metals with a non-ferrous filler metal using heat below 800°F and below the melting point of the base metals being joined. Also known as soft soldering or sweat soldering.

Solenoid: An electromagnetic coil used to control a mechanical device such as a valve or relay contacts.

Solid state: Having to do with semiconductors.

Specific heat: The amount of heat required to raise the temperature of one pound of a substance one degree Fahrenheit. Expressed as Btu/lb/°F.

Specifications: A document that describes the quality of the materials and work required. Specifications dictate the types of tubing, fixtures, hangers, etc., that must be used on a project.

Spud: A threaded metal device that screws into the gas manifold. It contains the orifice that meters gas to the burners.

Standard yoke (pipe) vise: A holding device used to hold pipe and other round objects. It has one movable jaw that is adjusted with a threaded rod.

Standing pilot: A gas pilot that is on continuously.

Starter: A magnetic switching device used to control heavy-duty motors.

Static pressure (s.p.): The pressure exerted uniformly in all directions within a duct system, as measured in in. w.c.

Stock: A tool used to hold and turn dies when cutting external threads.

Strap wrench: A tool for gripping pipe. The strap is made of nylon web.

Subcooling: Cooling a liquid below its condensing temperature.

Superheat: The measurable heat added to the vapor or gas produced after a liquid has reached its boiling point and completely changed into a vapor.

Swaged joint: A method of creating a pipe joint in which the diameter of one of the pipes to be joined is expanded using a special tool. The other pipe then fits inside the swaged pipe.

Sweating: A method of joining pipe in which solder is applied to the joint and heated until the solder flows into the joint.

Takeoffs: The process of itemizing and counting all the material and equipment needed for an installation.

Thermistor: A semiconductor device that changes resistance with a change in temperature.

Thermocouple: A device made of two different metals that generates electricity when there is a difference in temperature from one end to the other.

Ton of refrigeration: Large unit for measuring the rate of heat transfer. One ton is defined as 12,000 Btus per hour or 12,000 Btuh.

Total heat: Sensible heat plus latent heat.

Total pressure: The sum of the static pressure and the velocity pressure in an air duct. It is the pressure produced by the fan or blower.

Toxic: Poisonous.

Transformer: Two or more coils of wire wrapped around a common core. Used to raise and lower voltages.

Tubing: Thin-wall pipe; generally, pipe that can be easily bent.

Union: A pipe fitting used to join two lengths of pipe. It permits disconnecting the two pieces of pipe without cutting.

Unit: A definite standard measure of a dimension.

Vacuum: Any pressure that is less than the prevailing atmospheric pressure.

Variable: An element of an equation that may change in value.

Velocity: How fast air is moving. The rate of airflow usually measured in feet per minute.

Velocity pressure: The pressure in a duct due to the movement of the air. It is the difference between the total pressure and the static pressure in w.c.

Venturi: A ring or panel surrounding the blades on a propeller fan; used to improve fan performance.

Volt: The unit of measurement for voltage.

Voltage: A measure (in volts) of the electrical potential for current flow; also known as electromotive force (EMF).

Volume: The amount of space contained in a given three-dimensional shape; also, the amount of air in cubic feet flowing past a point in one minute (cfm).

Watts: The unit of measure for power consumed by a load.

Wet-bulb temperature: Temperature taken with a thermometer that has a wick wrapped around its sensing bulb that is saturated with clean, distilled water before taking a reading. The reading from a wet-bulb thermometer, through evaporation of the distilled water, takes into account the moisture content of the air. It reflects the total heat content of the air.

Wetting: A process that reduces the surface tension so that molten (liquid) solder flows evenly throughout the joint.

Figure Credits

Module 03101-07

Topaz Publications, Inc., 101SA01, 101SA03, 101SA04, 101SA05, 101SA06

Module 03102-07

Extech Instruments, 102SA06
Robinair, SPX Corporation, 102SA03, 102SA04, 102SA05
Topaz Publications, Inc., 102SA01, 102SA02, 102SA07, 102SA08

Module 03103-07

Bacharach Inc., 103F25
Charlotte Pipe and Foundry Company, 103F20
Legend Valve and Fitting, Inc., 103SA04
Ridge Tool Company (RIDGID®), 103F17
Topaz Publications, Inc., 103F03, 103SA01, 103F07, 103F11, 103SA02, 103SA03, 103F18, 103F19

Module 03104-07

Topaz Publications, Inc., 104F03, 104SA02, 104SA03, 104SA04, 104SA05, 104SA06, 104F07, 104SA07, 104SA08, 104SA09, 104SA10

Module 03105-07

Ridge Tool Company (RIDGID®), 105F09, 105F10, 105F12, 105SA03, 105F21 (left), 105SA05
Topaz Publications, Inc., 105SA01, 105F11, 105F14, 105SA02, 105F16, 105F17, 105F18, 105SA04
Tyco Valves & Controls, 105F25
Victaulic Company, 105F21 (right), 105F23, 105F24

Module 03106-07

Carrier Corporation, 106F27
National Oceanic and Atmospheric Administration/Department of Commerce, 106SA02
Topaz Publications, Inc., 106SA01, 106SA03, 106SA08, 106F14 (photo), 106F15 (photo), 106SA09, 106F18 (photo), 106SA10, 106SA11, 106SA12, 106F29, 106F30

Module 03107-07

Aprilaire, 107F41
Emerson Climate Technologies, 107SA13, 107F40
Extech Instruments, 107SA03
Honeywell International Inc., 107F39
Raytek Corporation, 107SA02
Robinair, SPX Corporation, 107F11, 107SA04
Courtesy of Snap-on Tools, www.snapon.com, 107F36
Swift Optics, 107SA01
Topaz Publications, Inc., 107F10, 107SA06, 107F18, 107F19 (photo), 107SA07, 107SA08, 107SA09, 107SA10, 107SA11, 107SA12, 107F33, 107F35, 107F37

Module 03108-07

Aprilaire, 108F12
Hearth, Patio & Barbecue Association, 108F01 (photo)
Topaz Publications, Inc., 108SA01, 108SA02, 108F07, 108F13, 108F26, 108F27, 108F28, 108F30 (photo), 108SA03, 108SA04, 108F37, 108F39, 108F40, 108F42

Module 03109-07

Bacharach, Inc., 109F55 (left)
Carrier Corporation, 109SA03, Appendix
Ductmate® Industries, Inc., 109SA02 (left)
Ductsox Corporation, 109SA01
Dwyer Instruments, Inc., 109F52, 109F53
Extech Instruments, 109F14, 109F50 (photo), 109F51
Hart & Cooley, Inc., www.hartandcooley.com, 109F29, 109F32, 109F35
Lockformer, 109SA02 (right)
Nailor Industries, Inc., 109F37
Northern Blower Inc., 109F08
TIF, SPX Corporation, 109F55 (right)
Topaz Publications, Inc., 109F07, 109F09, 109F27, 109F31, 109F33, 109F49

Figures are indicated with "*f*" following the page number.